ORGANIC CHEMISTRY LABORATORY

with Qualitative Analysis
Standard and Microscale Experiments
Third Edition

Charles E. Bell, Jr.
Old Dominion University

Douglass F. Taber
University of Delaware

Allen K. Clark
Old Dominion University

Harcourt
College Publishers

A Harcourt Higher Learning Company

Fort Worth   Philadelphia   San Diego
New York   Orlando   Austin   San Antonio
Toronto   Montreal   London   Sydney   Tokyo

Vice President/Publisher: John Vondeling
Vice President/Publisher: Emily Barrosse
Marketing Strategist: Pauline Mula
Developmental Editor: Nancy Lubars
Production Manager: Susan Shipe
Art Director: Cara Castiglio/Carol C. Bleistine
Cover Designer: Carol C. Bleistine
Cover Credit: © Mauritius, GMBH/Phototake

Bell, Charles E., Jr., Taber, Douglass F., and Clark, Allen K.

Organic Chemistry Laboratory with Qualitative Analysis: Standard and Microscale
Experiments, Third Edition
ISBN: 0-03-029272-7
Library of Congress Catalog Card Number: 00-30100

Address for domestic orders:
Harcourt College Publishers
6277 Sea Harbor Drive, Orlando, FL 32887-6777
1-800-782-4479
e-mail: collegesales@harcourt.com

Address for international orders:
International Customer Service, Harcourt, Inc.
6277 Sea Harbor Drive, Orlando, FL 32887-6777
(407) 345-3800
Fax (407) 345-4060
e-mail: hbintl@harcourt.com

Address for editorial correspondence:
Harcourt College Publishers
Public Ledger Building Suite 1250
150 S. Independence Mall West, Philadelphia, PA 19106-3412

Web Site Address:
http://www.harcourtcollege.com

Printed in the United States of America
1 2 1 2 3 4 5 6 7 8   066   10 9 8 7 6 5 4 3 2

Harcourt College Publishers

Where Learning Comes to Life

TECHNOLOGY

Technology is changing the learning experience, by increasing the power of your textbook and other learning materials; by allowing you to access more information, more quickly; and by bringing a wider array of choices in your course and content information sources.

Harcourt College Publishers has developed the most comprehensive Web sites, e-books, and electronic learning materials on the market to help you use technology to achieve your goals.

PARTNERS IN LEARNING

Harcourt partners with other companies to make technology work for you and to supply the learning resources you want and need. More importantly, Harcourt and its partners provide avenues to help you reduce your research time of numerous information sources.

Harcourt College Publishers and its partners offer increased opportunities to enhance your learning resources and address your learning style. With quick access to chapter-specific Web sites and e-books . . . from interactive study materials to quizzing, testing, and career advice . . . Harcourt and its partners bring learning to life.

Harcourt's partnership with Digital:Convergence™ brings :CRQ™ technology and the :CueCat™ reader to you and allows Harcourt to provide you with a complete and dynamic list of resources designed to help you achieve your learning goals. Just swipe the cue to view a list of Harcourt's partners and Harcourt's print and electronic learning solutions.

C 62 00 00 00 00 00 25 201

http://www.harcourtcollege.com/partners/

*This edition is dedicated to the memory of
Professor James A. Moore, the originator of this series of
textbooks for the organic laboratory course.*

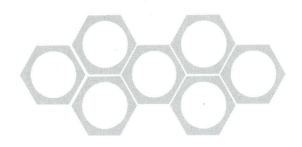

Acknowledgments

Many of the experiments in this text are direct descendants of *Experimental Methods in Organic Chemistry*, 3rd edition, written by James A. Moore, David L. Dalrymple, and Oscar R. Rodig, and previous editions by Moore and by Moore and Dalrymple. We are indebted to these colleagues, and to the many students and teaching assistants who worked with them.

We are grateful to Professor Allen K. Clark for his contributions to the first and second editions of this text. His attention to detail and insistence on clarity of text set an excellent example for us to follow in this edition.

Finally, we are grateful to the following people who reviewed and critiqued our manuscript:

- Dan Blanchard, Kutztown University
- Stephen E. Branz, San Jose State University
- Donald Byers, Johnson County Community College
- Philip Chenier, University of Wisconsin, Eau Claire
- Joseph M. Hornback, University of Denver
- James E. Hutchinson, University of Oregon
- Roy E. LaFever, California State University, Bakersfield
- Kirk Manfredi, University of Northern Iowa
- Aaron Monte, University of Wisconsin, LaCrosse
- Regina Zibuck, Wayne State University

Charles E. Bell, Jr.
Allen K. Clark
Old Dominion University

Douglass F. Taber
University of Delaware

July 2000

Contents

x   Contents

Preface

The purpose of the organic chemistry lab is to give students practical experience with operational organic chemistry. Our philosophy, expressed in this text, is that they will be more interested in the science—and ultimately will learn more—if the work is fun to do.

At the lowest level, this means that the experiments work as written. All new experiments included have been used in the laboratory course at the University of Delaware. Each can be carried out in one or two three-hour lab periods. Throughout, we have tried to make the experiments practical. In the Oxidation chapter, rather than prepare camphor, a volatile solid, we oxidize 4-nitrobenzyl alcohol to 4-nitrobenzaldehyde. Neither the starting material nor the product is volatile, and both are crystalline. They also both have strong UV chromophores, making TLC visualization and chromatographic purification easy.

At a higher level, fun means that the experiments involve more than making a product and taking the melting point. In addition to the procedure for preparing the Grignard reagent from bromobenzene, for instance, we have included a procedure for preparing the Grignard reagent from 4-bromo-N,N-dimethylaniline. Addition of this reagent to methyl benzoate provides, after hydrolysis, the dye Malachite Green. Use of diethyl carbonate in place of methyl benzoate leads to the dye Crystal Violet. Another illustration of this approach occurs early in the text. In addition to a procedure for preparing acetanilide from aniline, we have included a procedure for converting anthranilic acid to the N-acetyl derivative. In the latter case, the crystals obtained are triboluminescent—they flash when ground between two watch glasses.

At the highest level, fun means that the student gets to *reason* about chemistry. We have written the two introductory NMR chapters (12 and 13) to make them immediately accessible to the beginning student. We have found that even beginning organic students are able to deduce organic structures from simple data sets—and that they enjoy doing it! We have also included a chapter (14) on more advanced NMR, IR, and MS, with more advanced problems, for the students who have mastered the easier problems. We anticipate that the solving of spectroscopic problems (dry lab) will not be limited to a single laboratory period, but rather will be a continuing activity throughout the year. As a service to instructors adopting this text, Douglass F. Taber (taberdf@udel.edu) will supply ten fresh spectroscopy problems each year, to use for exams.

For those who asked for more syntheses, we have included three more, as outlined below. Space limitations precluded us adding any more to the text, so we have published additional syntheses on our web site (http://valhalla.chem.udel.edu/orglab.html). If you have a good one that you would like to share, send it in by e-mail (taberdf@udel.edu) and we will post that also!

Finally, a special note for microscale users: We have found that the students are frequently frustrated by the difficulties of crystallizing milligram quantities of pure products from crude microscale reaction mixtures. We have included (Chapter 7) instructions for a simple procedure for quick column chromatography, based on Flash Chromatography but using the simplest of equipment. It has been our experience that a quick silica gel column followed by crystallization is by far the easiest way to work up most microscale reactions.

New and Revised Experiments

The following experiments are new to this edition:

- Chapter 7, Thin Layer Chromatography: The preparation of acetylferrocene from ferrocene has been updated, and a procedure for separating acetylferrocene from ferrocene by silica gel chromatography has been added. This experiment is now an excellent one-period introduction to synthesis, TLC, and column chromatography that can be easily carried out using either microscale or standard glassware.
- Chapters 12, 13, and 14, Spectroscopic Structure Determination: As noted above, Chapters 12 and 13 are now simple and inviting introductions to ^{13}C and ^{1}H NMR, respectively. Chapter 14 covers more advanced applications, including the use of mass spectrometry in structure determination. Each of these chapters includes a new set of spectroscopic problems at the appropriate level.
- Experiment 24, Carboxylic Acid Mini-unknown: A procedure has been added for converting an unknown arene to the corresponding benzoic acid. The product acid is then titrated, and the structure of the starting arene is deduced from the melting point and calculated molecular weight of the product acid. Wet chemistry to deduce the structure of an organic unknown is an important part of the organic chemistry tradition, but it is increasingly difficult to make time for it in the curriculum. This experiment incorporates many of the desirable aspects of "qual organic," but can be accomplished in a single lab period.
- Experiment 26, The Friedel-Crafts Reaction: In the past, we have used only Friedel-Crafts alkylation. This chapter has been expanded to include Friedel-Crafts acylation, the preparation of a crystalline unsymmetrical benzophenone from two arenes. This experiment takes one lab period.
- Experiment 31, Photochemical Reactions: In the place of the previous experiment, we now have a procedure for the photochemical dimerizaton of benzophenone, and for the rearrangement of the product benzopinacol to benzopinacolone. The photodimerization proceeds smoothly with a commercially available (Fisher Scientific) UV display lamp or with strong direct sunlight. The rest of the experiment is easily accomplished in one lab period.

Web Features

This text is supported by a web page (http://valhalla.chem.udel.edu/orglabtext.html). The answers to the problems in Chapters 12, 13, and 14 are posted there. In addition, as noted above, several additional laboratory experiments are described there.

Also at this website, Data Report Sheets are provided for each of the chapters. The students can print these out directly from the Web, fill them in, and submit them. While many instructors may continue to require that the traditional laboratory notebook be submitted for grading, we have found that in a large pre-med organic laboratory course, grading of such notebooks takes an inordinate amount of time. We thank Professor Roger Murray of St. Joseph's University for this suggestion.

Microscale and Macroscale

We follow current practice by presenting microscale and macroscale procedures for many of the experiments. In general, the microscale experiments were developed using the minimal practical amounts of material. Our experience has been that the slight savings in chemical costs realized from working on still smaller scales does not justify the marked increase in frustration on the part of the students.

Safety

Safe laboratory practice and proper waste disposal are emphasized, both in Chapter 1 and in the Safety Notes associated with each experiment. Further advice on the safe dispensing of reagents and other general information on the experiments can be found in the accompanying Instructor's Manual.

In any laboratory course, the student will work more efficiently if he or she has thought the material through ahead of time. To this end, we have included prelaboratory questions for almost all of the chapters. The answers to these questions are in the Instructor's Manual. In addition, we have included questions of variable difficulty at the end of each chapter, which can best be answered after the experiments have been completed.

Introduction

Laboratory work is an integral and essential part of any chemistry course. Chemistry is an experimental science—the compounds and reactions that are met in lecture and classroom work have been discovered by experimental observations. Organic compounds exist as gases, liquids, or solids with characteristic odors and physical properties. They are synthesized, distilled, crystallized and chromatographed, and then transformed by reactions into other compounds. The purpose of laboratory work is to provide an opportunity to observe the reality of compounds and reactions and to learn something of the operations and techniques that are used in experimental organic chemistry and in other areas in which organic compounds are encountered.

A. LABORATORY SAFETY

Along with the opportunity to learn at first hand about the properties and reactions of compounds and the manipulation of laboratory equipment, there must be proper concern for safety. Most organic compounds are flammable, and they are toxic or irritating to a greater or lesser degree. Many organic reactions are potentially violent. Laboratory work in organic chemistry is not a dangerous occupation, however, nor is the laboratory a perilous place to be, provided that some simple precautions and safety rules are followed. There are some potential hazards that must be recognized and avoided; accidents can and do occur when these hazards are ignored. Throughout this book, **Safety Notes** are included for each experiment. They emphasize the specific hazards that may be encountered and must be kept in mind. In the following section, some general precautions are discussed, and a few rules are given that must be observed during any laboratory work.

RULES FOR PERSONAL SAFETY

Avoiding injuries is largely a matter of good sense. Carelessness can lead to accidents and injuries to yourself and others. The following rules cover some important general precautions and should be observed at all times.

1. **Eye Protection.** Approved eye protection must be worn at all times in the laboratory, regardless of what is being done. In many locations, chemical safety goggles are required by law. Safety glasses (either prescription or plain) with side shields to protect from splashes also offer good eye protection.
2. Never work in a laboratory without another person being present or within calling distance. Minor accidents can become disasters if help is not available.
3. Do not carry out any reaction that is not specifically authorized by the instructor.

4. Never taste a compound; never pipet a chemical by mouth; do not eat, drink, or smoke in the laboratory.
5. Avoid contact of the skin with any chemical. If a substance is spilled on your hands, wash them thoroughly with soap and water. Do not rinse them with a solvent, since this may cause more rapid absorption.
6. Long hair should be tied back. Shoes must be worn to prevent injury from spilled chemicals or bits of glass. Avoid loose-fitting sleeves and clothing that leave expanses of skin unprotected.
7. Never heat a flask or any apparatus that is sealed or stoppered (i.e., a closed system)— make certain that there is an opening to the atmosphere.
8. When inserting glass tubing into a rubber stopper or rubber tubing, lubricate it first with a drop of glycerin and protect your hands with a towel. The same is true for inserting thermometers into stoppers. Thin-walled transfer pipets should not be used as connectors for rubber tubing or stoppers; they are fragile and very easily crushed.
9. Some experiments require the use of a well-ventilated hood. These experiments should not be attempted in the open laboratory.
10. Peroxides. A major safety concern with certain organic substances is the buildup of potentially explosive peroxides. These can form by the light-catalyzed autoxidation of ethers, alcohols, aldehydes, alkenes, and also aromatic compounds having allylic or benzylic hydrogen atoms. The danger is particularly acute when these materials are used as reaction solvents or in extractions and then are concentrated in the workup. This is one reason why liquids should never be distilled to dryness. Such commonly used substances as diethyl ether, diisopropyl ether, tetrahydrofuran, cumene, tetralin, and 2-butanol (Fig. 1.1) should be checked for peroxides if they have been stored for an extended period of time, especially in a partially empty container. Particular care should be taken with diethyl ether because of its wide use as an extraction solvent.

    Peroxides can be detected with starch iodide paper or by adding 1 mL of the suspected material to 1 mL of glacial acetic containing 0.1 g of sodium or potassium iodide. A yellow-to-brown color indicates the presence of peroxides. A blank should be run to confirm the validity of the test.

    Peroxides can be removed from ethers by shaking with a 30% solution of aqueous ferrous sulfate or by percolating through a column of alumina. The latter method also removes traces of water. Additional information on peroxides can be found in the references cited at the end of the chapter.

Figure 1.1   Some substances that can form peroxides by autoxidation.

FIRST AID

1. **Emergency Equipment.** Learn the location of safety showers, eyewash fountains, and fire extinguishers, and know how this equipment is used.

2. **Chemical Spills on the Skin.** Immediately flush the skin with running water for several minutes; if the eyes or face are involved, use an eyewash fountain or the nearest faucet and wash for 15 minutes. For any serious burn or splashing of chemicals in the eyes, consult a physician as soon as possible after the initial thorough water flushing.

3. **Fire.** If clothing is ignited, immediately extinguish it in a safety shower or by rolling on the floor. If necessary, cover the victim with a coat or fireblanket. Note that it is no longer recommended to wrap the victim in a fire blanket while the victim is standing. The blanket can act as a chimney and encourage the fire. These blankets can be used to cover a shock victim. Do not allow a person with burning clothing to run; such action only fans the flames. Even standing should be avoided because of the danger of inhaling superheated fumes.

4. In the event of a severe chemical spill, burn, or cut, the affected person should be escorted to a physician or a hospital emergency room.

CHEMICAL TOXICITY AND CARCINOGENS

It has long been recognized that certain chemicals—for example, phosgene or the nerve gas isopropyl methylfluorophosphonate—are highly toxic substances that are lethal in extremely small amounts. In recent years there has been a considerable increase in awareness and concern about the toxicity of all chemicals encountered in laboratory and manufacturing environments. Major efforts are now being made to identify toxic chemicals and avoid exposure to them.

The National Institute of Occupational Safety and Health (NIOSH) has prepared a registry of a large number of compounds for which some data on toxic effects are available. Another government agency, the Occupational Safety and Health Administration (OSHA), issues regulations governing permissible limits of exposure to chemicals, with particular attention being placed on compounds that are commonly encountered as air contaminants. Many of the compounds for which limits have been set, such as diethyl ether and ethanol, have relatively low toxicity, but limits are nevertheless placed on prolonged exposure. On the other hand, compounds that may have high acute toxicity can be transferred and used in a laboratory experiment with simple precautions to avoid contact.

A major concern in recent years has been the carcinogenicity of organic compounds, that is, their ability to induce cancer. For certain compounds that were used industrially for many years, there is a clear link between exposure to the compound and the incidence of certain types of cancer in the workers who handled them. More recently, evidence has been found for the occurrence of tumors in experimental animals exposed to very large doses of a wide variety of other organic compounds. A much-publicized example is the artificial sweetener saccharin, which was used for many years in low-calorie beverages.

Government agencies, such as the Carcinogen Assessment Group (CAG) of the Environmental Protection Agency (EPA) and NIOSH, have compiled lists of suspected carcinogens, and laboratory chemical catalogs often identify these as "cancer suspect agents." The National Institutes of Health has published guidelines for the laboratory use of certain chemical carcinogens, and OSHA, in turn, has issued regulations on exposure to a number of carcinogenic substances, most of which are in commercial use.

A brief list of carcinogenic compounds that are sometimes encountered in laboratory work is given in Table 1.1. It should be pointed out that this list contains only a very small fraction of the compounds for which data on carcinogenic properties are known. Remember that carcinogenic activity is usually based on tests in animals at high doses for prolonged duration, and the risk from occasional brief exposure is unknown. The chief concern with carcinogens, as with other toxic substances, is with permissible levels for continuous exposure, not occasional use. The important consideration for laboratory work is to avoid any unnecessary exposure and to use these compounds only when essential, with due care and protection. In particular, benzene, chloroform, and carbon tetrachloride should not be used as solvents for extraction or column chromatography.

Table 1.1   Partial List of Chemical Carcinogens

acetamide	CH_3CONH_2	dioxane	$O(CH_2CH_2)_2O$
acrylonitrile	$CH_2\!=\!\!=\!CHCN$	ethyl carbamate	$NH_2CO_2C_2H_5$
aminobiphenyl	$NH_2C_6H_4C_6H_5$	hydrazine	NH_2NH_2
benzene	C_6H_6	methyl iodide	CH_3I
benzidine	$NH_2C_6H_4C_6H_4NH_2$	1-naphthylamine	$1\text{-}C_{10}H_7NH_2$
benzpyrene	$C_{20}H_{12}$	2-naphthylamine	$2\text{-}C_{10}H_7NH_2$
t-butyl chloride	$(CH_3)_3CCl$	4-nitrobiphenyl	$NO_2C_6H_4C_6H_5$
carbon tetrachloride	CCl_4	nitrosomethylurea	$NH_2CON(CH_3)NO$
chloroform	$CHCl_3$	phenylhydrazine	$C_6H_5NHNH_2$
chromic anhydride	CrO_3	thiourea	NH_2CSNH_2
diazomethane	CH_2N_2	*o*-toluidine	$CH_3C_6H_4NH_2$
dibromoethane	$BrCH_2CH_2Br$	trichloroethylene	$CHCl\!=\!\!=\!CCl_2$
dimethyl sulfate	$(CH_3)_2SO_4$		

In the experiments in this book, all compounds that have been implicated as carcinogens have been eliminated wherever possible; in the remaining few cases they have been identified as such. It must be emphasized that, with present knowledge, these compounds should *not* be considered as deserving any more concern than many others. Careless handling of simple acids or solvents normally presents a much greater safety hazard.

B. DISPOSAL

Recent laws and regulations have created a significant increase in public awareness of the dangers associated with improper disposal of hazardous waste. For many chemicals, the cost for disposal of residues greatly exceeds the original purchase price. This includes many of the chemicals used in organic teaching laboratories. Important new industries relating to waste disposal have arisen during the past ten years. Most academic laboratories contract with these companies to handle hazardous waste disposal. A major factor in the process is keeping records as to the exact identity of the substances being handled. The cost of disposal increases dramatically if the identity is not precisely known. Thus, for the sake of safety and keeping down costs, a well-regulated disposal procedure should be included in every experiment undertaken in the laboratory. This manual will include, for each chapter, information relating to the disposal of waste materials associated with the experiments. Special care should be taken to observe these disposal requirements. Many areas have very efficient state and local monitors which are authorized to invoke penalties for careless or deliberate improper handling of waste. A synopsis of the suggested methods for disposal is given below. Each chapter will contain specific instructions pertaining to the experiments contained therein. Consult local authorities if questions arise concerning the disposal of any chemical.

1. Small amounts of simple acids, such as hydrochloric, sulfuric, nitric, acetic, and propionic, or simple bases such as sodium or potassium hydroxide, should be neutralized before flushing down the drain with large amounts of water.

2. Small amounts of simple salts of mineral acids and ions, such as aluminum, calcium, magnesium, and ammonium, may be flushed down the drain with large amounts of water.

3. Organic solvents should be poured into a properly labeled waste container and stored in a properly ventilated place. Pouring volatile solvents down the drain may result in the vapors and residues remaining in the sink trap. This could cause a fire.

4. Small quantities of unreactive and nontoxic solid wastes can often be disposed of along with the paper trash. Hazardous solid waste should be disposed of in a properly labeled jar. The exact nature of the contents should be written on the label or an attached sheet.

5. Special care should be taken in disposing of suspected carcinogens. Consult *Prudent Practices in the Laboratory: Handling and Disposing of Chemicals,* National Academy Press, Washington, DC, 1995.

6. Chemicals that react violently with water, such as acid chlorides, alkali metals, or metal hydrides, should be decomposed in a hood in a suitable way, such as by controlled reaction with alcohol.

7. Harmless solids, such as silica gel or alumina, that are damp with volatile organic solvents present a safety as well as a monetary problem. Spread the solid out on a tray or other large container in the fume hood. Allow the solvent to evaporate and dispose of the solid in a nonhazardous waste container. The nonhazardous waste container can be dumped into the local trash, thereby eliminating a considerable expense for formal disposal.

8. Broken glass should be discarded in an earthenware crock or other nonmetallic waste container. Do not use a wastepaper basket.

9. Local authorities or college waste disposal officers should be consulted before disposing of anything down the drain other than "rinsate" from washing glassware. What may be legal in one locality may not be legal in another.

FIRE HAZARDS

Fire hazards are present in any organic laboratory because of the frequent use of volatile, flammable solvents. By far the greatest risk of fire is associated with gas burners for heating, and this form of heating should be avoided whenever possible. On the other hand, electric heating devices may contain exposed elements and are not totally immune from being fire hazards. Moreover, the low flash points of typical organic solvents present a continuing problem where electric sparks may result from thermostatic devices.

Organic vapors are heavier than air and flow downward; they diffuse rapidly and can be ignited several feet away from the source. The following precautions should always be observed:

1. Organic liquids should only be heated to boiling or distilled under a condenser. When refluxing a liquid, be certain that the condenser is tightly fitted to the flask.

2. Before lighting a flame, if one must be used, check to see that volatile liquids are not being poured or evaporated in your vicinity. Conversely, before pouring or evaporating a liquid, be certain that none of your neighbors is using a flame. Always turn off a burner as soon as you are finished using it—never leave it on unnecessarily.

3. Smoking creates an avoidable fire hazard and is not permitted in the laboratory.

C. LABORATORY EQUIPMENT AND TECHNIQUES

For successful laboratory work it is essential that, before beginning an experiment, you understand what you are going to do and why and how you are going to do it. Study the assigned experiment in advance and plan your operations. If required by your instructor, answer in writing the prelab questions at the end of the experiment. These questions provide background for the work and will acquaint you with points that can and should be understood before you actually do the experiment.

Experimentation in organic chemistry calls into play a number of operations and techniques and a rather large assortment of apparatus. Detailed instructions in the techniques and equipment used for various separation and purification methods are given in Chapters 2 through 10 in conjunction with actual experimental procedures. A few points of general practice that apply to nearly all experiments are covered in this section.

EQUIPMENT

Glassware

In this book you will learn two major laboratory techniques for carrying out chemical reactions. Those on a *macroscale* normally involve gram quantities of materials, and typical glassware used for this is shown in Figure 1.2. The glassware is equipped with standard-taper ground joints that permit quick and secure assembly of apparatus. It is expensive and must be handled with care. *Microscale* experiments are carried out on milligram amounts of material and generally demand more careful manipulation of reagents and products. Often these experiments can also be carried out with the glassware shown in Figure 1.2, but on a reduced scale.

Some examples of glassware designed specifically for microscale use are shown in Figure 1.3. This glassware contains joints with external threads and open-top caps that allow them to be screwed together. Since the pieces cannot pull apart, the assemblies require considerably less

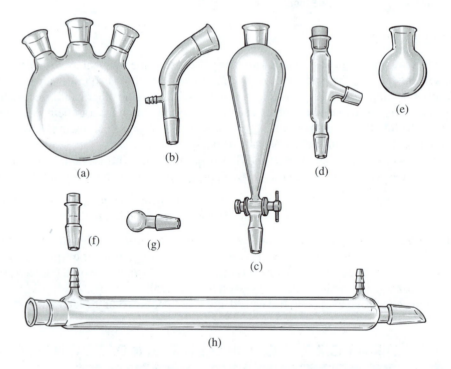

Figure 1.2   Standard-taper glassware. (a) three-necked round-bottom flask; **(b)** vacuum take-off adapter; **(c)** dropping funnel; **(d)** distillation head with rubber connector; **(e)** round-bottom flask; **(f)** tubing connector; **(g)** standard-taper stopper; **(h)** condenser.

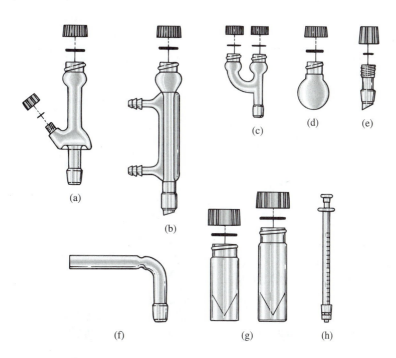

Figure 1.3 Microscale glassware. (a) Hickman still head; **(b)** condenser; **(c)** Claisen head adapter; **(d)** small round-bottom flask; **(e)** thermometer adapter; **(f)** drying tube; **(g)** 3- and 5-mL conical reaction vials; **(h)** 1-mL sample syringe.

clamping. In addition, the inner parts of the joints are also standard-taper ground glass that allow their attachment to other ground joint apparatus. Although convenient, such equipment is not required for experiments in this book.

Other equipment that will be needed includes Erlenmeyer flasks, beakers, and test tubes in various sizes, Büchner and Hirsch funnels, graduated cylinders, pipets, centrifuge tubes (preferably graduated), thermometers, spatulas, septums, syringes, and assorted clamps and other hardware.

For effective laboratory work it is most important that you develop good working habits, learn the proper equipment for a given purpose, and know how to use it. Maintain a well-organized locker or equipment drawer; keep your equipment clean and as conveniently located as possible. Make a practice of washing or rinsing glassware as soon as it has been used so that it will be clean and dry the next time you need it. Take enough time to clean up and store equipment properly before you leave the laboratory.

Reaction Setups

Many reactions involve the combining of two reactants at a controlled rate. Simple reactions on a small scale require only a pipet and centrifuge or test tube, which can be swirled by hand. In other cases, on both macroscale and microscale, the reactants are combined and heated in a simple setup, such as that shown in Figure 1.4. The temperature is controlled by a heating unit or by the boiling point of a reactant or solvent. Mixing, if needed, is provided by the turbulence of boiling. If the reaction is exothermic, cooling occurs by the heat being transferred to the condenser by the refluxing solvent, or, if necessary, instead of being heated the flask may be placed in an ice bath.

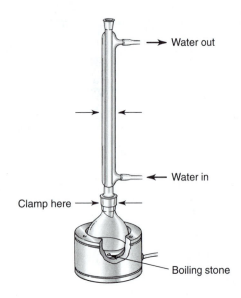

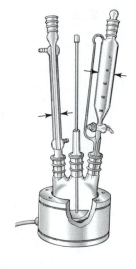

Figure 1.5   Reaction assembly with heating well, stirrer, and pressure-equalizing dropping funnel.

Figure 1.4   Simple reflux apparatus.

For reactions on a large scale or those requiring special conditions, a more elaborate setup such as that shown in Figure 1.5 may be needed. The separate necks for the dropping funnel, stirrer, and condenser permit flexible assembly and easy control of addition and mixing. Note: A Claisen head adapter may be used to convert a single-neck flask to one with two openings.

Stirring is important if a reactant must be added and dispersed at a steady rate or if the reaction must occur between separate phases. For relatively small flasks and for reaction mixtures that are not viscous, a bar magnet with an inert coating is placed in the flask and is spun by a motor-driven magnet below the flask (Fig. 1.6). Alternatively (as seen in Fig. 1.5), a rod with a small propeller or paddle, turning in a closely fitting sleeve, is driven by an electric or air-powered motor.

The atmosphere in a reaction may be critical. If the reactants or products are sensitive to water, a drying tube filled with desiccant attached to the condenser may be sufficient. For scrupulous removal of moisture or when oxygen must be excluded, a dry inert gas such as nitrogen is passed over the reaction mixture and out through a trap. With such a setup, the dropping funnel must be equipped with a pressure-equalizing arm (see Fig. 1.5).

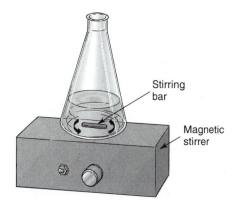

Figure 1.6 Magnetic stirrer.

Heating Sources

The sources of heat in most undergraduate laboratories are electric mantles or hot plates and steam baths. Bunsen burners are also used, but they are the most frequent cause of laboratory fires and must be used with great care, as noted in the section on Fire Hazards. When a flask must be heated over a burner, a piece of wire gauze placed under the flask will distribute the heat more evenly; one must be ever-vigilant about organic vapors.

When possible, the steam bath (Fig. 1.7) should be used for heating or evaporating an organic liquid. Place the steam bath on a hot plate and place the flask on the largest ring that will support it. Do not remove all of the rings or set the flask on the bottom of the steam bath. Be sure that the vent is facing away from your workplace.

Figure 1.7 Steam bath.

Electric heating elements, controlled by a variable transformer, are available in a variety of forms. In heating wells (see Fig. 1.4) the element is embedded in a hemispherical shell that fits around a spherical flask. Hot plates are convenient for heating a flat-bottomed vessel such as a beaker or an Erlenmeyer flask. Hot plates and heating wells may contain exposed contacts or elements and should not be used to vaporize flammable liquids without a condenser. The hot surfaces of the apparatus may exceed the flash point of the liquid and cause a fire.

For experiments on a microscale, a sand bath is a convenient heat source, and is readily made by filling a 100-mL heating well with sand.

A recently introduced heating method uses an aluminum block that is placed on a hot plate and contains various sized holes for flasks, test tubes, and a thermometer (Fig. 1.8a). Auxiliary aluminum blocks can also be placed around flasks and test tubes to improve further the heat transfer. Such an assembly is shown in Figure 1.8b using a conical reaction vial attached to a Hickman still head, an apparatus used to carry out microscale distillations.

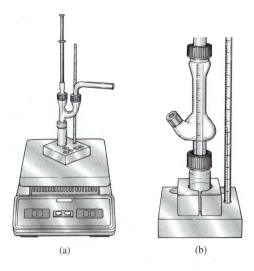

(a)   (b)

Figure 1.8 Aluminum heat transfer devices.

TECHNIQUES

Use of the Aspirator

An aspirator provides a convenient source of reduced pressure for vacuum operations. Water should be turned on to full capacity when an aspirator is used. Splashing may be a problem, but it can usually be avoided by attaching some rubber tubing, tying a piece of rag around the outlet, or wedging a piece of wire gauze in the trough below the stream. Heavy-wall rubber tubing should be used when connecting apparatus to the aspirator, since regular tubing will collapse and pinch off the system. If the water pressure drops, or the water is turned off while the aspirator is in use, water tends to be drawn back through the arm into the evacuated system. Most aspirators contain a check valve to prevent this, but a trap should always be connected between the aspirator and the evacuated vessel (Fig. 1.9) as a safety precaution. The vacuum should always be released *before* turning off the water.

Handling and Measuring Chemicals

In most preparative experiments, solid starting materials and reagents can be weighed on a beam or a top-loading balance to ±0.1g for the macroscale experiments and a top-loading

Figure 1.9   Aspirator and safety trap.

electronic balance (Fig. 1.10) with a 100-g capacity that is capable of weighing to ± 0.001g for the microscale experiments. A beaker is the most convenient vessel for handling more than a few grams. For weighing small quantities of solid reagents or products, weighing cups or glassine paper (not filter paper) should be used. The sample can then be placed in a vial or added to a solution simply by picking up the paper by opposite edges or corners and using it as an open funnel. Finely divided solids can be transferred quite completely by gently scraping with a spatula. Metal spatulas should be wiped clean after use and polished occasionally.

It is usually unnecessary to weigh certain reagents, such as activated carbon, salt, or drying agents; the appropriate amount often can be estimated by bulk. Small amounts of sodium hydroxide or potassium hydroxide normally need not be weighed. The pellets are fairly uniform and weigh roughly 0.1g each. If more accurate measurement of a small quantity is required, as in some microscale experiments, an aliquot of a solution of known concentration should be used. Never attempt to weigh hydroxide pellets on paper or a watch glass; a few inevitably roll away and liquify into a highly corrosive puddle.

Liquid reagents and starting materials are conveniently measured by volume in a graduated cylinder or sometimes with a graduated pipet or syringe, but these are accurate to only a few percent. For more accurate macroscale measurements, liquids should be weighed, particularly if they are rather viscous. For transfer of small volumes of liquids and very approximate measurements, thin-wall, soft-glass transfer (Pasteur) pipets are particularly useful. A supply of these should be maintained; they can be rinsed and reused repeatedly. Rubber bulbs of 1- and 2-mL capacity will fill these pipets to about half and full capacity, respectively; the drawn-out section of these pipets contains about 0.2 mL (see inside back cover). There will be many

Figure 1.10   A top-loading electronic balance.

occasions to obtain approximate volumes of liquid in this way. A detailed procedure for transferring small volumes of organic solutions is described in Chapter 4.

D. REPORTS AND RECORDS

Submission of Samples and Reports: Percentage Yield

In many of the described experiments, a sample of compound is to be turned in or a specific result may be called for in written form. Samples should be submitted in an appropriate-sized plastic-cap vial and neatly labeled with the student's name, the name of the product, its weight, percentage yield, melting point or boiling point, and any other data specified by the instructor.

The *yield* in a reaction is the amount of product of acceptable purity actually obtained. The percentage yield is the ratio of this amount to that theoretically obtainable $\times$ 100. In a reaction involving more than one starting material, the theoretical or 100% yield is determined by the reactant that is used in the smallest stoichiometric quantity called the "limiting reagent" or "yield limiting reagent." In calculating a percentage yield, therefore, one must first determine the molar amounts of each starting compound and then the theoretical amount of product that could be obtained from the limiting reagent.

As an example, consider the esterification of succinic acid in which 5.0 g of the acid is heated with 100 mL of ethanol and 6.2 g of ester is isolated.

$$(CH_2CO_2H)_2 \;+\; 2\,CH_3CH_2OH \;\xrightarrow{\;H+\;}\; (CH_2CO_2CH_2CH_3)_2 \;+\; 2\,H_2O$$

succinic acid   ethanol   diethyl succinate

The molar amounts are as shown.

SUCCINIC ACID	ETHANOL	DIETHYL SUCCINATE
5.0 g; MW 118 = 0.042 mole	100 mL; d, 0.79 g/mL = 79 g; MW 46 = 1.72 moles	6.2 g; MW 174 = 0.036 mole

The ethanol is present in large excess (0.084 mole required) and the yield of ester product is thus based on the acid. The theoretical amount of diethyl succinate is the same as that of acid, 0.042 mole, and the percentage yield is

$$\frac{0.036}{0.042} \times 100 = 86\%$$

Alternately, the percentage yield can be calculated from the grams of product divided by the theoretical yield in grams and multiplying by 100.

In the foregoing example, if 0.40 g of unreacted succinic acid were recovered from the reaction in usable form, the yield might alternatively be based on "unrecovered starting material," that is, $5.0 - 0.4 = 4.6$ g; the percentage yield on this basis would be 92%.

Records

Equally as important as the actual product or the final report is the **laboratory notebook.** An important objective of this course is to develop good practices and habits in keeping permanent notes of experimental work and in working in an orderly, systematic way.

The laboratory notebook must be bound with a hard cover and used only for experimental data in this course. It must, of course, be at hand at all times while you are in the laboratory. The notebook is the fundamental record of actual laboratory operations and observations. It should provide an account of *what* was done, *how* it was done, and *what happened.* The apparatus used, the sequence of steps, all measurements, significant time intervals, changes in appearance, and other relevant data should be recorded. It should be possible for someone else to repeat the experiment as *you* did it and obtain the same results. The notebook should *not* be a verbatim transcription of a procedure from the laboratory manual or any other source that you then purport to follow, nor should it be an after-the-fact scrapbook of recollections and miscellaneous jottings. It is impossible to reconstruct an accurate record from isolated numbers and memory; on the other hand, your experimental data cannot be recorded before they exist. It is necessary, therefore, to record the salient operations, measurements, and observations *as they are done or made,* insofar as possible. This procedure usually will not result in a flawless copybook record, but the notebook should be coherent and legible.

Each experiment should be dated (each day if protracted) and placed on a separate page or pages. It is a good idea to use only right-hand pages for the actual write-up; calculations and the like can be recorded on the facing page.

A common objection raised by students is "Why must I write down in a notebook all of the setups in a procedure that I am following in a lab manual?" The reason for doing this is to provide an orderly account into which you can incorporate your own data and observations. Moreover, it is a habit that must be acquired. In later experiments, you will be adapting a general procedure to your own situation; your own specific case is unique, and the record of what you do is vital.

An illustrative example of an actual notebook page that may serve as a guide to the type of record that should be kept in a typical experiment is shown in Figure 1.11.

QUESTIONS

1. Draw a sketch of the organic chemistry laboratory and the hallway outside, in which you clearly show the location of each of the following:
   a. the safety shower
   b. all fire extinguishers
   c. all fume hoods
   d. eyewash fountains
   e. your desk
2. Use the most recent *Handbook of Chemistry and Physics (CRC Handbook)* to find the following data:
   a. the refractive indices of acetone and hexane
   b. the boiling points of 1-propanol and 2-propanol
   c. the melting points of acetanilide and benzoic acid
   d. the densities of ethyl acetate and butyl acetate
3. Are organic vapors heavier or lighter than air?
4. Draw a simple reflux apparatus and identify each piece of glassware.
5. Calculate the percentage yield (page 12) for diethyl succinate if 10 g of succinic acid and 100 mL of ethanol were used to produce 8.7 g of diethyl succinate.

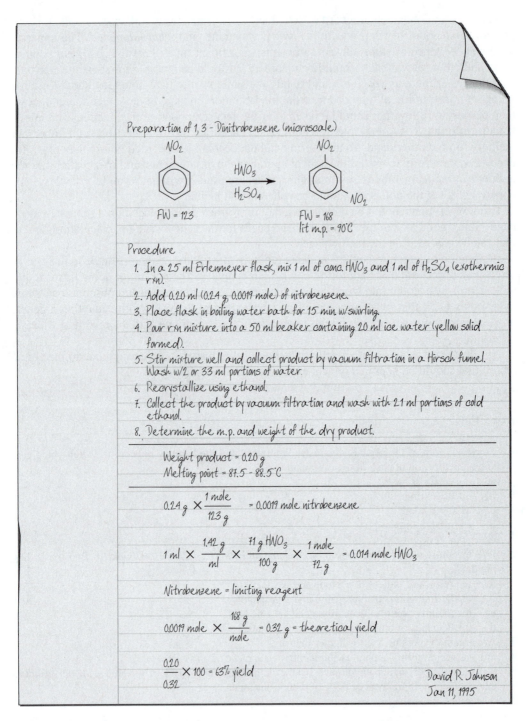

Preparation of 1, 3 - Dinitrobenzene (microscale)

$$NO_2 \quad \xrightarrow[H_2SO_4]{HNO_3} \quad NO_2 \quad NO_2$$

FW = 123   FW = 168
                  lit m.p. = 90°C

Procedure

1. In a 25 ml Erlenmeyer flask, mix 1 ml of conc. HNO_3 and 1 ml of H_2SO_4 (exothermic rxn).
2. Add 0.20 ml (0.24 g, 0.0019 mole) of nitrobenzene.
3. Place flask in boiling water bath for 15 min. w/swirling.
4. Pour rxn mixture into a 50 ml beaker containing 20 ml ice water (yellow solid formed).
5. Stir mixture well and collect product by vacuum filtration in a Hirsch funnel. Wash w/2 or 33 ml portions of water.
6. Recrystallize using ethanol.
7. Collect the product by vacuum filtration and wash with 2.1 ml portions of cold ethanol.
8. Determine the m.p. and weight of the dry product.

Weight product = 0.20 g
Melting point = 87.5 - 88.5°C

$$0.24\,g \times \frac{1\,mole}{123\,g} = 0.0019\,mole\;nitrobenzene$$

$$1\,ml \times \frac{1.42\,g}{ml} \times \frac{71\,g\;HNO_3}{100\,g} \times \frac{1\,mole}{72\,g} = 0.014\,mole\;HNO_3$$

Nitrobenzene = limiting reagent

$$0.0019\,mole \times \frac{168\,g}{mole} = 0.32\,g = theoretical\;yield$$

$$\frac{0.20}{0.32} \times 100 = 63\%\;yield$$

David R. Johnson
Jan 11, 1995

Figure 1.11   A typical notebook write-up.

REFERENCES

Laboratory Safety and Disposal of Waste

Armour, M. A. *Handbook of Hazardous Laboratory Chemicals;* CRC Press: Boca Raton, FL, 1991.

————. *Hazardous Laboratory Chemicals Disposal Guide;* 2nd ed.; CRC Press: Boca Raton, FL, 1996.

Documentation of the Threshold Limit Values for Substances in Workroom Air; American Conference of Governmental Industrial Hygienists: Cincinnati, OH, 1978.

Green, M. E.; Turk, A. *Safety in Work with Chemicals;* Macmillan: New York, 1978.

Handbook on Hazardous Materials Management, 4th ed.; Carson, H. T.; Cox, D. B., Eds.; Institute of Hazardous Materials Management: Rockville, MD, 1992.

Hazardous Waste Task Force. *Hazardous Waste Management at Educational Institutions;* National Association of College and University Business Officers: Washington, DC, 1987.

Hazards in the Chemical Laboratory, 4th ed.; Bretherick, L., Ed.; Royal Society of Chemistry: London (distributed by the American Chemical Society, Washington, DC), 1986.

Introduction to Safety in the Chemical Laboratory; Freeman, N.; Whitehead, J., Eds.; Academic Press: San Diego, 1983.

NIOSH Registry of Toxic Effects of Chemical Substances; Lewis, R. J., Sr., Ed.; National Institute for Occupational Safety and Health: Cincinnati, OH, 1978.

Lewis, R. J. *Sax's Dangerous Properties of Industrial Materials,* 9th ed.; Van Nostrand Reinhold: New York, 1994.

Lewis, R. J., Sr. *Rapid Guide to Hazardous Chemicals;* Van Nostrand Reinhold: New York, 1994.

Occupational Exposures to Hazardous Chemicals in Laboratories; Occupational Safety and Health Administration; *Federal Register,* January 31, 1990, vol. 55, pages 3300–3335.

Petersen, D. *The OSHA Compliance Manual,* rev. ed.; King Publications: Northbrook, IL, 1979.

Prudent Practices in the Laboratory: Handling and Disposing of Chemicals; National Academy Press: Washington, DC, 1995.

Safe Storage of Laboratory Chemicals; Pipitone, D. A., Ed.; John Wiley: New York, 1984.

Safety in Academic Chemistry Laboratories; American Chemical Society: Washington, DC, 1994.

Sax, N. I.; Lewis, R. J. *Carcinogenically Active Chemicals;* J. Wiley & Sons: New York, 1999.

————. *Hazardous Chemicals Desk Reference;* Van Nostrand Reinhold: New York, 1991.

Woodside, G. *Hazardous Material and Hazardous Waste Management;* J. Wiley & Sons: New York, 1999.

Laboratory Equipment and Techniques

Elvidge, J. A.; Sammes, P. G. *A Course in Modern Techniques of Organic Chemistry,* 2nd ed.; Butterworth: London, 1966.

Kanare, H. M. *Writing the Laboratory Notebook;* American Chemical Society: Washington, DC, 1985.

Lodwig, S. N. The Use of Solid Aluminum Heat Transfer Devices in Organic Chemistry Laboratory Instruction and Research; *J. Chem. Ed.,* 1989, *66,* 77.

Vogel, A. I., et al. *Practical Organic Chemistry,* 5th ed.; Longman/Wiley: New York, 1989.

Melting Points

The melting point of a solid is defined as the temperature at which the solid and liquid phases are in equilibrium. The time necessary to obtain such an equilibrium value is not practical for organic chemists; therefore, the melting-point range of temperatures between the first sign of melting and the complete melting of the solid is taken. A *narrow range* indicates high purity of the sample, whereas a *broad range* usually indicates an impure sample.

To determine a melting-point range, a small sample of the solid in close contact with a thermometer is heated in an oil bath or metal heating block so that the temperature rises at a slow, controlled rate. As the thermal energy imparted to the substance becomes sufficient to overcome the forces holding the crystals together, the substance melts. The rate of heating should be controlled so that the melting range is as narrow as possible. The temperature is recorded when the first melting appears and when the last solid disappears. A sharp melting point is generally accepted to have a range of 1 to 2°C. Impurities will usually cause the melting-point range to become wider and melting to occur at lower temperatures than that of a pure compound. A familiar example of this is the lowering of the melting point of ice by the addition of salt. This phenomenon results from the fact that both the liquid and solid are in equilibrium with the vapor. An impurity dissolved in the substance lowers the vapor pressure of the liquid, causing the solid to melt and restore the equilibrium among the three phases. The amount of lowering will depend on several factors, among which are the molal freezing point lowering constant (K_f), the concentration, and whether the solute is ionic or not. A typical curve for the lowering of the melting point of substance A by added amounts of substance B is shown in Figure 2.1.

The melting point of pure A is 120°C and that of B is 110°C. The upper curves connecting points *A* with *E* and *E* with *B* are the boundary above which mixtures of A and B of any composition are completely melted. The lower horizontal line through point *E* is a boundary representing the temperature (99°C) below which the sample is completely solid at any composition. In the areas between these phase boundaries are mixtures of one or the other pure solids plus liquid. A sample of composition M, for example, will show a melting-point range between the temperature at *M* (112°C) and that at line *E* (99°C). Point *E,* where the lines intersect, is called the **eutectic point.** At this composition (40% B) the mixture would melt sharply at 99°C.

The phase diagram in Figure 2.1 represents a simple or limiting case in which no solution of B in A or A in B occurs in the solid phase. More complex phase diagrams are possible, in which a solid solution does occur or in which there are two eutectic points with a maximum between them, corresponding to the formation of a complex or compound between the two components. Even in these cases, the melting point of the pure compound is generally lowered by the presence of a second component.

The number of exceptions to the rule is small, and chemists are generally safe in using melting point determinations as criteria for purity of known compounds. However, the use of melting points as criteria of purity for unknown substances must be tempered by the knowledge that exceptions do exist.

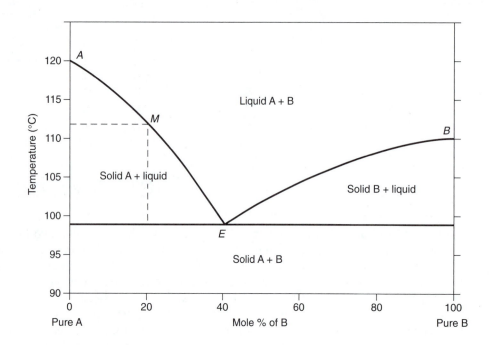

Figure 2.1   Solid-liquid phase diagram for a mixture of two compounds.

MELTING POINT BEHAVIOR

Many solids undergo some degree of decomposition or unusual behavior prior to melting. There may be changes in appearance of the sample, such as loss of luster or darkening, before the sample actually begins to melt. Some compounds melt with decomposition, as evidenced by bubbling or formation of a dark char. In such cases, the observed melting point (or decomposition point) frequently depends on the rate of heating.

Often compounds begin to soften, shrink, or appear moist before melting. These changes are not the beginning of melting and are sometimes referred to as sintering; the actual melting begins when the first drop of liquid is visible and is completed when the last solid disappears.

If a compound is appreciably volatile, determination of the melting point may be accompanied by sublimation, wherein the solid vaporizes and disappears before it melts. With a capillary tube, sublimation can be prevented by sealing the top of the tube after filling it. On an open block, the sample will seem to shrink into an increasingly small circle and may disappear before melting occurs. It may be possible to observe the melting point by using a larger than normal sample and placing it on the block at a temperature not too far below the expected melting point.

MIXTURE MELTING POINTS

In identifying an unknown, a useful practice is to take the melting point of the unknown and, at the same time, to take that of a mixture of the unknown and a known compound suspected to be the same as the unknown. If the known and unknown samples are the same substance, there will be no depression of the melting point; however, if they are different, a depression of the melting point will occur. To make sure that the mixture was not one of the exceptions mentioned above, another ratio of known to unknown should be melted.

To prepare a sample for determination of mixture melting point, place approximately equal amounts of the two compounds on a piece of glassine paper or a watch glass and mix them together. Then crush the crystals to a powder and grind them thoroughly with a spatula. When the pile of crystals is spread out to a thin layer, scrape the powder together and grind again. Fill the melting point capillary in the usual way and fill other capillaries with samples of the two individual compounds that have been ground to the same degree of fineness. Place all three samples together, with the mixture in the middle, in the bath or block and observe the melting points simultaneously. The temperature can be raised rapidly to about 20° below the expected melting point, then more slowly.

APPARATUS

A thermometer and a means of heating the sample in close contact with the thermometer bulb at a steady, controlled rate are required for the determination of the melting point. In the most commonly used method, the sample is enclosed in a glass capillary tube that is sealed at one end. Commercially available 2-mm capillaries are somewhat thicker than necessary but are satisfactory for most purposes. Thinner-walled tubes of smaller diameter, which permit better heat transfer, can be pulled from a soft-glass disposable pipet.

The capillary tube containing the sample is heated either in a liquid bath or on a metal block. The bath is usually in the form of a vertical tube with a side loop to provide convection (Thiele–Dennis tube), shown in Figure 2.2. A 150-mL round-bottom flask also can be used. The bath contains a high-boiling mineral oil or silicone fluid that can be heated to 250°C or higher for brief periods without decomposition. The capillary is held on the thermometer with a small ring sliced from a piece of rubber tubing, with the sample as close to the thermometer bulb as possible. Thermometer and capillary are clamped in the bath or held in place by a cork with a wedge cut out to permit the thermometer to be seen and to prevent pressure from building up. The bulb and sample should be just below the upper arm of the loop. The bath is heated on the loop with a microburner; a current of heated fluid circulates upward in the loop and down the tube to provide mixing and an evenly rising temperature.

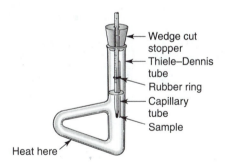

Figure 2.2   Thiele–Dennis tube melting-point bath.

A typical melting-point block for use with capillary tubes is illustrated in Figure 2.3. The block is heated by electrical resistance, with a variable voltage control to regulate the rate of heating. A slot in the front of the block can hold three capillary tubes adjacent to the thermometer. The samples are illuminated and can be observed through a magnifying lens.

In another method for determination of melting point, the sample is sandwiched between thin cover glasses, which are then placed directly on an electrically heated block (Fig. 2.4). The

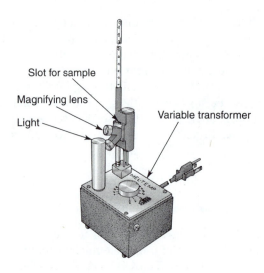

Figure 2.3   Mel-Temp capillary melting-point apparatus. (Courtesy of Laboratory Devices, Holliston, MA)

sample is illuminated and viewed through a lens above the block. A smaller sample (only a few tiny crystals) can be used with this type of block than with a capillary tube.

Either of the melting-point blocks shown in Figure 2.3 or 2.4 can be used for temperatures up to 300°C. A block is preferable to a liquid bath for any melting point above 250°C.

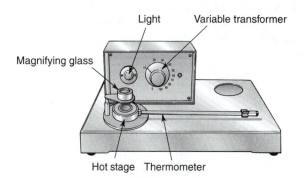

Figure 2.4   Fisher–Johns melting-point block. (Courtesy of Fisher Scientific Company)

EXPERIMENTS

A. MELTING POINT OF A PURE KNOWN COMPOUND

Fill a melting-point capillary tube (2 × 90 mm) with a small portion of sample (a column approximately 5 mm long is ample). This can be accomplished most readily by pressing the open end of the capillary into a small mound of the sample, inverting the capillary so that the closed end is down, and tapping it gently on the bench top or by drawing the flat side of a triangular file across the upper end of the tube. Another method involves dropping it and its contents through a piece of 8- to 10-mm glass tubing 100 cm long onto a hard surface. The tube will bounce off the surface and, in the process, will pack the sample uniformly. Repeat this packing several times, and note the number of times for future reference. All subsequent packing should be done in exactly the same manner. Insert the capillary into the melting point apparatus, and apply heat at a rate that will cause the temperature to rise about 10°C per minute. Watch the sample carefully for any sign of melting. Record the temperatures when melting first occurs and when the sample is all liquid; this is the melting-point range. Repeat the procedure with a fresh sample and tube; however, this time when the temperature of the sample reaches 15°C below the first temperature you previously recorded, slow the heating rate to about 1 or 2°C per minute. This slower heating rate is important to allow time for a better equilibrium to be established among sample, heating source, and thermometer. Record the temperatures as before. Duplicate results should agree within 1 to 2°C; the range should also be within a 1 or 2°C span. If you do not obtain results within these parameters, check with your instructor.

B.   MELTING POINTS AS A THERMOMETER CALIBRATION

The thermometer used for melting points should be calibrated by determining the melting points of three or four standards listed in Table 2.1 that melt near 50°, 100°, 150°, and 200°C. Determine the melting points as accurately as possible (within 1 degree) and plot the literature value (x-axis) *vs.* experimental value (y-axis). The resulting graph will allow you to correct any subsequent temperature value that you determine in this course. The same thermometer should be used for all melting-point determinations.

Table 2.1   Melting-Point Standards

COMPOUND	MP (°C)	COMPOUND	MP (°C)
p-Dichlorobenzene	53	Salicylic acid	159
m-Dinitrobenzene	90	Succinic acid	188
Acetanilide	114	3,5-Dinitrobenzoic acid	205
Benzamide	130	*p*-Nitrobenzoic acid	241

C.   MELTING POINT OF AN UNKNOWN COMPOUND

Obtain an unknown from your instructor and determine its melting-point range. The unknown will be one of the substances in Table 2.2. Determine the melting points of mixtures of your unknown with known samples of comparable melting-point ranges and use this information to identify the unknown. You should strive to have results in agreement within 1 or 2°C. One determination of a melting point is usually insufficient evidence for a positive identification. Before you report your results you should run duplicate samples that agree. Capillary samples

Table 2.2   Compounds for Melting-Point Unknowns

COMPOUND	MP (°C)	COMPOUND	MP (°C)
Benzhydrol	68	*o*-Anisic acid	100
Phenyl benzoate	69	Phenanthrene	101
Biphenyl	70	*o*-Toluic acid	105
Phenylacetic acid	78	Acetanilide	114
Naphthalene	80	Fluorene	114
Vanillin	81	Cholesteryl acetate	116

should not be used more than once because decomposition may occur during the heating process; always use a fresh sample.

QUESTIONS

1. What effect would each of the following have on the observed melting-point range of a sample?
    a. filling the melting-point capillary two-thirds full.
    b. not packing the sample tightly in the capillary.
    c. the presence of an insoluble impurity such as glass in the sample.
2. Why should the heating rate during melting-point determinations be as slow as 1 or 2°C per minute?
3. Suppose your sample melts before you were ready to read the temperature. Should you cool the capillary and start again or prepare a fresh sample? Explain.
4. Suppose you have an unknown X that melts at 85–86°C and you do a mixture melting point with a known compound A that melts at 85-86°C. What results would be observed if:
    a. X is identical with A.
    b. X is not identical with A.

PRELABORATORY QUESTIONS

1. Why is a melting-point range determined rather than the actual melting point of a compound?
2. What does a broad melting-point range indicate about the purity of a sample?
3. Gram for gram, table salt lowers the freezing point of water much more than table sugar. Explain.
4. If a volatile compound sublimes and disappears before it melts, how can its melting point range be determined?
5. If a compound is pure, how narrow should its melting point range be expected to be?

REFERENCE

Physical Methods of Chemistry; Weissberger, A., Rossiter, B., Eds.; Wiley-Interscience: New York, 1971; Vol. I, Part V.

Recrystallization

Of all the techniques studied in the organic chemistry laboratory, the purification of a solid by recrystallization from a solvent is the one most widely employed and at the same time the most universally misused. Students should master this skill early in their laboratory experience so that they will not be hampered by it when they should be concentrating on other aspects of subsequent experiments.

In the process of recrystallization, an impure sample is dissolved in an appropriate solvent that will differentiate between the sample and the impurities. Insoluble impurities may be removed by filtration, whereas the more soluble impurities remain in solution when the purified sample crystallizes. The exact process will depend upon the following criteria:

1. The nature of the differences in structure of the sample and the impurities (if known).
2. Differences in solubility of the sample and impurities in various hot and cold solvent systems. For ease of operation, the sample should be more soluble in the hot solvent than in the cold solvent; the impurities should be either highly soluble or extremely insoluble.
3. The solvent should be readily volatilized after collection of the purified sample crystals. The solvent should also be unreactive towards the sample.
4. The solvent may be a pure substance, such as diethyl ether or ethyl alcohol, or it may be a mixture of two or more substances that give the desired solubility characteristics. "Ready-mixed" solvents may be available in the laboratory, or the proper solvents can be mixed as needed by adding one of the solvents to the other (in portions) until the desired characteristics are achieved.

For crystallization to occur, the concentration of a dissolved compound must exceed the equilibrium solubility; that is, the solution must be supersaturated at the given temperature. With few exceptions, the solubility of a solid increases with an increase in temperature, often manyfold over a 30° to 40°C temperature increase. Thus, a sample is dissolved in a minimum of hot solvent, and crystallization occurs on cooling. For a proper solvent system, the sample's solubility should increase rapidly as the melting point is approached. Some typical temperature versus solubility curves are shown in Figure 3.1.

The solubility of crystalline organic compounds in a solvent depends on two major factors: (a) the relative polarity of the solvent and solute and (b) the energy of the crystal lattice of the solute. The effect of polarity can be summed up by the statement "like dissolves like." Thus, compounds that contain one or more polar groups, such as $-OH$, $-NH_2$, $-CO_2H$, or $-CONH_2$, are usually more soluble in hydroxylic solvents (e.g., water or alcohols) than in hydrocarbons, such as toluene or hexane. Conversely, the latter more easily dissolve compounds of low polarity. Within a series of compounds of the same type, the usual relationship is that the higher the melting point (high crystal lattice energy), the lower the solubility in a given solvent. This can be seen in a comparison of the melting points and solubilities of the isomeric

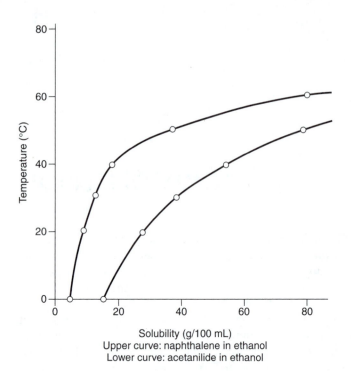

Figure 3.1 Typical temperature-solubility curves.

Solubility (g/100 mL)
Upper curve: naphthalene in ethanol
Lower curve: acetanilide in ethanol

nitrobenzoic acids (Table 3.1). The *para-* isomer, having a highly symmetrical structure, can pack into a more stable lattice (strong attraction) than can the others and consequently has a much higher melting point and corresponding lower solubility.

In selecting a solvent for the recrystallization process, the main requirement is suitable solvating power, that is, one that provides high solute solubility in the hot solvent and a significantly lower solubility at room temperature or below. Frequently, this requirement is met by using a mixture of two solvents in which the solute has quite different solubilities. Solvents that are generally useful for crystallization are listed in Table 3.2 in order of decreasing polarity, together with the solubilities of two representative compounds.

Table 3.1   Melting Points and Solubilities of Nitrobenzoic Acids

	ISOMER		
	ortho-	*meta-*	*para-*
MP (°C)	147	141	242
Solubility, g/100 mL at 10 to 12°C			
Methanol	43	47	9.6
Ethanol	28	33	2.2
Ether	21	25	0.9

Water is a convenient solvent for recrystallizing some moderately polar compounds with melting points above 100°C. Mixtures of water and alcohols are widely used. The low boiling points of diethyl ether and methylene chloride (dichloromethane) make them attractive for recrystallizing low-melting solids. Mixtures of low-boiling alkanes (petroleum ether, bp 35 to 50°C, and ligroin, bp 130 to 145°C) have wide utility and are commercially available. These generalizations are illustrated with the two compounds mentioned in Table 3.2. Acetanilide is much too soluble in methanol and ethanol to permit crystallization from these solvents. The solubility in water is much lower, and since the solubility at 100°C is about ten times higher than at 25°C, water is a suitable solvent, although a rather large volume is needed. On the other hand, naphthalene is very soluble in hot ethanol, but the solubility drops to 5% at 0°C. Thus, ethanol is a practical solvent; adding a small amount of water would probably be even better.

The general procedure for recrystallization involves the following steps:

1. Selection of a suitable solvent through experiment or from data on solubility.
2. Dissolution of the material in the hot solvent (near the boiling point).
3. Filtration of the hot solution to remove insoluble impurities or impurities adsorbed on activated carbon. (This step is sometimes omitted—see below.)
4. Crystallization of the solute from the cool solution.
5. Collection of the purified crystals.
6. Washing and drying the product.

A polar solvent system (e.g., ethanol/water) will retain impurities that are more polar than the substance being purified. Conversely, a nonpolar solvent (e.g., hexane) will retain impurities that are less polar than the substance being purified. Thus, one can often most efficiently purify a crystalline substance by alternating solvent systems.

Table 3.2 Useful Recrystallization Solvents

SOLVENT	BOILING POINT (°C)	SOLUBILITY AT 25°C, g/100 mL SOLUTION	
		Acetanilide (Polar)	Naphthalene (Nonpolar)
Water	100	0.53	0.002
Methanol	65	48	9.9
Ethanol	78	30	11.8
Methylene chloride (CH$_2$Cl$_2$)	40	(17)*	(55)*
Diethyl ether	35	2.8	57
Hexane	68	0.03	20
Pentane	36	<0.01	

* These values are for chloroform (CHCl$_3$), which has similar solvent properties but is more toxic than CH$_2$Cl$_2$.

FILTRATION

Insoluble matter is removed from the hot solution by gravity filtration (Step 3). This step may be omitted if the solution is clear and obviously free of insoluble material. Hot filtration must be carried out with precaution to avoid premature crystallization in the funnel or on the filter paper because of the cooling of the solution. Thus, a conical funnel with a short stem or no stem should be used. The funnel is placed in a beaker or flask containing a small amount of the boiling solvent (Fig. 3.2). This arrangement allows the funnel to be bathed in the hot solvent vapors during the process. The solution to be filtered should be maintained close to the boiling temperature, and small amounts of additional solvent should be added as necessary. Rapid filtration can be achieved using fluted filter paper, which provides a maximum amount of surface area for the solvent to pass through. Details for proper folding to flute the filter paper are shown in Figure 3.3.

Note that the above procedure works well on samples as small as 100 mg. Smaller amounts should be handled by using a cotton plug or specially adapted pipet. A small cotton wad the size of a pea stuffed into the funnel stem very often results in fast filtration and eliminates the losses on the filter paper. This method is particularly convenient for microscale procedures to filter a few milliliters of solution or less. A disposable (Pasteur) pipet may serve as the funnel

Figure 3.2   Filtration of a hot solution.

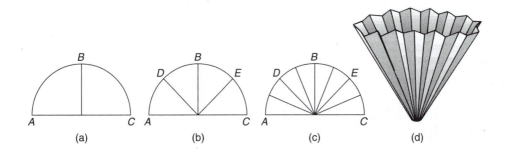

Figure 3.3   Folding fluted filter paper.
On a hard surface, fold a piece of 11- or 15-cm circular filter paper in half and then in quarters. Open the second fold like a book **(a),** and fold corners A and C to meet point B. Reopen it to a semicircle **(b)** and fold corner A first to point D and then to point E. Similarly, fold corner C to D and E, reopening it to the semicircle after each fold. The paper should now look like **(c),** with the folds bending the semicircle into a partially open cone. In each of the eight segments, make a fold in the center, alternating with and opposite to the direction of the previous folds. This accordionlike arrangement will open into a fluted funnel **(d).**

(Fig. 3.4). With the tip of a second pipet, a very small pellet of *loose* cotton is pushed through the top and wedged gently in the constriction in the pipet. Snip off the narrow part of the pipet to a length of about 2 cm and transfer the solution to be filtered through the top with another pipet. (Fig. 3.4b)

Another useful adaptation of the Pasteur pipet is the filter tip pipet, which can be used to remove solid from small amounts of liquid. A very small piece of cotton is rolled into a ball and placed in the large end of a Pasteur pipet; it is then pushed down to the narrow tip of the pipet with a piece of wire (see Fig. 3.4c). Some practice is needed not to get the cotton plug too loose or too tight. This filter tip pipet also helps to prevent "squirting" when volatile organic liquids are transferred.

A filter aid such as Celite is used to remove very small particles that otherwise would pass through normal filters. The filter aid retains the fine particles more effectively than the paper alone. The procedure involves the laying down of a thin layer (about 0.5 cm) of filter aid over the filter paper as follows: A piece of filter paper is placed in a Büchner or Hirsch funnel attached to a vacuum flask. Solvent is added until the funnel is approximately half filled and sufficient dry filter aid is added. The mixture is stirred until the filter aid is evenly suspended and suction is applied slowly. When the solvent has drained, the suction is broken and the flask removed, cleaned, and reattached to the funnel. The funnel is now ready for use. If the filtration becomes slow because of the buildup of a coating on the surface of the Celite, fresh surface can be exposed by gently scraping with a spatula.

Highly colored contaminants may be removed from solution by adding activated carbon (Norit, Darco, or other trade-name samples of finely divided carbon particles, or an easily filtered granulated form, which works especially well with microscale experiments) to the warm (not hot) solution, heating the solution to boiling and filtering while hot. Polar substances with chromophores that are usually responsible for imparting a color to the solution are strongly adsorbed on the surface of the carbon. The use of carbon is by no means a panacea for removing impurities; sometimes it is ineffective and can reduce the yield of product because of adsorption. If the compound to be purified is known to be colorless but the solution is dark, it will often be worthwhile to try the effect of carbon on a small portion. Activated carbon is usually most effective in polar solvents, such as water or alcohol.

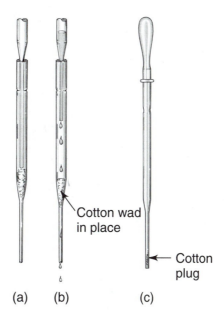

Figure 3.4   **(a)** Cotton stuffed in pipet to be used as funnel. **(b)** Filtration in progress. **(c)** Filter tip pipet.

CRYSTALLIZATION

In the optimum case, crystallization occurs spontaneously when the solution is cooled to room or ice-bath temperature; however, in some cases, crystallization has to be induced. Sometimes the volume of solvent is too large and the solution must be concentrated. Crystal growth often can be initiated by scratching the inside wall of the flask with a glass rod or spatula or by removing a few drops of the solution and rubbing it with a spatula on a watch glass as the solvent evaporates. Crystals obtained in this way can then be used as a "seed" to induce crystal growth in the main solution.

The size of crystals depends on the rate of formation and the number of particles present. Rapid cooling and shaking or stirring the solution lead to the rapid formation of small crystals. This is normally desirable, since it helps to eliminate solvent molecules entrapped in the crystal lattice. In delicate selective crystallization, it may be crucial to allow crystals to form slowly and without agitation.

COLLECTION AND WASHING OF CRYSTALS

When a crystallization is complete, the solid is scraped loose from the wall of the flask and collected in a porcelain Büchner or Hirsch funnel using vacuum (Figs. 3.5 and 3.6). If a small volume of liquid is to be filtered and retained, a test tube can be placed inside the filter flask or side-arm test tube and the stem of the funnel directed into the tube (see Fig. 3.6). These porcelain funnels are fitted with a circle of filter paper that should lie flat and cover the perforations, which permits liquid to drain rapidly into the evacuated flask. The paper is wet with a few drops of solvent and the flask is connected to suction. This will seat the filter paper firmly on the funnel.

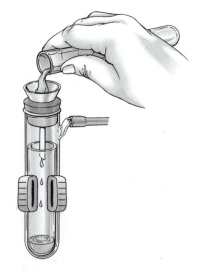

Figure 3.5   Büchner funnel and suction flask.

Figure 3.6   Hirsch funnel and side-arm test tube.

After transferring the mixture to the funnel, any solid clinging to the wall of the flask is scraped loose and rinsed into the funnel with a small portion of the filtrate (mother liquor) or cold solvent. In microscale crystallizations in a test tube, crystal growth may occur entirely on the walls, thereby permitting the mother liquor to be decanted cleanly without the necessity of transferring the solid. Alternately, crystallization can be performed in a centrifuge tube, and the crystals centrifuged to the bottom of the tube for easy removal of the supernatant mother liquor with a pipet. Normally, the crystals are washed free of the mother liquor using small portions of chilled solvent. This insures the removal of impurities that might be carried along in the mother liquor. If a more thorough washing to remove oily mother liquor is needed, the suction should be broken and the crystals stirred well on the funnel before again applying suction and allowing the liquid to drain through. If a solvent with a high boiling point has been used, a final rinse with a more volatile solvent, such as ether or pentane, will facilitate rapid drying of the crystals.

The crystal mass may be compacted with a cork or spatula to hasten the removal of solvent, and it is often possible to remove traces of solvent by drawing air through the funnel using suction. The crystals are transferred to a watch glass, filter paper, or glassine paper for further drying in air or in an oven with the temperature well below the melting point of the sample.

EXPERIMENTS

A.   RECRYSTALLIZATION OF BENZOIC ACID

(Microscale)
Place 60 mg of benzoic acid and approximately 2 mL of water in a 10-mL Erlenmeyer flask, add one boiling stone, and heat to a gentle boil (simmer) on a hot plate; continue to add boiling water in small increments until all of the benzoic acid is dissolved (be patient; organic compounds often do not dissolve instantaneously). Record the total amount of water used. Remove the flask from the heat and allow it to cool to room temperature. If crystals do not appear, induce their formation by scratching the inside of the flask with a glass rod or spatula or chill in an ice bath. Filter the crystals with suction using a Hirsch funnel, side-arm test tube, and aspirator (see Fig. 3.6). Allow the sample to air dry, and weigh and record the amount of benzoic acid recovered. Calculate the percent recovery. The aqueous filtrate may be diluted with water and flushed down the drain.

B.   CRYSTALLIZATION OF IMPURE ACETANILIDE

(Macroscale)
Weigh 2.0 g of impure acetanilide into a 100-mL Erlenmeyer flask and add approximately 20 mL of water and a boiling stone (used to prevent superheating of the solvent that might result in a violent eruption of the solution). Place the flask on a hot plate and slowly heat the contents to boiling. Stir the contents occasionally, and when the solvent starts to boil, add small portions of water (about 5 mL each) until you cannot detect any further disappearance of the solute. Keep the solution boiling; you may add as much as a total of 50 mL. *Note:* Take into account that if the impure sample contained insoluble impurities you would never get a complete solution. Cool the solution slightly and add a small amount of activated carbon to remove any highly colored impurities. (*Note:* Do not add the carbon to the hot solution; it may boil over when the carbon is added.) Reheat to boiling and filter the solution into a second Erlenmeyer flask through a double layer of fluted filter paper in a funnel heated with boiling solvent (see Fig. 3.2). If the mixture is still gray or highly colored, you may have to repeat the filtration process after reheating the solution. Set the flask aside to cool. When it has cooled to room temperature, chill it in an ice bath to induce crystallization. When crystallization appears to be complete, filter the mixture using suction. Consult your instructor

on disposal of the filtrate. Partially dry the crystals on the filter by drawing air through the vacuum system for several minutes, place them on a large watch glass or glassine or filter paper, and allow them to dry in air for approximately an hour. Determine the melting-point range; when the correct melting point is reached (an indication that the sample is dry), the crystals can be weighed and stored in a properly labeled sample vial. Calculate the percentage recovery and submit the sample and required data to your instructor.

C.   PREPARATION OF ACETANILIDE

(Microscale)

$$NH_2 \quad + \quad (CH_3CO)_2O \quad \longrightarrow \quad NHCOCH_3 \quad + \quad CH_3CO_2H$$

aniline	acetic	acetanilide	acetic acid
C_6H_7N	anhydride	C_8H_9NO	
	$C_4H_6O_3$		
	(d. 1.08 g/mL)		

[3.1]

In the late 1800s, acetanilide was introduced as an antipyretic under the name antifebrine and was widely used for this purpose and as an analgesic. It has now been replaced for these purposes by such compounds as aspirin and acetaminophen. Acetanilide is an intermediate in the production of sulfa drugs and dyes. In this experiment, aniline is acetylated and the product, acetanilide, recrystallized from water.

SAFETY NOTE

Aniline is a very toxic compound; take care to avoid contact with your skin. Acetic anhydride is irritating to the eyes and nasal membranes. If either compound is spilled, wash the affected area thoroughly with soap and cold water. If either is splashed into the eyes, treat it as a genuine emergency and seek a physician's help immediately.

PROCEDURE

Using a medicine dropper, place 0.100 to 0.125 g of aniline (about 4 drops) in a tared 13 × 100-mm test tube and determine the weight to the nearest milligram. Add 2 mL of water to the test tube and then 6–7 drops of acetic anhydride, again using a medicine dropper; stir 3–4 minutes until solid forms.

The product is recrystallized in the same test tube. Add 5 mL of water and heat the test tube in a hot water bath with occasional stirring until all the solid has dissolved. Set the test tube aside to cool for 3–5 minutes and then chill it in an ice bath. When crystallization is complete, collect the product by vacuum filtration using a small Hirsch funnel (see Fig. 3.6). Allow the sample to air dry until the next class period. Weigh the dry product, determine its melting point, and calculate the percent yield. Transfer the product to a labeled vial and submit it for grading. The aqueous filtrate may be flushed down the drain.

D.   PREPARATION OF N-ACETYLANTHRANILIC ACID: A TRIBOLUMINESCENT MATERIAL

(Macroscale)

Anthranilic acid (2-aminobenzoic acid) may be acetylated with acetic anhydride to give N-acetylanthranilic acid. In the hot acetic acid-acetic anhydride environment, the compound loses water to form 2-methylbenzisoxazinone. Addition of water and brief heating hydrolyzes this compound back to N-acetylanthranilic acid (2-acetamidobenzoic acid).

The product crystals, once isolated and dried, should exhibit the property of triboluminescence. It is left to you to find out what this property is and to suggest what may cause it. Record this information in your laboratory notebook write-up.

SAFETY NOTE

Acetic anhydride is irritating to the eyes and nasal membranes. If it is spilled on the skin, wash the affected area thoroughly with soap and cold water. Particular care should be taken in the handling of hot solutions of acetic anhydride and acetic acid.

PROCEDURE

Place 2.00 g of anthranilic acid (MW = 137 g/mole) in a 25-mL Erlenmeyer flask along with 6 mL of acetic anhydride; the mixture may solidify. Add a carborundum boiling chip and, in a hood, warm the mixture to a gentle boil on a hot plate for 15 minutes. Allow the reaction mixture to cool, then add 2 mL of distilled water. Again warm the mixture just to boiling, then allow it to cool very slowly to room temperature. Crystals of the product, N-acetylanthranilic acid (MW = 179 g/mole), should form during this cooling period. Generally, the slower the cooling, the better the crystal growth.

Isolate the crystals by vacuum filtration, wash them with a small amount of chilled ethanol, and suck them thoroughly dry. Determine the weight and melting point of the product (lit. mp 185°C) and calculate the percent yield. The filtrate should be neutralized with $NaHCO_3$, diluted with water, and flushed down the drain.

In a dark room, crush crystals of the product between two watch glasses, looking for evidence of triboluminescence. Give your eyes time to adjust to the darkness to better observe the effect. Record your observations. Submit the product to your instructor in an appropriately labeled container.

QUESTIONS

1. During recrystallization, an orange solution of a compound in hot alcohol was treated with activated carbon and then filtered through fluted paper. On cooling, the filtrate gave gray crystals, although the compound was reported to be colorless. Explain why the crystals were gray and describe steps that you would take to obtain a colorless product.
2. The solubility of compound A in ethanol is 0.8 g per 100 mL at 0°C and 5.0 g per 100 mL at 78°C. What is the minimum amount of ethanol needed to recrystallize a 12.0-g sample of compound A? How much would be lost in the recrystallization, that is, would remain in the cold solvent?
3. Describe how you would purify *p*-nitrobenzoic acid from a mixture of 9.0 g of *p*-nitrobenzoic acid and 2.0 g of *o*-nitrobenzoic acid using recrystallization (see Table 3.1). Tell what solvent you would use and how the components would be separated.
4. Assume that 3.0 g of aniline and 4.5 mL of acetic anhydride are used in the preparation of acetanilide (Eq. 3.1). What is the limiting reagent? What is the theoretical yield of acetanilide? What is the percent yield if 3.3 g of acetanilide is obtained?
5. Compound B is quite soluble in toluene, but only slightly soluble in petroleum ether. How could these two solvents be used in combination in order to recrystallize Compound B?

PRELABORATORY QUESTIONS

1. Describe the requirements for a "good" recrystallizing solvent.
2. How are insoluble impurities removed in a recrystallization?
3. What is the purpose of adding activated carbon to the warm solution during a recrystallization?
4. In most cases, how does the solubility of a solid in a solvent change with increasing temperature?
5. What are boiling stones? Why are they used?

REFERENCES

Erickson, J. *J. Chem. Ed.,* 1972, *49,* 688.

Tipson, R. S. In *Techniques of Organic Chemistry,* 2nd ed.; Weissberger, A., Ed.; Interscience: New York, 1956; Vol. III, Part I, Chapter 3.

Extraction

O ne of the main concerns of experimental organic chemistry is the separation of mixtures and the isolation of compounds in as pure a form as needed for subsequent use. Crystallization is a useful method for purification and isolation, but it is restricted to solids, and as indicated in Chapter 3, it is a relatively inefficient way to separate a mixture of very similar compounds.

Several more general separation methods are described in this and the following chapters. All depend in some way on the *partitioning* of the compounds to be separated *between two distinct phases.* By choosing the phases so that the different compounds are *unequally distributed* between them, **fractional separation** of the compounds is effected. The various methods are mechanically quite different, but they all depend on this common principle, and it should be thoroughly understood. The principle is most easily illustrated by the general method of extraction, and this method is therefore covered in some detail in this chapter.

THEORY OF EXTRACTION

Extraction is the general term for the recovery of a substance from a mixture by bringing it into contact with a solvent that preferentially dissolves the desired material. The initial mixture may be a solid or liquid, and various techniques and apparatus are required for different situations. After an organic reaction, the reaction product is frequently obtained as a solution or a suspension in water along with inorganic and other organic by-products and reagents. By shaking the aqueous mixture with a water-immiscible organic solvent, the product is transferred to the solvent layer and may be recovered from it by evaporation of the organic solvent.

The extraction of a compound from one liquid phase into another is an equilibrium process governed by the solubilities of the substance in the two solvents. The ratio of the solubilities is called the *distribution coefficient,* $K_d = C_1/C_2$, and is an equilibrium constant with a characteristic value for any compound and pair of solvents at a given temperature.

Let us consider the following situation: A 100-mL aqueous solution contains 1 g each of compounds A and B, the solubilities of which in water and ether are given below.

COMPOUND	SOLUBILITY IN WATER C_{Water}	SOLUBILITY IN ETHER C_{Ether}	$K_{Ether/Water}$
A	10 g/100 mL	1 g/100 mL	$\frac{1}{10} = 0.1$
B	2 g/100 mL	10 g/100 mL	$\frac{10}{2} = 5.0$

By the foregoing definition, the distribution coefficients of A and B (C_{ether}/C_{water}) are 0.1 and 5, respectively. If the aqueous solution is shaken with 100 mL of ether, the amount of each compound transferred to the ether phase, X, can be calculated as follows:

<table>
<tr><td align="center">*Compound A*</td><td align="center">*Compound B*</td></tr>
<tr>
<td align="center">$\dfrac{C_{ether}}{C_{water}} = \dfrac{X_A/100}{(1 - X_A)/100} = 0.1$</td>
<td align="center">$\dfrac{C_{ether}}{C_{water}} = \dfrac{X_B/100}{(1 - X_B)/100} = 5$</td>
</tr>
<tr>
<td align="center">$X_A = 0.091$ g in ether
$1 - X_A = 0.909$ g in water</td>
<td align="center">$X_B = 0.833$ g in ether
$1 - X_B = 0.167$ g in water</td>
</tr>
</table>

If the same aqueous solution is extracted with the same amount of ether, but in four 25-mL portions, we have for the first extraction:

$$\frac{X_A/25}{(1 - X_A)/100} = 0.1 \qquad\qquad \frac{X_B/25}{(1 - X_B)/100} = 5$$

(1) $X_A = 0.0244$ g in ether, $X_B = 0.556$ g in ether
       leaving 0.9756 g in water   and 0.444 g in water

For the second 25-mL extraction:

$$\frac{X_A/25}{(0.9756 - X_A)/100} = 0.1 \qquad\qquad \frac{X_B/25}{(0.444 - X_B)/100} = 5$$

(2) $X_A = 0.0238$ g in ether, $X_B = 0.247$ g in ether
       leaving 0.9518 g in water   and 0.197 g in water

After the third and fourth extractions:

(3) $X_A = 0.0232$ g in ether $X_B = 0.109$ g in ether
(4) $X_A = 0.0226$ g in ether $X_B = 0.049$ g in ether

Totals:

$$X_A = 0.094 \text{ g A in ether} \qquad\qquad X_B = 0.961 \text{ g B in ether}$$
$$1 - X_A = 0.906 \text{ g A in water} \qquad 1 - X_B = 0.039 \text{ g B in water}$$

It can be seen from these values that even with a relatively small distribution coefficient ($K_d = 5$) virtually complete extraction of compound B can be effected and that several extractions with small volumes of extractant are more efficient than a single extraction with the same total volume in one portion.

In the extractions described above, a significant amount of compound A was also transferred to the ether layers. If the ether were removed from the solution at this point, the residue would be 1.055 g of material that is only 92% pure B. However, if the ether is shaken (back-extracted) with 50 mL of water before evaporation, the amounts (Y) of A and B removed from the ether solution can be calculated as follows:

$$\frac{C_{ether}}{C_{water}} = \frac{(0.094 - Y_A)/100}{Y_A/50} = 0.1 \qquad\qquad \frac{(0.961 - Y_B)/100}{Y_B/50} = 5$$

$Y_A = 0.078$ g in water, $Y_B = 0.087$ g in water
leaving 0.016 g in ether   and 0.874 g in ether

Evaporation of the ether now leaves a residue of 0.890 g that is more than 98% pure B. Some B was lost in this process, to be recovered by subsequent extractions, but that which remains may be pure enough for its intended use.

APPLICATIONS OF EXTRACTION

LIQUID–LIQUID EXTRACTION

As described in the preceding section, extraction of one liquid with another is a standard operation that is used for a number of purposes. Simple extractions on a macroscale are carried out in a separatory funnel, which is a conical vessel with a stopcock at the bottom and a stopper at the top (Fig. 4.1). The two layers are mixed by shaking to permit transfer from one layer to another; after they separate, the lower layer is drained out through the stopcock.

In the most common applications of liquid–liquid extraction, such as the recovery of an organic compound from an aqueous solution containing inorganic acids, bases, or salts, the distribution coefficient is very favorable. In this case, one or two extractions with an organic solvent, followed by the back-extraction or "washing" of the solvent layer with water, is often sufficient. If the procedure does not specify the number of extractions, it is normally understood by organic chemists to be three, followed by one back-extraction.

Figure 4.1   Separatory funnel.

If the distribution coefficient is unfavorable, that is, if the compound has a high solubility in water, sufficient extraction with many small portions of solvent becomes impractical. For such extractions, an apparatus such as that shown in Figure 4.2 is used. The extracting solvent is continuously distilled into the condenser, from which the condensed solvent returns in a stream that passes down through the solution. The extracted material collects and becomes concentrated in the flask containing the boiling extraction solvent.

A common and important use of extraction is in the separation of acidic, basic, and neutral organic compounds. An organic acid, RCO_2H, or a base, such as RNH_2, is usually much more soluble in organic solvents than in water. However, the salts of these compounds, for example, $RCO_2^- Na^+$ or $RNH_3^+Cl^-$, have much higher solubilities in water, since they are ionic substances. To separate an acid and a neutral compound, for example, the mixture is dissolved in ether and extracted with an aqueous solution of a base such as sodium hydroxide or sodium bicarbonate. The acid is converted to the salt (Eq. 4.1) and is extracted into the water layer. After separation of the water layer, the acid is recovered by reacidification of the aqueous solution (Eq. 4.2).

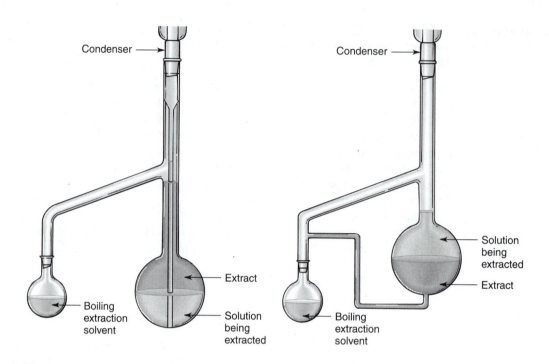

Figure 4.2   Continuous liquid–liquid extraction apparatus. *Left,* lighter-than-water solvent. *Right,* heavier-than-water solvent.

$$RCO_2H + NaHCO_3 \rightarrow RCO_2^-Na^+ + CO_2 + H_2O \qquad (4.1)$$

$$RCO_2^-Na^+ + HCl \rightarrow RCO_2H + Na^+Cl^- \qquad (4.2)$$

By the same principle, a base can be separated from neutral compounds by extraction with aqueous acid (Eq. 4.3). After separation of the aqueous solution of the salt, the addition of NaOH yields the organic base (Eq. 4.4).

$$RNH_2 + HCl \rightarrow RNH_3^+Cl^- \qquad (4.3)$$

$$RNH_3^+Cl^- + NaOH \rightarrow RNH_2 + H_2O + NaCl \qquad (4.4)$$

The separation scheme is outlined in the following diagram.

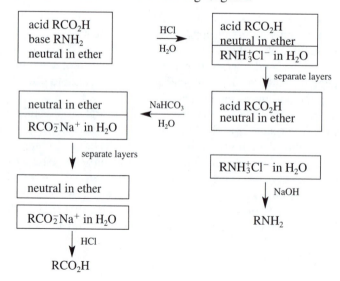

SOLID–LIQUID EXTRACTION

Another important use of extraction is in the isolation of a compound from a solid. In the study of naturally occurring compounds in plants, for example, the first step is the extraction of ground leaves, bark, or wood with solvents to remove all soluble compounds before further separation. An extraction of this type is illustrated by the isolation of leaf pigments in Chapter 7. A very familiar example is the brewing of coffee or tea by extraction of the flavors, caffeine, and other substances with hot water. If these compounds from the coffee bean or tea leaf are desired, they can be isolated from the aqueous brew by further extraction using organic solvents (see Chapter 7).

STEPS AND TECHNIQUE OF EXTRACTION

USE OF THE SEPARATORY FUNNEL

A separatory funnel is expensive and fragile, and when full, it is top-heavy. The funnel should be supported on a ring of the proper size at a convenient height; it should not be propped up on its stem. Before each use, the stopcock should be checked to be sure that it is seated and rotating freely. If the stopcock is glass, a clip or leash should be used to prevent the stopcock from falling out if it is accidentally loosened. A *very light* film of stopcock lubricant should be applied around the stopcock in bands on each side of the hole. Excess grease is to be avoided, since it will be washed away by organic solvents and contaminate the solution. (Teflon stopcocks require no lubricant.)

The separatory funnel should be filled to no more than approximately three fourths of the total depth, so that thorough mixing is possible. After filling the funnel (check first that the stopcock is closed!), it is stoppered with a properly fitting plastic, glass, or rubber stopper. (Before the funnel is used for the first time, it is good practice to shake it with a few milliliters of liquid to make sure that stopcock and stopper are tight.)

Holding the funnel with the stopcock end tilted up, the stopper is kept in place securely with the heel of one hand and the stopcock end is supported in the other hand (Fig. 4.3). As soon as the funnel is inverted, the stopcock is opened to release any pressure. The stem must be pointed away from you and your neighbors, since a small amount of liquid may spurt out. The stopcock

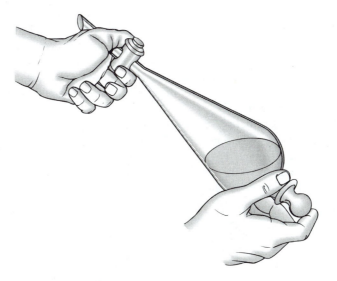

Figure 4.3   Use of a separatory funnel.

is then closed and the funnel is shaken gently in a horizontal position for about 30 seconds. The process is stopped periodically and the stopcock is opened slowly to vent any pressure that may have built up; this step is particularly important in extractions with bicarbonate, when CO_2 pressure may develop during the extraction. The funnel is replaced in the ring and the stopper is loosened immediately. When the phases have separated completely, the lower layer is drawn off through the stopcock.

In the extraction of an aqueous solution, the solvent may be either less dense (e.g., ether) or more dense (e.g., methylene chloride) than water. In case of doubt as to which layer is the aqueous phase, several drops of the lower layer can be drained into a test tube and checked for miscibility with water.

When the solvent is less dense than water and forms the upper phase, as in an extraction with ether, the aqueous layer must be drained into a receiver (usually the flask in which it was originally contained) and the ether layer transferred to a second flask over drying agent. The aqueous phase is then returned to the separatory funnel for further extraction with second and third portions of solvent. The (now dry) first organic extract is transferred to a round-bottom flask for evaporation, and the second and third organic extracts are used sequentially to rinse the drying agent, then added to the first organic extract in the round-bottom flask. If a second aqueous wash is called for, the two aqueous phases are not combined. Rather, the first organic extract is washed with the first and second aqueous, which are kept separate. Then, the second portion of organic solvent is shaken sequentially with the two aqueous washes. Finally, the third portion of organic solvent is shaken sequentially with the two aqueous washes. With a solvent denser than water, the aqueous solution is simply retained in the funnel and shaken with further portions of fresh solvent.

In either case, the organic layers are usually then combined, returned to the funnel, and shaken with a small volume of water to remove traces of the original aqueous phase that are suspended in the organic layer (back-extraction). When the organic layer is less dense (e.g., an ether or pentane layer), it is good practice to swirl the funnel while the aqueous layer is drained. Droplets of water clinging to the funnel are swept down and collected at the bottom of the funnel. The residual water is drained off before the organic layer is collected. The organic layer may be poured out the top of the funnel.

A difficulty that is sometimes encountered in extraction is the formation of an **emulsion** between one layer and the other. Mayonnaise is an example of an emulsion of oil and water. Emulsions prevent a sharp boundary between layers and can be frustrating to deal with. They are caused by the presence of colloidal impurities or surfactant materials in the solution being extracted and are very common in the extraction of an aqueous solution of plant or animal tissue. The formation of emulsions is favored by vigorous mixing of the layers, and if emulsification is anticipated, it can often be avoided by mixing the layers with gentle swirling for a longer time than is usually used for shaking. Sometimes emulsions can be cleared or broken up by the addition of an electrolyte, such as saturated salt solution, or by adding aqueous acid. Another measure that can be effective is to filter the emulsified portion of the mixture through filter paper or through Celite (see p. 27).

It is important to note that an aqueous solution to be extracted can contain a water-miscible solvent such as alcohol. If the amount of alcohol is significant, an excessive volume of extracting solvent (such as ether) is required to form two liquid phases, and the ether layer will contain a substantial fraction of the third solvent as well as water. If this situation arises, it may be possible to remove some of the alcohol before extraction by distillation or evaporation. Otherwise, a larger-than-normal amount of ether must be used in the extraction, and the ether phase is then thoroughly back-extracted with water.

In many cases the separatory funnel is used simply as a means of recovering a small amount of an insoluble organic product, such as precipitated acid, from a large amount of water with minimum mechanical loss. A small volume of ether or other solvent is added to the mixture to permit a sharp separation of layers. Even though the organic product may have a negligible solubility in water, a second portion of solvent should be used to rinse the aqueous layer and the separatory funnel.

DRYING AGENTS

After any of these extraction processes, the organic solution, as received, is saturated with water; therefore it is desirable to dry the solution before evaporating the solvent. Water is an impurity and should often be removed before a crystallization procedure is carried out or before a liquid is distilled. A number of salts that form hydrates can be used for this purpose. The efficiency of a drying agent depends on the completeness of drying (**intensity**), the degree of hydration (**capacity**), and the rate at which the salt absorbs water. A few of the more commonly used agents are as follows:

1. Potassium carbonate is the agent most often used for drying neutral and basic compounds; it has intermediate intensity and capacity.
2. Sodium sulfate has high capacity but low intensity; it can be used only at room temperature or below.
3. Magnesium sulfate has high capacity and intermediate intensity. It is inexpensive and rapid in action, but somewhat acidic.
4. Calcium chloride has high intensity but generally is useful only for hydrocarbons or halides, because of complex formation with most compounds containing oxygen or nitrogen.
5. Calcium sulfate (Drierite) has very high intensity and is rapid, but it has low capacity; it is most often used as the drying agent in dessicators.
6. Molecular sieves are complex silicates with a porous structure that selectively entraps water molecules. They are useful for obtaining rigorously anhydrous liquids but are not usually used for organic solvents.

It is difficult to specify how much drying agent to use in any one case but a rough rule of thumb is that for solutions the amount of drying agent usually should be about one twentieth of the liquid volume; a smaller amount is used for a pure liquid.* After allowing the solution to stand over the drying agent for 10 to 20 minutes, with occasional swirling, the dried solution is decanted and the residual drying agent is washed with a little fresh solvent, but are not usually used for organic extracts.

EVAPORATION OF SOLVENT

An important operation in extractions and other procedures such as column chromatography (Chapter 7) is the removal of a solvent to permit recovery of a relatively nonvolatile residue. Ether, methylene chloride, and hydrocarbon solvents are flammable and toxic. If more than a few milliliters of solvent is to be removed, provision must be made to avoid the discharge of vapor into the laboratory. Where permitted by safety codes, a convenient way to accomplish this is to sweep the vapor into an aspirator as it distills. This can be done by inserting a piece of glass tubing into the rubber tubing leading to the aspirator and clamping it vertically above the flask on the steam bath, with the glass tubing extending a short distance into the neck. The pressure in the system is not reduced, but most of the vapor will enter the aspirator, where it is diluted with water and discharged into the drain. This technique is inadequate for large volumes of solvent, which must be removed by distillation through a condenser.

For rapid evaporation of small amounts of solvents with minimum heating, the solution is placed under reduced pressure in a round-bottom flask or test tube. This is very easily arranged by fitting a large one-hole rubber stopper on the mouth of the flask (Fig. 4.4a). Atmospheric pressure holds the stopper in place, but it can be quickly released; an alternate procedure, particularly useful for microscale experiments, utilizes the ground glass equipment shown in Figure 4.4b. When this method is used, the flask or test tube must be no more than 10 to 15% full, and it must be agitated constantly, with the pressure controlled by placing a thumb on the tubing, or with the stopcock, to prevent bumping and frothing. This technique requires a little practice, but it will be found to be a great time saver and well worth learning.

---

* The drying agents tend to clump when they pick up water. Thus, a useful technique is to add the agent in small increments with intermittent swirling and standing until added portions remain free-flowing.

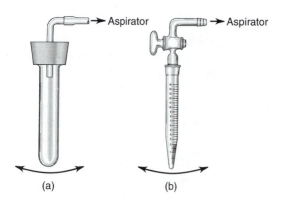

Figure 4.4   (**a** and **b**) Evaporation of solvent at reduced pressure.

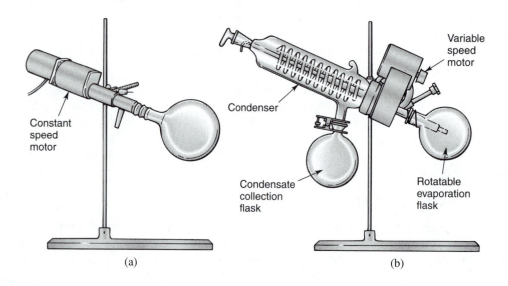

Figure 4.5   Vacuum rotary evaporators.

For rapid evaporation of larger volumes of solvent, rotary evaporators (of the types shown in Fig. 4.5) are standard equipment in most advanced laboratories. These evaporators are operated at reduced pressure, some with a very efficient condenser (Fig. 4.5b). The flask is held at an angle and rotated in the heating bath; this action spreads the liquid in a film on the inside wall and provides a large surface for evaporation.

The correct size and type of glassware for solvent removal deserves comment. There must be room for boiling and agitation, whether under reduced pressure or not, and the flask should never be more than one-fourth to one-third full. Much time can be wasted by gingerly evaporating a solution from a flask that is too full. Evaporation of solvent should never be done in a beaker. Although it has a wide mouth, a beaker is actually a very inconvenient vessel from which to remove a crystalline residue or a liquid, since solvent tends to creep up the walls, and rinsing is less efficient than with a flask or test tube; moreover, a beaker cannot be stoppered, swirled, or evacuated. Vessels containing liquids to be dried or stored should always be securely stoppered with corks to prevent evaporation and contamination. A beaker is really useful only as a container for weighing solids or as a bath for cooling or heating. For practically all other purposes, Erlenmeyer flasks, round-bottom flasks, or test tubes are preferable.

SAFETY NOTE

Only a *spherical* or *cylindrical* vessel (e.g., a test tube) should be placed under reduced pressure unless it is made of heavy-wall glass, such as a filter flask. A flat-bottom surface is not built to withstand the force exerted by nearly 1 atmosphere of pressure if the flask is evacuated. In a round-bottom flask or test tube, the pressure is distributed over a continuously curved surface. Most modern glassware is being made with fairly thick walls, but it is *extremely* hazardous to evacuate a large Erlenmeyer flask, because of the danger of implosion and resultant flying glass.

It is frequently necessary to weigh a residue after removing a solvent or decanting a liquid from a solid; if this is the case, the empty weight, or **tare,** of the vessel should be determined in advance.

TRANSFER OF LIQUIDS

A few points of technique are important in manipulating small volumes of organic liquids and solutions. It is usually necessary to transfer a solution to a smaller vessel after solvent removal is about 80% complete, to facilitate removing the last trace solvent and to minimize losses in recovering a crystalline residue.

A small volume of organic liquid should be transferred with a pipet (pp. 26–27). An attempt to pour it will result in a significant fraction of the residual compound being spread over the inside wall, the lip, and the outer surface of the neck of the flask because of the low surface tension of the organic solvent. An excessive amount of solvent is then required for rinsing. In rinsing a relatively large flask with ether or methylene chloride, a useful technique is to add a small amount of solvent and then warm the bottom of the flask so that the condensing vapor rinses down the walls; then, the rinse is concentrated in a small pool for transfer.

EXPERIMENTS

A. DETERMINATION OF A DISTRIBUTION COEFFICIENT

(Macroscale)

In this experiment a K_d value will be determined and the efficiency of single and multiple extractions compared. A solution containing a known amount of benzoic acid ($C_6H_5CO_2H$) in water is extracted with methylene chloride (CH_2Cl_2). The amount of acid remaining in the aqueous layer is determined by titration with base, and the amount of acid in the organic layer is obtained by difference. From these two values, K_d is calculated. Another extraction is then carried out with the same amount of CH_2Cl_2 divided into two portions, and the amount of acid extracted is compared.

SAFETY NOTE

Methylene chloride (CH_2Cl_2) is toxic. Do not get it on your skin or inhale its vapors.

Disposal. All aqueous solutions may be poured down the drain provided they have been previously neutralized. Any methylene chloride or its solutions should be poured into a container labeled as "halogenated hydrocarbon waste."

Benzoic Acid Solution. Weigh out 0.61 g (5.0 mmoles) of benzoic acid and place the acid in a 250-mL Erlenmeyer flask. Add about 150 mL of water and heat the mixture on a hot plate or burner with swirling until the acid is dissolved. Pour the solution into a 250-mL volumetric flask or graduated cylinder and rinse several times with water to transfer all the acid. Cool the solution and adjust the volume to 250 mL to give a 0.020 M solution of benzoic acid.

NaOH Solution. Dissolve one pellet of NaOH in 100 mL of water and rinse and fill a buret with this solution. Alternately, use a stock solution of 0.02 M NaOH. Place a 10.0-mL sample of the 0.02 M benzoic acid in a small flask and add one drop of phenolphthalein solution. Titrate to a pink end point. Repeat with a second 10.0-mL sample, average the buret readings, and calculate the molarity of the NaOH.

Distribution Coefficient

Place 50 mL of the benzoic acid solution in a separatory funnel, and using a volumetric pipet, add 10 mL of methylene chloride to the funnel. Stopper the funnel and shake vigorously for 30 seconds. Remove the stopper, allow the layers to separate, and drain off the lower (CH_2Cl_2) layer. Swirl the funnel and after a minute or two, a small additional amount of lower layer can be removed.

Pour the aqueous layer into a 250-mL Erlenmeyer flask, rinse the funnel with a few mL of water, and add the rinse to the flask. Add a drop of phenolphthalein solution and titrate with the NaOH solution to a pink end point. To get the most accurate value of the titration, it is good practice to pull up some of the solution in a pipet resting in the flask. Titrate the rest of the solution to get an approximate end point, then mix in the solution held in reserve and titrate carefully to a precise end point. Repeat... Calculate the number of mmoles of acid in the water and, by difference, in the CH_2Cl_2 layer, and from these values and the volumes, calculate K_d as described on page 41.

Multiple Extraction

Place another 50-mL sample of the benzoic acid solution in the separatory funnel and extract with 5.0 mL of CH_2Cl_2, drain off the lower layer, and extract with a second 5.0-mL portion of CH_2Cl_2. After the second extraction, remove and titrate the aqueous layer and calculate the amounts of acid in the two CH_2Cl_2 layers. Compare the amount of acid removed by 10 mL of CH_2Cl_2 in one and two portions.

As a check on the accuracy of your experimental values, calculate the amount of acid that should be removed by two 5-mL extractions, using the value of K_d obtained in the 10-mL extraction, and compare it with the amount found.

B.   SEPARATION OF ACIDIC, BASIC, AND NEUTRAL COMPOUNDS

(Macroscale)

The separation of acids, bases, and neutral compounds (described earlier) by conversion of the acid and base to water-soluble ionic forms and extraction of these into an aqueous layer are illustrated in this experiment. In this procedure, an ether solution of an unknown that may contain an acid, base, or neutral compound, or some combination of these, is first extracted with acid to remove the base. This aqueous extract is then made basic with hydroxide to liberate any organic base. Extraction of the unknown solution with aqueous NaOH removes any acid, which is then recovered by reacidification. Finally the neutral compound, which remains unextracted, is recovered by removal of the solvent.

Disposal. All aqueous solutions may be poured down the drain provided they have been neutralized. Methyl *tertiary*-butyl ether solutions should be poured into a properly labeled container and a notation made as to the nature and amount of waste disposed. The ether chosen for these experiments, methyl *tertiary*-butyl ether (MTBE), can be replaced by diethyl ether (common name "ether"); however, ether is more toxic and flammable than MTBE and can form peroxides.

SAFETY NOTE

None of the unknowns is particularly toxic or irritating, nevertheless, care should be taken to avoid spilling ether solutions on the hands. If this occurs, wash your hands thoroughly with soap and water. Ethers are highly flammable, and there should be *no flames* of any kind in the laboratory.

PROCEDURE

Obtain 1.0 g of an unknown mixture, dissolve it in 50 mL of MTBE, and place it in a separatory funnel. Extract the solution with three 15-mL portions of 5% aqueous HCl, draining each portion successively into the same 125-mL Erlenmeyer flask. Extract the combined acid extracts with a 15-mL portion of MTBE (this is called "back-extraction"), and dispose of the MTBE extract as directed. Add 10% NaOH solution to the combined acid extracts until the solution is basic to indicator paper. If a basic compound is present, it will separate at this point. If it is a solid, collect it by filtration on a small Büchner or Hirsch funnel (see Figs. 3.5 and 3.6), wash it on the filter funnel with a small amount of cold water, and set the filter paper with the product aside to dry.

If the base is an oil, extract the aqueous solution with fresh MTBE (3 × 15 mL), combine the MTBE layers, and dry them over about 2 g of anhydrous sodium sulfate or other suitable drying agent for 15 to 30 minutes. (If sodium sulfate is used, some granular, uncaked drying agent should remain after the drying time. If this is not the case, add an additional small amount of sodium sulfate and let the solution stand for another 10 to 15 minutes.) Remove the drying agent by gravity filtration through filter paper or a cotton plug and evaporate the MTBE on a steam bath to obtain the basic component of the mixture.

To isolate the acid component of the mixture, extract the original MTBE solution with a 5% aqueous NaOH solution (3 × 15 mL), and combine the extracts in a 125-mL Erlenmeyer flask. Back-extract the combined NaOH extracts with 15 mL of fresh MTBE and dispose of the MTBE as directed. To recover the acidic component from the combined NaOH extracts, slowly add 10% HCl solution, shaking after each small portion is added. Collect any solid acid that precipitates by suction filtration, wash with cold water, and air dry. (If the acid separates as an oil, it is recovered by MTBE extraction in the same way as described previously for a liquid basic component.)

Back-extract the ether solution with water. Dispose of the water as directed. Pour the remaining ether solution from the top of the separatory funnel into an Erlenmeyer flask, rinse the funnel with a few milliliters of ether, and dry the ether solution with Na_2SO_4 as described earlier. Filter through a cotton plug into a dry flask and evaporate the ether on a rotary evaporator. If a neutral compound is present, it will remain as a crystalline or oily residue.

Table 4.1 Possible Unknowns

ACIDS	MP (°C)	BASES	MP (°C)	NEUTRAL	MP (°C)
o-Toluic acid	104	*p*-Chloroaniline	72	Benzil	95
Benzoic acid	122	Ethyl *p*-aminobenzoate	89	Phenanthrene	101
trans-Cinnamic acid	133	2-Aminobenzophenone	106	Fluorene	114
m-Nitrobenzoic acid	140	*p*-Phenylenediamine	140		
m-Bromobenzoic acid	155	4-Aminoacetanilide	162		
p-Toluic acid	180				

In each fraction in which a crystalline compound is obtained, spread the substance out on glassine paper and finely divide it to dry. Determine the weight and melting point and identify it from the list of unknowns in Table 4.1. If an identification is questionable, it may be desirable to recrystallize the compound or to determine a mixture melting point with an authentic sample, or both. If a residue is noncrystalline, crystallization should be attempted by using the technique described in Chapter 3.

When working with very small quantities of material, the use of a separatory funnel is impractical. Instead, separations are carried out in a centrifuge tube (Fig. 4.6) and transfers are made with a Pasteur pipet (see inside back cover).

(a)   (b)

Figure 4.6   Two different types of centrifuge tubes. Type **(b)** has a glass stopper and is also called a distillation receiver.

C.   SEPARATION OF ACIDIC, BASIC, AND NEUTRAL COMPOUNDS

(Microscale)
Disposal

All aqueous solutions may be poured down the drain provided they have been neutralized. Methyl *tertiary*-butyl ether solutions should be poured into a properly labeled container and a notation made as to the nature and amount of waste disposed.

In a centrifuge tube, dissolve 100 mg of an unknown in 2 mL of methyl *tertiary*-butyl ether. Add 1 mL of 5% HCl, cap or stopper the tube, and shake gently for about 1 minute. Allow the layers to separate and carefully draw off the bottom layer with a Pasteur pipet,* depositing it in a second (properly labeled) centrifuge tube. Repeat the procedure with two additional 1-mL portions of 5% HCl, combining these with the first portion. Back-extract the combined HCl layers with 1 mL of fresh methyl *tertiary*-butyl ether by shaking gently for about 1 minute,* and then carefully draw off the methyl *tertiary*-butyl ether layer with a clean pipet. Add this methyl *tertiary*-butyl ether layer to the original ether solution.

The combined HCl layers are treated with 10% aqueous NaOH until basic, at which point a precipitate indicates the presence of a basic component in the original unknown. If the

---

* An alternate procedure involves the mixing of the layers by using the pipet to draw up and squirt out the mixture several times. The layers are then allowed to separate and the bottom layer drawn off as described.

precipitate is a solid, it can be isolated by direct filtration using a Hirsch funnel, washed with a small amount of water, and spread out on glassine paper to dry. Alternatively, the aqueous layer can be carefully drawn off with a pipet into which a small amount of cotton or glass wool has been inserted in the tip (Fig. 3.4C) to prevent the removal of solid. The solid is washed by adding a small amount of water, and as much of the water removed as possible with the pipet. The centrifuge tube containing the remaining wet solid is then attached to an aspirator and dried under vacuum (with gentle heating on a steam bath, if necessary). When the solid is dry, it is weighed and the melting point determined.

If the base precipitate is noncrystalline, the mixture is extracted with 3×1 mL of ether (in this case the ether layer is carefully removed each time with the pipet). The combined ether extracts are dried by adding a small amount of Na_2SO_4 or other suitable drying agent, separated from the drying agent by transferring to a clean centrifuge tube with a pipet containing cotton or glass wool in the tip, and the ether removed under vacuum with gentle heating on a steam bath and shaking to prevent bumping. If the residue continues to be an oil, crystallization should be attempted by adding a small amount of suitable solvent and following one of the procedures outlined in Chapter 3 under "Crystallization."

An acidic component in the unknown is separated in similar fashion. Extract the original ether solution with 3×1 mL of 5% aqueous NaOH, each time transferring the NaOH layer with a pipet to a properly labeled centrifuge tube. The combined NaOH extracts are back-extracted with 1 mL of fresh methyl *tertiary*-butyl ether and then acidified by cautiously adding 10% aqueous HCl. A precipitate indicates the presence of an acidic component. Separate and dry the solid using one of the procedures discussed previously in the basic component isolation step. Weigh, and record the melting point. If the acid separates as an oil, it is recovered and treated as described earlier for a noncrystalline basic component.

To isolate a neutral component, wash the original ether solution with 1 mL of water, dry the methyl *tertiary*-butyl ether layer with a drying agent, and after separating the drying agent, remove the ether solvent under vacuum with gentle heating on a steam bath and constant agitation (gentle shaking) to prevent bumping.

QUESTIONS

1. The distribution coefficient, $K = C_{ether}/C_{water}$, is 10 for compound Y. What weight of compound Y would be removed from a solution of 4.0 g of Y in 100 mL of water by a single extraction with 100 mL of ether?
2. Reconsider Question 1 above; what weight of compound would be removed by two extractions using 50 mL of ether each time?
3. Describe how you would separate a mixture containing benzoic acid and a neutral substance, benzophenone. Both compounds are insoluble in water and soluble in ether.
4. Assume that during an extraction experiment you have lost track of which layer is the aqueous layer. How could you determine which layer is which by the use of a simple test?
5. The compounds isolated during Chapter 4, Experiments B and C may be impure or wet. What effect will this have on the melting points?
6. Why must the stopper be removed from the separatory funnel before the lower layer is removed?
7. Most phenols (ArOH) do not react with sodium bicarbonate solution but do react with sodium hydroxide solution. Describe how to separate, by extraction, a mixture of a water-insoluble phenol (ArOH) and a water-insoluble carboxylic acid (RCOOH).

PRELABORATORY QUESTIONS

1. Explain why it is more efficient to make several extractions using small portions rather than one extraction using the same total volume.

2. Define the following terms:
    a. drying agent intensity
    b. drying agent capacity
3. Select the upper layer in each pair of mixed solvents (see Appendix C):
    a. water and benzene
    b. water and carbon tetrachloride
    c. water and ether
4. Write an equation for the reaction of a benzoic acid with aqueous sodium bicarbonate.
5. Complete the following table:

compound	capacity	intensity
$MgSO_4$	_____	_____
Na_2SO_4	_____	_____
$CaCl_2$	_____	_____
K_2CO_3	_____	_____

REFERENCE

Craig, L. C.; Craig, C. In *Techniques of Organic Chemistry,* 2nd ed.; Weissberger, A., Ed.; Interscience: New York, 1956; Vol. III, Part I, Chapter 2.

Distillation

Distillation is the process of heating a liquid to its boiling point, continuing the heat input to convert liquid to vapor and to force the vapor to another portion of the apparatus, cooling the vapor there to condense it, and collecting the liquid condensate. In this way, a liquid can be recovered from nonvolatile contaminants, or mixtures of liquids with different volatilities can be separated. When vaporization occurs from a solid, the process is called **sublimation.** The term **evaporation** is used when the residue rather than the distillate is of importance and the volatile liquid is allowed to escape uncondensed into a hood or an aspirator. The term **reflux** means boiling the liquid with return of the condensate to the original flask.

Distillation depends on the fact that at a given temperature, some molecules of a liquid have sufficient kinetic energy to escape from the surface and create a vapor pressure. This tendency for vaporization becomes greater as the kinetic energy is increased by raising the temperature. When a liquid is heated to the temperature at which the vapor pressure equals that of the surrounding atmosphere, the liquid boils, and this temperature, at a given pressure, is called the **boiling point.** Two or more components of a liquid that have different boiling points may be separated by distillation under favorable circumstances. This process is known as **fractional distillation** and involves the distillation of one component (or fraction) at a time.

Consider a mixture of two volatile liquids. The question might be asked, "What is the composition of the vapor that is distilled from this liquid?" The expectation would intuitively be that the vapor would be richer in the more volatile (lower boiling) component; thus, it might be supposed that one could heat the liquid to a temperature between the boiling points of the two components and cleanly distill off the lower boiling component. Such is not the case. For example, it is not possible to heat a mixture of benzene (bp 80°C) and toluene (bp 111°C) to a temperature of 81°C and distill off pure benzene. Consideration of some general relationships will explain why this is so.

In a mixture of two compounds, A and B, the total vapor pressure is the sum of the partial pressures of each component (**Dalton's Law**). Furthermore, the partial pressures are proportional to the amount of each component in the mixture. For an ideal solution of two compounds, in which the molecules do not interact, the vapor pressures P_A and P_B of each compound are defined by **Raoult's Law** as the vapor pressure of the pure compound multiplied by the mole fraction of that compound.

$$P_A = \eta_A \bullet P_A^{\circ} \qquad \text{and} \qquad P_B = \eta_B \bullet P_B^{\circ}$$

where

P_A = vapor pressure of component A
η_A = mole fraction of A
P_A° = vapor pressure of pure A at the temperature in question

Combining Dalton's Law and Raoult's Law gives

$$P_{\text{total}} = P_A + P_B = \eta_A \cdot P_A^{\circ} + \eta_B \cdot P_B^{\circ}$$

Consider a mixture of equimolar amounts of A and B where the mole fraction = 0.5 for each and component A is the more volatile, with the higher vapor pressure and the lower boiling point. When the mixture boils, the vapor in equilibrium with the liquid will be richer in the more volatile component A ($\eta_A \cdot P_A^{\circ} > \eta_B \cdot P_B^{\circ}$).

The distillation of such a two-component mixture is most readily described in terms of a phase diagram in which equilibrium curves for vapor and liquid are plotted with temperature versus composition (Fig. 5.1). The boiling points of pure A and pure B are T_A and T_B respectively. The upper curve represents a boundary above which there is only vapor, and the lower curve is the boundary below which there is only liquid. The lower curve represents the boiling points of various mixtures of components A and B, going from pure A on the left to pure B on the right. At any point on the curve, a horizontal line drawn across to the upper curve shows the composition of the vapor in equilibrium with the liquid.

In an equilibrium mixture of components A and B, corresponding to 0.5 mole fraction A in Figure 5.1, the boiling point of the liquid is T_1. At this temperature, however, the vapor has a composition of 0.8 mole fraction of component A, much richer in A, as seen from Raoult's Law.

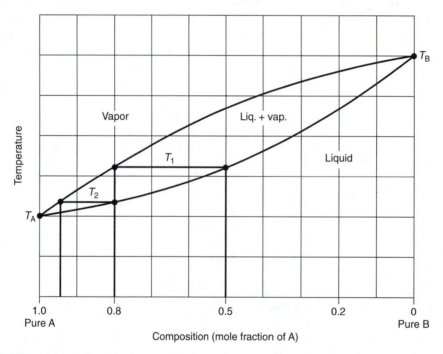

Figure 5.1 Vapor–liquid phase diagram.

If a sample of this vapor is removed, the composition of the remaining liquid shifts to the right on the diagram; that is, the relative amount of component B increases and the mixture has a higher boiling point. Continued distillation from the mixture provides distillate progressively enriched in B, but each fraction of the distillate will contain both components. The net result is that a simple distillation of two liquids, whose boiling points differ by less than about 50°C, effects relatively little separation, and the temperature will rise continuously during the distillation (dashed line in Fig. 5.2).

How the separation efficiency of a distillation can be increased can be illustrated by further study of Figure 5.1. If the condensed vapors of 0.8 mole fraction A are reequilibrated at tem-

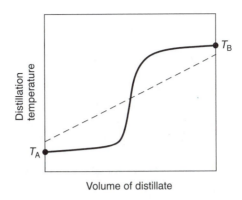

Figure 5.2   Distillation curves. Dashed line equals simple distillation; solid line equals fractional distillation with efficient column.

perature T_2, the vapors will now have a composition of 0.95 mole fraction A. If the equilibration can be repeated several times, it is possible to obtain a final distillate that is nearly pure A. This process, termed **fractional distillation,** is accomplished by distilling through a column with a large surface area, at which repeated exchange of molecules between the liquid and vapor phases occurs.

Each of the steps seen in Figure 5.1 (from 0.5 to 0.8 to 0.95 mole fraction A) represents in principle a simple distillation step, with one equilibration between vapor and liquid. In fractional distillation, successive distillation steps occur as vapor moves up the column and liquid flows back down. The extent of separation by a column depends on the number of theoretical distillation stages, usually referred to as "theoretical plates." The more efficient the column, the larger is the number of plates in a given length, or the smaller the height equivalent to one theoretical plate (HETP). The HETP is thus an index of the *efficiency* of a column; highly efficient columns have HETP values of less than 1 cm per plate. For a mixture of two compounds with boiling points differing by 10°C, a column of at least 25 theoretical plates is required for a reasonably efficient separation, such as that shown by the solid curve in Figure 5.2.

The essential function of a fractionating column is to allow vapor to condense and revaporize so that the more volatile material continues to progress upward while the less volatile component flows back down to the distilling flask. This is accomplished by providing ample surface for condensation and revaporization as well as sufficient length for the necessary temperature gradient. The column may be packed with glass beads, wire-mesh sponge, glass wool, or other inert material. A problem with a large amount of packing is that there is a rather large "holdup" of material. This term refers to the amount of condensed vapor that remains in the fractionating column when the distillation is completed. One of the most efficient types of column is a so-called spinning band column, in which a spiral of metal or Teflon on a central shaft extending down through the column is rotated at high speed. This action provides rapid exchange of vapor with a continuously flowing film of liquid over the spiral band.

To achieve high efficiency of separation, a column must be heated to a temperature intermediate between that of the distilling flask ("pot temperature") and the still head. Also the distillation must be carried out slowly, with only a small amount of the distillate reaching the condenser being removed; the remainder is allowed to flow back down the column for further equilibration.

AZEOTROPES

In the earlier discussion of fractional distillation, the mixture was assumed to be an ideal solution; that is, it obeyed Raoult's Law. However, this is not always the case. For example, with some dissimilar compounds such as an alcohol and a hydrocarbon, association between molecules of one compound may be greater than association between unlike molecules. The vapor pressure of the mixture at certain compositions is greater than the vapor pressure of the more volatile compound alone, and there is a minimum in the vapor–liquid phase diagram. Such a mixture is called a **minimum-boiling azeotrope** (azeotropes are also called constant boiling mixtures). The mixture at composition X shown in Figure 5.3 distills completely at temperature T_X with no separation of the components. Fractional distillation of a mixture at any other composition will provide distillate of composition X until only one component remains in the liquid. **Maximum boiling azeotropes** also occur but are less common. In these cases, fractional distillation of a mixture of two components yields one pure substance until the azeotropic composition is reached, and the remaining material then distills as the azeotrope.

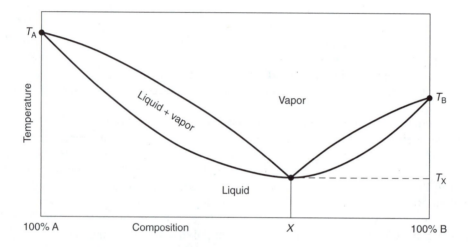

Figure 5.3   Phase diagram for minimum-boiling azeotrope.

SUBLIMATION

Sublimation is very effective for purification of volatile solids. In this technique, a solid is placed in a vessel that can be evacuated and heated, and in which a cold finger is positioned a short distance above the material to be sublimed. A typical all-glass sublimer is shown in Figure 5.4. A very simple sublimer can be fashioned out of ordinary glassware, as shown in Figure 5.5. A 125-mL filter flask serves as the subliming vessel and an 18×150-mm test tube as the cold finger. The test tube is fitted with a rubber stopper with a 16-mm hole. Lubricate the rubber stopper with glycerin, slip it over the test tube, and adjust the position so that when the apparatus is tightly sealed, the cold finger is approximately 1 cm above the bottom of the flask or compound to be sublimed. Be sure to wash off the excess glycerin with water, wipe the test tube clean, and dry thoroughly before use.

Add the solid to be sublimed to the filter flask and assemble the apparatus. Using an aspirator, evacuate the flask and add crushed ice to the test tube. Warm the flask gently

Ice and water

Figure 5.4   All-glass sublimer.   **Figure 5.5**   Filter flask–test tube sublimer.

until sublimation is complete. Remove the water (the melted ice) from the inside of the tube. Gently break the vacuum, remove the cold finger, and scrape the sublimed solid onto a tared watch glass.

APPARATUS AND TECHNIQUE

The basic elements of a distillation apparatus are a boiling flask, a column through which the vapor rises, a thermometer, a condenser in which the vapor is cooled and liquefied, and a receiver. Standard assemblies for distillation are illustrated in Figures 5.6 and 5.7. The distilling flask is always supported by a heating mantle or ring and wire gauze. The flask with liquid (and a boiling stone) is clamped in place and the column and/or distilling head put in place. The condenser is clamped on a second ring stand with height and angle adjusted so that the joints fit without stress; none of the clamps should be tightened until the apparatus is carefully aligned. The rubber condenser tubing is connected with the inlet at the bottom and the outlet at the top. An adapter and suitable receiver (Erlenmeyer flask, round-bottom flask, distillation receiver, or vial) are attached and are supported by a third ring stand if necessary. The receiver should not be propped on a makeshift support. The adapter should be fastened to the condenser with a rubber band or a clamp.

In assembling ground-glass apparatus, it is initially important that the joint is clean, free of grit, and properly aligned. A very thin film of stopcock grease may be used to ensure a snug fit. A common error is to use too much grease; this results in contamination of the liquid that comes in contact with the joint. Ground joints may be used without grease and will seldom stick if the contents of the apparatus are not alkaline and if the apparatus is taken apart before it has completely cooled.

It is important to select the proper size flask for a distillation. Too large a flask results in loss due to "holdup" of condensing vapor on the walls. In too small a flask, the liquid will expand and fill the flask to the point where there is insufficient room for boiling to occur without forcing liquid up into the column or distilling head. As a general rule, the flask should be one-third to two-thirds full initially. A boiling stone (small lump of carborundum or other porous material) is essential to maintain constant formation of bubbles (ebullition). Without a boiling stone to provide a steady stream of gas bubbles, superheating and bumping of the liquid can occur. This is particularly true when a liquid on a steam bath is evaporated without a condenser. If the boiling stone is forgotten until the liquid is hot and possibly superheated, the flask must be cooled before the boiling stone is added or a violent and dangerous eruption of the liquid may occur. Since the pores fill with liquid as soon as boiling ceases, a boiling stone cannot be reused and a fresh one must be added if the distillation is interrupted.

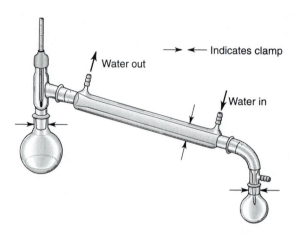

Figure 5.6   Simple distillation apparatus.

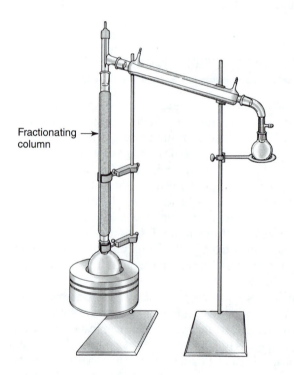

Figure 5.7   Fractional distillation apparatus.

In order to reduce loss due to "holdup," a "chaser" may be used. This is a high-boiling, inert liquid that is added to the sample and serves as a heat transfer medium, thus allowing most of the lower boiling sample to be distilled.

Distillation assemblies similar to that shown in Figure 5.6 may be used to distill liquid samples as small as 1 to 2 mL in volume by simply using smaller-scale apparatus. For samples smaller than this, different apparatus and techniques must be used (see Cheronis, 1954). A Hickman still (Fig. 5.8) may be used for simple distillations of 200 to 900 μL of liquid or for the fractional distillation of similar volumes.

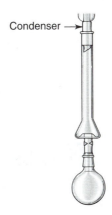

Figure 5.8 Hickman still.

Another method of distillation on a microscale (50 to 500 μL) utilizes the Kugelrohr apparatus (Fig. 5.9). The name comes from the German meaning "bulb tube," which describes the glass distillation tube. The material to be distilled is placed in the terminal bulb with a pipet, and the tube is then inserted into the glass heating jacket so that all other bulbs are outside the jacket. The apparatus provides for rotation and distillation under vacuum should these features be desired. As the material distills, it condenses in the cooler bulb outside the jacket. By inserting this receiver bulb into the jacket, the material can be distilled again (at the same or different temperature) to the next bulb. In this manner, a number of distillations can be carried out without removing the material from the tube.

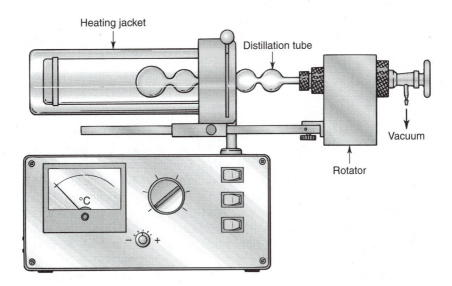

Figure 5.9 Kugelrohr distillation apparatus.

Safety Note

A distillation must *always* be stopped before the flask becomes completely dry. Without the absorption of heat due to vaporization, the flask temperature can rise very rapidly. Many liquids, particularly alkenes and ethers, may contain peroxides that become concentrated in highly explosive residues (see page 50).

EXPERIMENTS

A.   DISTILLATION OF A PURE COMPOUND

(Macroscale)

Assemble a simple distillation apparatus (see Fig. 5.6) using a 50-mL boiling flask that contains 20 mL of 1-propanol (or other liquid as designated by your instructor) and a boiling stone. Be sure the thermometer is correctly positioned with the top of the thermometer bulb just below the bottom of the distilling head side arm, the water flow is up the condenser, and the assembly is protected from drafts by aluminum foil. Heat the flask gently with a mantle or other source until the liquid begins to boil, and then adjust the heat to distill the liquid at a rate of about 2 mL per minute (2 drops every 3 seconds) using a graduated cylinder or distillation receiver as a receptacle. Record the temperature when the first drops of distillate are collected in the receptacle and after each 2 mL of distillate is obtained. Always stop a distillation before the distilling flask becomes dry; in this case you should be able to collect 15 to 17 mL of distillate. Make a plot of temperature versus volume of distillate.

B.   SIMPLE DISTILLATION OF A MIXTURE

(Macroscale)

Place 20 mL of a mixture provided by your instructor in a 50-mL flask with a boiling stone. Assemble the flask in a simple distillation apparatus and conduct a distillation as described in Experiment A, again noting the temperature as every 2 mL of distillate is collected. Make a plot of temperature versus volume of distillate. Save the total 20 mL of liquid for use in Experiment C.

C.   FRACTIONAL DISTILLATION

(Macroscale)

Place 20 mL of the mixture from Experiment B or provided by your instructor in a 50-mL flask with a boiling stone. Assemble a distillation apparatus with a fractionating column (large-diameter condenser, such as a Liebig condenser, packed with a stainless steel spring [1cm diameter], glass beads, or a wire-mesh sponge) between the distillation flask and the still head (see Fig. 5.7). Use a 10-mL graduated cylinder or a distillation receiver as a receptacle and have four clean dry test tubes available (label these fractions 1, 2, 3, and 4). Distill the mixture at the rate of about 1 or 2 mL per minute and record the temperature every 2 mL and at both the beginning and end of each fraction collected. Collect an initial fraction of 2 mL and quickly empty it into the first labeled test tube. Then collect separate fractions during the distillation of the low-boiling component, the period of rapid temperature rise, and the distillation of the high-boiling component.

   Plot the boiling point versus the volume of distillate. The fractions may be examined by gas chromatography (Chapter 6) or refractive index to determine the relative amounts of the two compounds present in each fraction.

The mixture provided by your instructor will contain two components listed in Table 5.1 that differ in boiling point by at least 25°C. From your data, determine the boiling points of the two components, their identities, and their relative ratio in small, whole numbers such as 1:1, 1:2, 2:1, and so on. Confirm your results by Gas Chromatography of the fractions you collect (see Chapter 6, Experiment B).

Table 5.1   Unknown Components

COMPOUND	BP
hexane	69.0°C
cyclohexane	80.7
toluene	110.6
p-xylene	138.3

D.   FRACTIONAL DISTILLATION

(Microscale)

Obtain an unknown mixture from your instructor that consists of two compounds from Table 5.1 (see last paragraph of Experiment C). Place 3.5 mL of this mixture in a 5-mL flask and conduct a fractional distillation using a fractionating column with a low holdup (2 to 3 in. of wire mesh sponge). Do not use a cooling condenser, but connect the still head directly to the take off adaptor. Carefully distill the mixture at not more than 2 to 4 drops per minute into a 10-mL graduated cylinder or distillation receiver cooled in an ice bath.

Plot the boiling point versus the volume of distillate to determine the boiling point of the two components, their identities, and their relative ratio.

E.   BOILING POINTS

Reflux Boiling Point Method

When sufficient liquid is available, the best method for determining the boiling point is distillation. This allows the boiling range to be determined and gives an indication of the purity. The distilled (purified) material may then be used to determine other physical constants such as the refractive index. Often one wishes to determine the boiling point of a small sample of a liquid. This can be done by the reflux method or the micro boiling point method.

Place about 0.6 mL of the liquid along with a boiling stone in a 15 × 125-mm test tube that has been clamped in a vertical position and supported on a wire gauze (see Figure 5.10). Suspend a thermometer in the test tube so that the bulb is about 10 mm above the surface of the liquid and so that no part of the thermometer touches the test tube. Heat the liquid gently with a microburner until it begins to boil and the ring of condensate rises about 1.5 inches up the thermometer (about one-half way up that portion of the thermometer that is in the test tube). Don't boil the liquid out of the top of the tube but adjust the rate of heating so that the ring of condensate remains essentially stationary; the thermometer will reach a constant temperature that is the boiling point of the liquid sample.

F.   MICRO BOILING POINT DETERMINATION (THIELE–DENNIS TUBE METHOD)

Seal one end of a piece of 7- or 8-mm glass tubing using a Fisher burner. Allow the glass to cool and then cut the tube using a triangular file to make a "boiler tube" about 3 to 4 inches long.

Place the liquid whose boiling point is to be determined into the boiler tube so that the level of the liquid is approximately 1 cm from the bottom. Break off a melting point capillary 1 inch

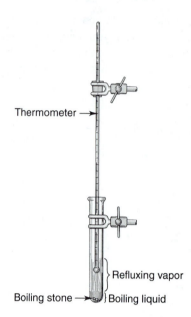

Figure 5.10 Reflux boiling point method apparatus.

from the closed end and drop it in the boiler tube with the closed end up. Attach the boiler tube to a thermometer using a small rubber band cut from 8- or 10-mm rubber tubing, and place the thermometer with attached boiler tube in a Thiele–Dennis tube (Fig. 5.11). Slowly heat the Thiele–Dennis tube until a rapid stream of bubbles comes out of the capillary. Remove the heat and allow the apparatus to cool. When the bubbles just cease and liquid is drawn up into the capillary tube, observe the temperature. This is the boiling point of the liquid. Carry out a micro boiling point determination with your unknown or with 1-propanol if you were not assigned an unknown. It may be helpful to practice this technique by carrying out a micro boiling point determination using water.

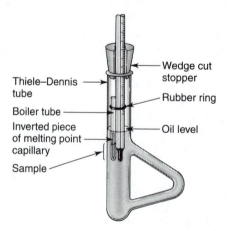

Figure 5.11 Micro boiling point apparatus.

G.   MICRO BOILING POINT DETERMINATION (CAPILLARY METHOD)

This method for determination of a boiling point of a small sample is based on the method reported by Siwoloboff (A. Siwoloboff, Ber., *19* [1886]). An accurate determination can be made on as little as a few microliters of liquid. A standard melting point capillary and a melting point apparatus such as a Mel-Temp are used.

1. Prepare a tiny sealed capillary (one that will fit inside of a standard melting-point capillary) by heating the middle of a Pasteur pipet in a small blue flame. Hold the pipet by the ends and when the glass starts to sag in the flame, remove it from the flame and pull the two ends until about 5 inches of a capillary of the desired diameter is made. Allow the pipet to cool then break off approximately 2 inches of the fine capillary and seal both ends. Break this 2-inch capillary in half for use.

2. Use the very fine tip of the Pasteur pipet that you just melted to introduce a small amount (about ½″) of the liquid sample into a melting-point capillary. Add one of the (now 1-inch) capillaries to the melting-point capillary, open end DOWN (see Fig. 5.12). Be sure there are no air bubbles in the melting-point capillary. If air bubbles are present, tap the melting-point capillary with a pencil or with your finger until they disappear. Alternately, warm the closed end of the melting point capillary tube over a small flame, remove, and insert the open end into the liquid in order to suck up 1 cm of the sample as it cools.

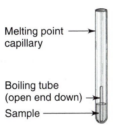

Melting point
capillary

Boiling tube
(open end down)

Sample

Figure 5.12   *Capillary boiling point apparatus.*

3. Place the melting point capillary in a standard melting-point apparatus and heat it gently. Bubbles (mostly expanded air) will start to come out of the bottom of the tiny capillary. Occasionally observe the temperature on thermometer. When a *stream* of bubbles (mostly vapors from the boiling liquid) start to come from the bottom (open end) of the tiny capillary, turn off the heater. The temperature of the sample tube will start to cool. When the boiling point (condensation point) is reached, the vapor will suddenly condense and a vacuum will be formed. This causes the liquid to be suddenly sucked up into the tiny capillary. Read the temperature at which this occurs. This is the "BOILING POINT."

Note: Before you try an unknown liquid, try two or three knowns such as water, isopropyl alcohol, or hexane. You should not be satisfied with more than 2 degrees error.

QUESTIONS

1. The normal boiling point of toluene is 110.6°C. What is its vapor pressure at this temperature?
2. What effect does an increase in atmospheric pressure have on the boiling point of a liquid? A decrease?
3. A distilling flask should never be filled to more than two-thirds of its capacity. Explain.
4. Compound A has a vapor pressure of 400 mm Hg at 95°C and Compound B has a vapor pressure of 250 mm Hg at 95°C. If A and B are miscible, what is the vapor pressure of a mixture of 3 moles A:1 mole of B at 95°C?
5. What causes the liquid to be drawn up the capillary tube during a micro boiling point determination?

PRELABORATORY QUESTIONS

1. Define the term "distillation."
2. Write an equation to exemplify Dalton's Law.
3. Maximum boiling azeotropes differ from minimum boiling azeotropes in what significant way?
4. Define the term "reflux."
5. Draw a sketch of a fractional distillation setup. Be sure to show where external support is necessary.

REFERENCES

Cheronis, N. D. In *Techniques of Organic Chemistry;* Weissberger, A., Ed.; Interscience: New York, 1954; Vol. VI.

Techniques of Organic Chemistry, 2nd ed.; Weissberger, A., Ed.; Interscience: New York, 1965; Vol. IV.

6

Gas Chromatography

The term **chromatography** refers to a general method of separation in which a mixture is partitioned between a stationary phase and a moving phase. The moving phase may be a vapor or a liquid; the stationary phase is a solid or a liquid film coated onto a solid. As the mobile phase flows over the stationary phase, compounds in the mixture are continuously equilibrated between the two phases according to their distribution coefficients. The greater the affinity of a compound for the moving phase, the greater will be the proportion of it in that phase, and therefore the faster it will be transported along the stationary phase. As a result, individual compounds become separated into zones or bands as the molecules are carried along in the moving phase. Eventually, if the resolution is sufficient, these bands emerge from the chromatograph as discrete fractions.

In **gas chromatography** (GC), sometimes called vapor phase chromatography (VPC), the mobile phase is a stream of inert gas and the stationary phase is a high-boiling liquid film supported on a solid and packed in a heated glass or metal column of about 2 to 3 meters in length. In capillary gas chromatography, the liquid may be affixed directly to a quartz capillary tube which may be as long as 50 meters in length. A sample of the mixture is injected with a syringe into a vaporization chamber and from there it enters the gas stream. The sample then passes through the column, where the partitioning process and separation take place. From the column, the components travel past a detector that delivers a signal to an electronic recorder. If desired, the compounds can be collected in a cold trap. A schematic diagram of the apparatus is shown in Figure 6.1.

The separation process in gas chromatography is illustrated in Figure 6.2 for a mixture of two compounds, A and B. As the sample is carried along the column by the gas flow, the faster-moving compound B becomes separated from A and passes the detector first. The recorder chart displays a series of peaks resulting from the detector response to each component in the mixture. In a typical gas chromatogram (Fig. 6.3), while the earlier peaks can be quite narrow, the width becomes progressively broader for later peaks. The position of each peak, expressed as the time required for the compound to pass through the column, is called the **retention time, R_t.**

For a given compound in the same column and exactly the same operating conditions, the retention time is a characteristic value and is independent of the presence of other compounds that may be present in the mixture. For qualitative analysis, if a peak in a gas chromatogram of a mixture is suspected to arise from a certain compound, a sample of the mixture can be "spiked" with an authentic sample of the compound in question. The absence of an additional peak in the chromatogram of the spiked sample is strong evidence that the compounds are identical. For positive identification, the compounds should be compared using several columns packed with different stationary phases. (In general, it is not acceptable to make an identification of any substance solely on the basis of one physical measurement, whether it be melting point, refractive index, density, or one retention time.)

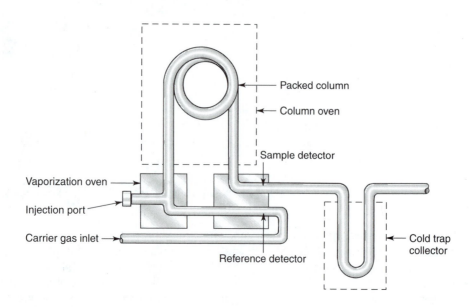

Figure 6.1   Schematic diagram for gas chromatograph.

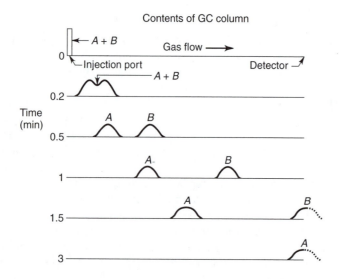

Figure 6.2   Separation of compounds by gas chromatography (GC).

The power of the method of gas chromatography lies in its high resolution and sensitivity, permitting a quantitative analysis of complex mixtures with less than a milligram of sample. The area of a peak in the chromatogram trace is proportional to the amount of the compound present in the sample, and as described below, the composition of the mixture can be determined by very simple measurement.

For analytical work the column is often a bent metal tube. The length is usually 2 to 3 meters (5 to 10 feet), and the diameter 2 to 5 mm. In general, the resolution of peaks increases with increasing length and decreases with increasing diameter. In current research settings where

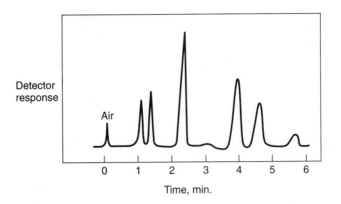

Figure 6.3   Typical gas chromatogram.

high resolution is necessary, capillary silica columns of very small diameter (e.g., 50 m long with a 0.2-mm diameter) are used.

Gas chromatography also can be used for preparative scale separations and purification. Small amounts of material can be separated by repeated injections and collection on an analytical-type column. In general, the maximum amount per injection without overloading the column is less than 20 μL. For larger samples, special columns with a diameter of 2 cm or larger are used. With these "preparative" columns, samples up to 1 mL in size can be applied.

The column packing used for the stationary liquid phase is a crucial factor in gas chromatography. Separation depends on the distribution of compounds between the vapor and liquid and is therefore affected by both the vapor pressures of the compounds to be separated and their solubilities in the liquid phase. The vapor pressure is controlled by the temperature of the oven surrounding the column, which is usually set at or a little below the average boiling point of the mixture. To obtain adequate solubilities in the liquid phase, it is chosen to be similar in polarity to the compounds in the mixture.

Several types of liquid, with numerous minor variations, are used for most gas chromatography. An important requirement is that the liquid phase be nonvolatile, so that it is not swept along and out of the column during use at elevated temperatures. For this reason the liquid is usually a polymeric substance. For nonpolar mixtures, various long-chain hydrocarbon greases (e.g., squalane or Apiezon) or silicone oils (e.g., SE-30, DC-200, or OV-101) are used. More polar liquid phases are polyethylene glycol (e.g., Carbowaxes, Durabond-Wax, or UCON) and polyesters such as polydiethyleneglycol succinate (DEGS).

$$HO \xrightarrow{\quad} (CH_2CH_2O)_n H \qquad\qquad \left[\begin{array}{cc} CH_3 & CH_3 \\ | & | \\ Si-O-Si-O \\ | & | \\ CH_3 & CH_3 \end{array} \right]_n$$

polyethylene glycol   silicone oil

$$\xrightarrow{\quad} (CH_2CH_2OCH_2CH_2O\overset{\overset{O}{\|}}{C}CH_2CH_2\overset{\overset{O}{\|}}{C}\xrightarrow{\quad} O)_n$$

diethyleneglycol succinate

These liquids are supported on finely divided, uniform particles of crushed firebrick or diatomaceous earth, usually in amounts of 5 to 15%. To provide an even coating, the stationary phase is prepared by evaporating a solution of the liquid dissolved in a low-boiling solvent in the presence of the support. Most commercial gas chromatographs come equipped with several standard columns. For most of these columns, the maximum operating temperature is 200 to 275°C. At higher temperatures the liquid phase will slowly "bleed off" and damage the detector.

Detectors used in gas chromatography are of two general types: thermal conductivity and flame ionization. **Thermal conductivity detectors,** which are normally used in undergraduate laboratory equipment, contain heated filaments or thermistor beads whose electrical resistance varies with temperature. One of these elements is placed in the gas stream coming out of the column and another in a reference stream of the pure carrier gas. The gas used with a thermal conductivity detector is helium, which has a very high thermal conductivity because of its high diffusion rate. When the stream containing a compound from the sample passes the detector, the rate of cooling by the gas stream decreases because of the lower thermal conductivity of the organic compound. This change in temperature results in a change in resistance and voltage, which is detected by the recorder.

Table 6.1   Response Factors for Thermal Conductivity Detectors

COMPOUND	WEIGHT FACTOR (W_F)	MOLAR FACTOR (M_F)
Benzene	1.00	1.00
Toluene	1.02	0.86
Heptane	0.90	0.79
Ethyl acetate	1.01	0.90
Ethanol	0.82	1.39

In a **flame ionization detector,** the stream from the column passes through a hydrogen flame, and ions formed by the combustion of the compound are captured by a grid to produce a current. Flame ionization is much more sensitive than thermal conductivity, and nitrogen, which is less expensive than helium, can be used as a carrier gas. In many undergraduate laboratories the detector of choice is a mass spectrometer using a relatively inexpensive combination of Gas Chromatography and Mass Spectrometry.

Since the thermal conductivity or extent of ionization varies somewhat from one compound to another, particularly with compounds of different types, the response of the detector per milligram or micromole of sample differs. For accurate work, corrections or response factors must be applied to take account of these differences. Factors for a few compounds are given in Table 6.1 for thermal conductivity detectors, relative to benzene, which is taken as 1.00. These factors can be expressed on a weight basis or on a molar basis by multiplying by the ratio of molecular weights. For closely similar compounds, such as isomeric alkanes or alkyl halides, the response factors are usually very nearly the same.

The measured areas in a chromatogram are multiplied by the factors shown in Table 6.1 to give the actual amounts of the compounds. For example, in a chromatogram of a benzene-ethanol mixture, let us say the peak areas are benzene, 50 mm^2, and ethanol, 70 mm^2. The composition on a weight basis is as follows:

benzene: 50 mm^2 × 1.00 = 50;   ethanol: 70 mm^2 × 0.82 = 57.4
total corrected area = 107.4
% benzene = 50/107.4 × 100 = 46.6%
% ethanol = 57.4/107.4 × 100 = 53.4%

OPERATING TECHNIQUE

The major variables in carrying out gas chromatography are column size and packing, column temperature, flow rate of carrier gas, sample size, and an instrument setting called the attenuation. The selection of a column packing depends on the factors discussed earlier; in undergraduate laboratory work the choice of column is usually limited to either a polar or nonpolar stationary phase.

Temperature and flow rate are interdependent, and both affect the resolution and time required for the chromatogram. The higher the temperature or the flow rate, the faster the mixture will pass through the column and, thus, the smaller the differences in retention times of the components. If either temperature or flow rate is too high, resolution may fall off to the point that separation is incomplete. If the temperature is too low, on the other hand, the retention times will become very long, and also, the peaks in the chromatogram will be broadened. The temperature is usually set at or a little below the boiling point of the sample, and the flow is then adjusted (around 20 to 30 mL/minute) to give the optimum performance.

The amount of sample depends on the column size and is usually 1 μL or less. Too large a sample overloads the column, and resolution is lost. The **attenuation** setting controls the sensitivity of the instrument and the amplitude of the peaks on the chart (higher attenuation leads to lower sensitivity and lower amplitude). This is set so that the largest peak in the sample registers full scale. If the sample contains a low-boiling solvent, the solvent peak often goes off scale.

The sample is injected by means of a syringe with very small bore. The sample, usually about 1 μL or less, is drawn up into the barrel of the syringe and then 1 to 2 μL of air is drawn in to empty the needle. This is done to avoid vaporization of sample from the needle before injection of the main sample. The air also provides a peak on the chromatogram, with very short R_t that can be used as a marker to measure retention times of other peaks. To make the injection, the needle is inserted through a soft rubber septum on the injection port to a depth of 2 to 3 cm. The plunger on the syringe is then pushed in with a quick, smooth motion to place all the sample on the column in the shortest time possible.

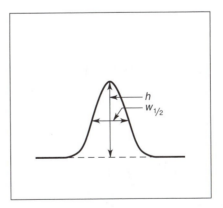

Figure 6.4   Measurement of peak area.

For measurement of the peak areas in the chromatogram, some instruments have **integrators,** which record the instrument response for each peak. If an integrator is not available, the areas can be obtained by multiplying the height of each peak (h) by the width of the peak at half the height ($w_{1/2}$) as shown in Figure 6.4. Alternatively, if the peak is asymmetric or very broad, the peaks can be cut out of the chart and weighed on an analytical balance. The area under the recorded curve is proportional to the amount of material causing the peak. As mentioned earlier, detector response differs from one compound to another; therefore, for accurate work the area-quantity relationship should be determined by calibration of the chromatograph. With calibration, the area under the curve can be used in the quantitative analysis of mixtures. For most routine analyses, the approximate ratios of components can be estimated reasonably well using only the chromatogram of the analysis sample.

Safety Note

The liquids used in this experiment are relatively harmless; however, avoid excessive inhalation of the vapors. Exercise care with the microsyringe as you would any sharp object.

EXPERIMENTS

Your laboratory instructor will demonstrate the use of the gas chromatograph and discuss retention times, sample size, peak areas, and other factors affecting this experiment.

A. EXAMINATION OF AN UNKNOWN MIXTURE

PROCEDURE

1. You will be given an unknown mixture made up of two to four liquids (these may be a series of hydrocarbons or alcohols; see Table 6.2). Samples of the pure liquids will be available for your use. Determine the retention times of each of the four liquids and run a gas chromatogram of the unknown. By comparison of the retention times of the known compounds with the unknown, determine which liquids are in your unknown. (Record temperature, chart speed, gas flow rate, and column used.)

Table 6.2   Unknown Components

	COMPOUND	BP (°C)
alcohols	methanol	65
	1-propanol	97
	1-butanol	118
	1-pentanol	137
alkanes	pentane	36
	hexane	69
	heptane	98
	octane	126

2. Determine the approximate percentage composition of your unknown by comparison of the areas under the peaks. (With a thermal conductivity detector, assume the peak areas are directly proportional to moles of sample present.)
3. Prepare a mixture that corresponds to the estimated percentages of your unknown, determine its chromatogram, and compare it with the chromatogram of your unknown.
4. Label your chromatograms and affix them in your notebook.

As an alternate introductory experiment in gas chromatography, fractions from the fractional distillation experiment, Chapter 5, can be examined to evaluate the separation efficiency of the distillation.

B.   GC OF FRACTIONS FROM FRACTIONAL DISTILLATION

PROCEDURE

1. Determine the gas chromatograms of three fractions from the distillation mixtures in Chapter 5, Distillation.
2. Determine the areas under each curve and the relative amounts of each component of the sample.
3. From the retention times of the knowns and unknowns, determine which components were present in your original sample for Chapter 5.
4. Report the percentage composition of each component of each fraction you took.

   Disposal. All distillates and residues not required for further use in the gas chromatography experiment should be discarded in the properly labeled organic waste container.

QUESTIONS

1. At what temperature would you set a GC oven in order to separate a mixture of heptane (bp 98°C) and octane (bp 126°C)?
2. A mixture of heptane (bp 98°C) and 1-propanol (bp 97°C) is not fully resolved on a column of SE-30 but is easily separated on a column of DEGS. Explain.

PRELABORATORY QUESTIONS

1. Define the term "retention time."
2. What are the requirements for a mobile phase in gas chromatography?
3. What volume of sample is appropriate in the running of an analytical gas chromatograph?
4. Which would you expect to have the shorter retention time, hexane (bp 69°C) or toluene (111°C)? Explain.
5. How is an air peak often useful in gas chromatography?

REFERENCES

Crippen, R. C. *Identification of Organic Compounds with the Aid of Gas Chromatography*; McGraw-Hill: New York, 1973.

Perry, J. A. *Introduction to Analytical Gas Chromatography: History, Principles, and Practice;* Dekker: New York, 1981.

Schupp, O. E., III. In *Techniques of Organic Chemistry;* Weissberger, A., Ed.; Interscience: New York, 1968; Vol. XIII.

Column Chromatography

In liquid phase chromatography, compounds are separated by partitioning them between a moving liquid and a finely divided solid phase. The process is closely analogous to gas chromatography (see Chapter 6), and the same basic principle applies. A mixture is distributed between the two phases according to differences in affinity of the components for liquid and solid. The compounds that spend more time in the solvent move more quickly, and are eluted first. The solid phase can be in the form of a thin layer on some kind of backing or packed into a vertical column. In either case, the sample to be separated is placed at one end of the layer or column, and solvent is then allowed to flow over the sample and through the solid.

The term **column chromatography** is usually reserved for the practice of separating compounds by partitioning them between a mobile liquid phase and a stationary solid phase that is packed in a glass column. The mixture is applied at the top of the column and is washed through the column by a pure solvent or combination of solvents. As the mixture moves over the solid phase, the compounds move apart into discrete zones. The separated compounds exit the column and are collected or detected. For preparative work the solvent is evaporated and the compounds are further purified and identified.

In most chromatography involving liquid and stationary solid phases, molecules are **adsorbed** on the solid, which has a very high surface area. The most common solid used in column chromatography is silica gel (SiO_2), the surface of which can be modified by chemical bonding to organic moeties (see **Reverse Phase** in Chapter 9).

In **straight-phase chromatography**, the surface provides a highly polar environment, and compounds with polar groups will be adsorbed more strongly than less polar ones. Hydrogen bonding to the surface is an important effect and causes a compound with an OH group, for example, cholesterol (Fig. 7.1), to be adsorbed more strongly than similar compounds lacking the OH group, such as the ketone, cholest-4-en-3-one.

cholesterol   cholest-4-en-3-one

Figure 7.1   Alcohols are more polar than ketones, and so run more slowly on silica gel.

Separation in adsorption chromatography depends on the competition between adsorption on the solid surface and desorption by the solvent needed to elute it. As a corollary, it follows that the more polar the solvent, the more rapidly the compounds will move with the liquid phase. A list of a few common chromatographic solvents, in order of increasing polarity, is given in Table 7.1. Relatively nonpolar solvents, such as hydrocarbons and chlorinated hydrocarbons, are most commonly used, but solvents such as acetone or moist ether may be required for elution of very strongly adsorbing compounds.

A typical chromatography column is shown in Figure 7.2. The weight of silica gel ("flash," 40–60 μm) is usually about 20 to 30 times the amount of pure compound in the crude sample. Usually a column height of about 10 cm of silica gel is sufficient to effect separation. For 5g of silica gel, a 1.4 cm interior diameter column is about right. For larger-scale separations, a wider, not a longer, column is used. It is necessary to have some free volume in the tube above the adsorbent for a supply of solvent. A separatory funnel mounted above the column can be used as a convenient solvent reservoir.

Table 7.1   Chromatographic Solvents

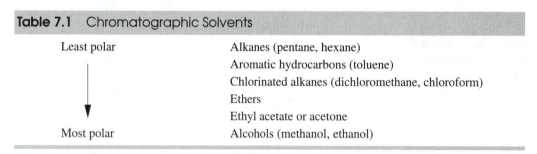

Least polar	Alkanes (pentane, hexane)
	Aromatic hydrocarbons (toluene)
	Chlorinated alkanes (dichloromethane, chloroform)
	Ethers
	Ethyl acetate or acetone
Most polar	Alcohols (methanol, ethanol)

The column is prepared in the following manner:

1. With a long wire (a straightened coathanger works well), tamp a small wad of glass wool into the bottom of the column. Gently pour a layer of sand over the glass wool, remove the wire, and tap the column to settle the sand. The sand retains fine particles and also provides a flat horizontal base for the adsorbent column.

2. Add the weighed amount of dry silica gel in a fine stream. Tap the tube gently to dislodge air bubbles and to level the top of the silica gel.

3. Dissolve your sample in a volatile solvent—methylene chloride works well. Add to the resulting solution an amount of silica gel that is one-tenth of the weight of silica gel that you are using in your column. Evaporate the solvent—the rotary evaporator works well. Add the resulting dry powder to the top of the silica gel in your column, and tap to settle this new layer.

4. Add a layer of sand (both the top and the bottom sand layers should be about 0.5 cm thick) to prevent disturbance of the surface when solvent is added. Tap again to settle the new layer.

Choosing a solvent:

Find a solvent combination that moves your desired product with an R_f of about 0.3 on thin layer chromatography (TLC–Chapter 8). This same solvent mixture should give good separation on the column. For a very close separation of a mixture of two or more components, gradient elution is often effective. If, for instance, the solvent mixture that gives an R_f of about 0.3 is 20% MTBE/petroleum ether, one might elute the column with first 5% MTBE/petroleum ether, then 10% MTBE/petroleum ether, then 20% MTBE/petroleum ether.

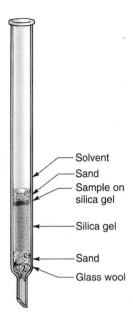

Solvent
Sand
Sample on
silica gel

Silica gel

Sand
Glass wool

Figure 7.2 Preparation of a chromatography column. Note that the silica gel is packed into the column as a dry powder.

To elute the column:

1. Add solvent, gently at first, so as to not disturb the sand on top of the column. When the solvent starts to drip out of the bottom of the column, begin collecting fractions—test tubes in a test tube rack work well. Keep adding solvent to the top of the column so it does not go dry. If the column runs slowly, gentle air pressure (5–15 psig) may be applied.

2. Check each fraction by TLC. Note that more than one fraction can be spotted on a plate. Combine in a tared flask the fractions that contain only pure product, then evaporate the solvent.

EXPERIMENTS

SAFETY NOTE

Care should be taken with the volatile solvents used. Diethyl ether, MTBE, and petroleum ether are flammable. Acetylferrocene and caffeine are toxic.

A. SEPARATION OF FERROCENE AND ACETYLFERROCENE

(Microscale)

Acetylferrocene is prepared from ferrocene (Fig. 7.3) by a Friedel-Crafts reaction, and the product is readily separated from starting material by column chromatography. Since both materials are colored, it is easy to follow their separation.

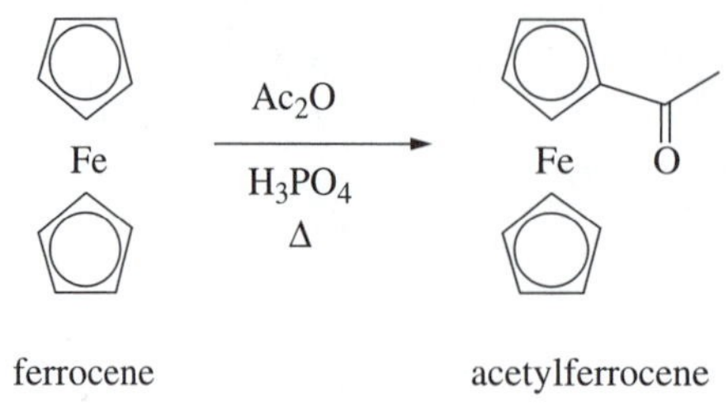

Figure 7.3   Preparation of acetylferrocene.

PROCEDURE

Preparation: Combine 0.4 g of ferrocene, 2 mL of acetic anhydride, and 0.5 mL of 85% phosphoric acid in a test tube. Stir the mixture well, then heat it in a 60 to 80°C water bath for 5 minutes. Add an equal volume of crushed ice, stir until the ice has melted, then extract the aqueous suspension with methylene chloride. Wash the methylene chloride extract with 5% aqueous NaOH. Dry the methylene chloride solution over anhydrous Na_2SO_4, and decant the dried solution into a clean flask. Check this solution by TLC, using 4:1 petroleum ether/MTBE. You should observe an orange spot for the product at $R_f = 0.31$, and perhaps a yellow spot for the starting material at $R_f = 0.79$. Add 0.5 gram of silica gel to the rest of the solution, and evaporate to give a dry orange powder.

   Purification: Prepare a silica gel column as outlined above, using five grams of silica gel in a 1.9 cm i.d. $\times$ 30 cm column. Tap to settle the silica gel, then add the dry orange powder prepared above. Tap to settle, then add a layer of sand. Have fifteen test tubes, each of which can hold at least 5 mL, ready in a test tube rack.

   Add 75 mL of 4:1 petroleum ether/MTBE in portions to the column. As the solvent moves down the column, you will probably observe a yellow band followed by an orange band. Collect the effluent from the column in 5 mL fractions, and check each of the fractions by TLC (remember, you can often spot three or even four fractions on the same TLC plate). Combine in a tared flask the fractions that have the pure product spot, evaporate the solvent, and record the weight of the product acetylferrocene. Record the melting point and calculate the yield of the reaction. The product may be recrystallized from petroleum ether or hexane. The literature mp's of ferrocene and acetylferrocene are 173–174°C and 85–86°C respectively.

B. ISOLATION OF CAFFEINE

Tea and coffee contain caffeine (Fig 7.4), an alkaloid that is a central nervous system stimulant. Several commercial pain relievers also have caffeine added to counteract the drowsiness produced by the analgesics that they contain. In the following procedure caffeine is extracted from tea, which contains 3 to 4% caffeine, and then purified by column chromatography.

PROCEDURE

Obtain 40 mL of brewed tea, which contains approximately the extract of two tea bags. Alternately, place two tea bags in a 125-mL Erlenmeyer flask with 45 to 50 mL of distilled water and bring the mixture to a gentle boil. After two minutes, remove the heat and allow the solution to cool. Squeeze out the tea bags and discard them. Dilute the tea with 20 mL of 5% aqueous HCl to help prevent emulsion formation, and extract it with four 10-mL portions of methylene chloride. Do not shake the separatory funnel; just swirl it with moderate vigor so an emulsion might

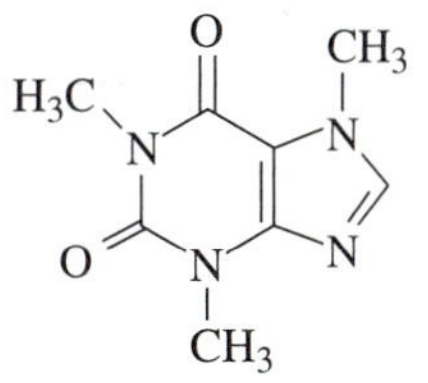

caffeine

Figure 7.4   Caffeine is the stimulant that is found naturally in coffee and tea, and that is added to some soft drinks.

be avoided. The combined extracts are dried over Na_2SO_4 and filtered through a small cotton plug, silica gel (0.2 g) is added, and the solvent is evaporated on a steam bath or rotary evaporator to give a dry powder.

The chromatographic column is packed according to the instructions using an approximately 10×150-mm column and 0.7 g of silica gel. Be sure to have your solvents and collecting test tubes ready before applying the solvent! Elute the column with three 5-mL portions of methylene chloride, then two 5-mL portions of 5% ethyl acetate in methylene chloride, then three 5-mL portions of pure ethyl acetate. Eluate from the column is collected in eight 13×100-mm test tubes held in a test tube rack (collect approximately 5 mL in each test tube); number these and keep them in order.

The contents of each test tube may be checked by TLC with comparison to an authentic sample of caffeine using silica plates and 1-butanol/ethyl acetate/methyl isobutyl betone (2:9:9) as the solvent. Spot each tube of eluate several times in one place to make sure that enough sample is applied. If the column has worked properly, early yellow fractions that smell like tea will be followed by fractions that contain caffeine. Combine the fractions that contain caffeine in a tared round-bottom flask, evaporate the solvent, and weigh the fluffy white residue. The literature melting point of caffeine is 238°C.

QUESTIONS

1. Suppose that your mixture showed good TLC separation with methylene chloride, but you inadvertently eluted the column with methanol. What would happen?
2. What would be the effect on the results of a column chromatogram if
    a. the top layer of sand was not included in making the column?
    b. solvent was not maintained above the adsorbent, and a crack developed in the column?
    c. you do not number the test tubes in which you are collecting fractions?
3. List the similarities between gas chromatography and column chromatography.

PRELABORATORY QUESTIONS

1. Why is the glass wool included in the construction of the chromatography column?
2. Which would elute more quickly from a silica gel column, a ketone or an alcohol?
3. Which is more polar, ether or petroleum ether?
4. If the components of a mixture are colored, it is a simple matter to follow the components as they come down the column and collect them separately as they exit the column. How would you detect different components that were not colored?
5. Draw the structure of ferrocene.

REFERENCES

Bobbitt, J.M. *Introduction to Chromatography*; Van Nostrand Reinhold: New York, 1968.

Chromatography: A Laboratory Handbook of Chromatographic and Electrophoretic Methods; Heftmann, E., Ed.; Van Nostrand Reinhold: New York, 1975.

Perry, S.G.; Amos, R.; Brewer, P.I. *Practical Liquid Chromatography*; Plenum Press: New York, 1972.

Snyder, L.R.; Kirkland, J.J. *Introduction to Modern Liquid Chromatography*, 2nd ed.; Wiley-Interscience: New York, 1979.

Still, W.C.; Kahn, M.; Mitra, A. *J. Org. Chem.,* 1978, *43,* 2923.

Taber, D.; Hoerrner, R.S. *J. Chem. Educ.,* 1991, *68,* 73.

Thin Layer Chromatography

8

Thin layer chromatography (TLC) is a primary tool for rapid qualitative analysis and is extremely effective and convenient for this purpose. The chromatography is carried out on a plate that is covered on one side with a thin (200 micrometers to 1 mm) coating of silica gel.

Very small amounts of samples are applied to the adsorbent surface as small spots in a row near one end of the plate (Fig. 8.1). The plate is then placed, sample-end down, in a widemouth jar containing a shallow pool of solvent. As the solvent rises over the adsorbent layer by capillary action, the compounds in the samples move to varying heights on the plate. The individual compounds can then be detected as separate spots on the plate.

The chief uses of TLC are to determine the number of components in a sample, to detect a given compound or compounds in a very crude mixture, and to serve as a preliminary trial in finding conditions before running a chromatography column. Since tiny amounts of material are exposed on an open surface, the use of TLC is limited to relatively nonvolatile substances; in this respect it is complementary to gas chromatography. Neither the sensitivity nor the resolving power, however, are as high as in gas chromatography (GC), and the detection methods do not lend themselves readily to quantitative determinations.

The easiest method for visualizing spots is illumination of the plate with an ultraviolet (UV) lamp. The TLC adsorbent layer contains a trace of fluorescent dye. Compounds that are fluorescent will show up as bright spots on a light background; compounds with UV chromophores will appear as dark spots, since they adsorb the UV light and prevent fluorescence of the dye.

Alternatively, the developed TLC plate can be treated with a general reagent, such as iodine vapor. Nearly all compounds adsorb iodine or react with it to form violet or brown spots on the slide. However, the relative intensity of the spots is not an accurate indication of the amounts of the compounds present, since the extent of reaction varies. Other widely used general reagents are phosphomolybdic acid, ceric sulfate in sulfuric acid, and vanillin in acidic ethanol.

The distance that a compound travels on the TLC plate is expressed as the **R_f value**, which is the ratio of the distances from the starting line to the compound and to the solvent front. Under exactly the same experimental conditions, the R_f for a compound is a characteristic value. However, R_f may be affected by small changes in the coating thickness, solvent, temperature, the amount of sample applied, and the presence of other compounds. In practice, it is usually not possible to control all of these conditions completely. For this reason, a comparison of two samples suspected of being the same compound should always be carried out by spotting both samples on the same plate, one on one side of the plate, the other on the opposite side of the plate, with both spotted together on the same spot in the middle.

A typical example of the use of TLC is the examination of a reaction mixture as illustrated in Figure 8.1. A sample of the crude mixture before the isolation of products, spotted in the left lane, shows the presence of one main component and one minor component, with R_f values of 0.38 and 0.16 respectively. A small dark spot at the starting line indicates some polymeric tar. Comparison with the right-hand lane, in which the starting material of the reaction is spotted,

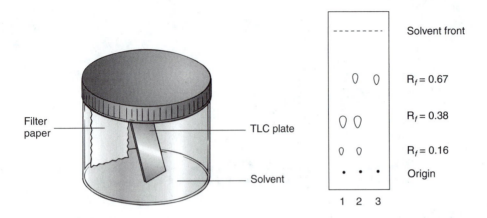

Figure 8.1 How to develop a TLC plate. (1) Reaction mixture. (2) Co-spot. (3) Starting material.

shows that the starting material has been consumed. The reaction mixture and the starting material are together in the center lane.

APPLICATION OF THE SAMPLE

The sample is applied (Fig. 8.2) with a *very fine* capillary. This should be about one-tenth the diameter of a melting-point capillary and can be prepared conveniently by softening a 1-cm section in the middle of an open-ended melting-point capillary with a microburner and drawing it out to about 4 to 5 cm. The thin portion is then broken to make two TLC capillaries. A larger quantity of capillaries can be made from a soft-glass transfer pipet; the pipet is drawn out to the appropriate thickness (about a 2-foot length) and the capillary section is broken into 3-inch pieces.

The sample to be applied can be dissolved in any volatile solvent; acetone or methylene chloride is convenient. Make a roughly 5 to 10% solution of the sample in a 10 × 75-mm test tube or 10-mL Erlenmeyer flask. With a sharp pencil, mark a light, straight line approximately 1 cm from the bottom edge of the plate. Do not dig into the silica gel surface. Dip a capillary into the solution and then touch the end very lightly at a spot on the pencil line (Fig. 8.2). It is a good idea to practice sample spotting first on a scrap plate to check your technique and to

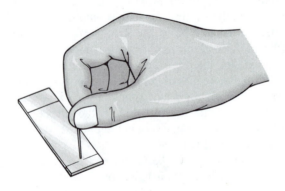

Figure 8.2 How to spot a TLC plate. Note that it is possible to put four or even five spots side by side on the same TLC plate.

make sure that the capillary does not deliver too large a spot. The spot should spread to a diameter of no more than 1 mm. If it is necessary to apply a larger amount, gently blow the first spot completely dry and then touch the capillary again at the same place. When working with UV fluorescent or adsorbing substances, a useful technique is to carry out the spotting under a UV lamp. This makes it considerably easier to see the size of the spot and the amount of sample applied. It is important that no crystals from the sample be transferred to the plate, since this causes streaking during development.

Two common errors are allowing the spot to become too broad and applying too much sample. For a preliminary investigation, it is useful to make several spots with one, two, or three applications to determine the proper volume of solution to apply. Three, four, or even five spots can easily be placed side-by-side on a microscope slide.

Care should be taken not to touch the adsorbent surface with the fingers. Always hold the TLC plate by its edges, as you would a CD!

DEVELOPMENT

The choice of solvent for development depends on the nature of the compounds to be separated. For relatively nonpolar compounds, the solvents shown in the upper part of Table 7.1 are appropriate. To obtain the proper polarity, a few percent of a polar solvent such as MTBE or acetone can be added to a less polar one, such as hexane. A third or even fourth component is sometimes used to further adjust the polarity and R_f values.

The best separation of closely similar compounds by TLC is usually achieved when the R_f values are 0.3 to 0.6, since the spots run together as they move farther up the plate. Ten to 30% MTBE in hexane is a generally useful TLC solvent for a fairly wide range of compounds. If a less polar mixture is desired, 10 to 30% methylene chloride in hexane works well. If a more polar solvent is necessary, 10 to 30% acetone in methylene chloride is often suitable.

Widemouthed screw-cap or snap-cap bottles are useful chambers for small-scale TLC work. To maintain equilibrium conditions during the development of the chromatogram, it is important that the space in the developing chamber above the solvent remains saturated with solvent vapor. This is accomplished by fitting a piece of filter paper around the inner wall of the jar to act as a wick; the wick must not be in contact with the TLC plate during development and must be placed in such a way so at least the upper part of the plate will be visible. Solvent is added to the jar to a depth of about 0.5 cm, saturating the wick; the solvent level must not be above the sample spots on the plate. Wait a few minutes for equilibration and then lower the plate into the chamber, spotted end down, with the silica gel surface facing the filter paper, and cap the jar.

When the solvent has risen to about 0.5 cm from the top of the slide, remove the plate, immediately mark the solvent front with a sharp pencil, and allow the excess solvent to evaporate. Visualize the chromatogram by viewing with a UV light or placing the plate for a few minutes in a jar containing a few crystals of iodine. Mark the outline of the spots with a sharp pencil (spots visualized with I_2 often fade in a few minutes).

If the R_f values are too low, that is, if the compounds have moved only a short distance, the chromatogram can be developed a second (and third) time, allowing it to dry between each development. Another possibility is to repeat the chromatogram with a freshly spotted plate, using a more polar solvent mixture.

EXPERIMENTS

A. TLC OF NITROANILINES

This experiment is a simple demonstration of TLC to provide practice in the basic techniques of the method.

SAFETY NOTE

Nitroanilines are toxic; avoid contact with the skin. Avoid breathing of methylene chloride vapors.

PROCEDURE

Obtain eight TLC plates from your instructor, and prepare six to eight TLC capillaries. Obtain, in labeled 10 × 75-mm test tubes or vials, very small samples of *ortho-*, *meta-*, and *2, 4-*dinitroaniline (about 5 mg, or just enough to cover the bottom of the test tube, is sufficient). Dissolve the samples in approximately 0.5 mL of acetone; the entire sample must be in solution.

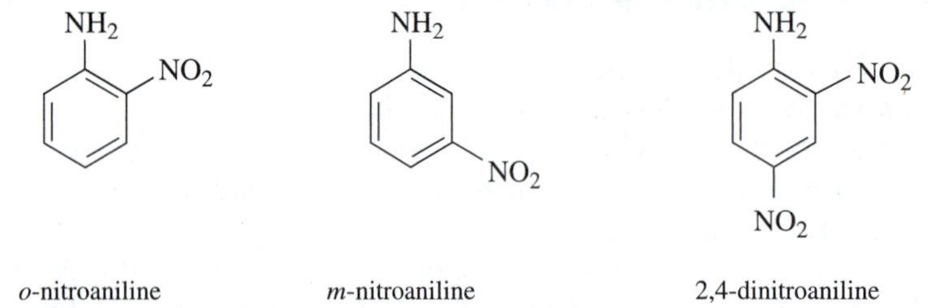

| *o*-nitroaniline | *m*-nitroaniline | 2,4-dinitroaniline |

Figure 8.3   The three nitroaniline derivatives separate easily on TLC.

On one plate, spot samples of each of the three nitroanilines. Develop the plates with methylene chloride. If the spots are not distinct or are too large or have run together, repeat the chromatogram. Sketch the appearance of the developed plates in your notebook, and then develop again. If you feel it may be useful, develop a third time. The spots should be visible on the dry plate, but placing them in an iodine vapor jar or viewing them under UV light may bring them out more distinctly. (It is not practical to redevelop a plate that has been exposed to iodine, even if the color disappears, since some chemical change may have occurred.)

On other plates, spot mixtures of two or three of the compounds in various combinations in the same lane; apply the second solution right over the spot of the first, taking care to avoid overloading. Spot the individual compounds in other lanes on the same plates. Develop and record the appearance of each plate.

When satisfactory results have been obtained with single compounds and known mixtures, obtain samples of unknowns from your instructor and determine which compound(s) is present in each.

B. TLC OF ANALGESIC DRUGS

Analgesics (compounds that relieve pain) range from aspirin, which is consumed at the rate of many millions of pounds per year, to morphine and related narcotics. In addition to aspirin, several other chemically similar compounds are widely used in nonprescription or proprietary analgesic tablets. Among these are acetaminophen and ibuprofen. Caffeine is sometimes added to these formulations to overcome drowsiness. In addition to the active ingredients, the tablets of these drugs contain starch, lactose, and other substances that act as binders and permit rapid solution. Inorganic bases are sometimes also included.

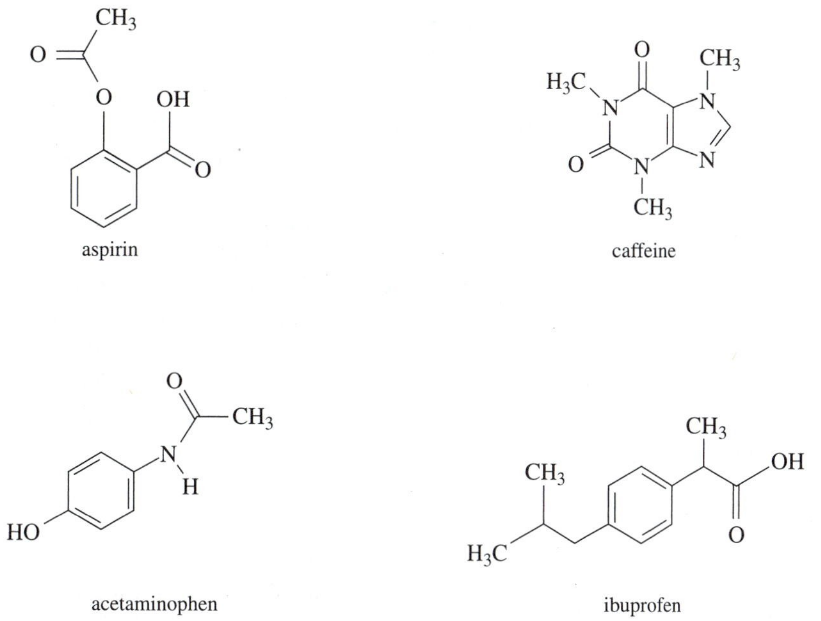

Figure 8.4   The four components of over-the-counter analgesics.

In this experiment, you will obtain as an unknown a proprietary analgesic drug. The objective is to identify the unknown drug by TLC comparison with several known compounds. The unknown will be one of those listed in Table 8.1. The amounts of ingredients in some cases are given in *grains* per tablet (grain [gr] is an apothecary unit; 1 gr = 64 mg).

Pure samples of the compounds found in the analgesics are used as reference standards. These are aspirin (acetylsalicylic acid), acetaminophen (4-acetamidophenol), caffeine, and ibuprofen.

Several solvent systems have been recommended for the separation of these analgesic drugs, and the relative merits of these solvents have been compared (see Roswell, D. F., and Zaczek, N. M.).

The solvent system you will use is a mixture of petroleum ether, methylene chloride, and ethyl acetate in a 1:1:2 ratio by volume. The effect of slight changes in polarity can be observed by comparing ethyl acetate alone and with added ethanol and acetic acid, using a combination of the reference compounds and drug samples.

The proper choice of solvent system depends somewhat on the plates and the conditions used for the experiment; it is best to experiment in advance with several systems to find the one that is most effective.

Table 8.1   Common Analgesic Drugs

DRUG (BRAND NAME)	INGREDIENTS
Anacin	aspirin, caffeine
Motrin	ibuprofen
Tylenol	acetaminophen
Vanquish	aspirin, acetaminophen, caffeine

PROCEDURE

Place approximately 10 mg of each of the four reference compounds in labeled vials or 10 × 75-mm test tubes and dissolve the samples in a few drops of methanol. In a fifth tube place a quarter of an analgesic tablet and add 1 mL of methanol. Crush the tablet with a rod, stir well, and allow the insoluble material to settle. Note that aspirin is not stable in methanol, so fresh solutions must be made each lab period.

TLC the samples as outlined above. Mark the solvent boundary with a pencil mark or small scratch. Examine the chromatogram under a UV lamp and sketch the appearance of the plate in your notebook, indicating the location and approximate size of the spots and any distinctive colors. After this examination, place the plate in a jar of iodine vapor for approximately 30 seconds, remove, and again record the appearance. Identify the spots in the chromatogram, including as many of the spots in the unknown lane as possible. From the identity and number of the spots in the unknown and the composition of the possible unknowns, deduce the identity of your unknown.

QUESTIONS

1. Explain why it is important to develop a TLC plate in a closed container and to have the chamber saturated with the vapor of the developing solvent.
2. Calculate the R_f values for Compound A and Compound B from the following data:

SUBSTANCE	DISTANCE TRAVELED
Solvent	12.8 cm
Compound A	4.0 cm
Compound B	7.5 cm

3. What will be the result of the following errors in TLC technique?
   a. too much sample applied
   b. solvent of too high polarity
   c. solvent pool in developing jar too deep
   d. forgetting to remove the TLC plate when the solvent has reached the top of the plate
4. Explain why the presence of crystals of the sample at the starting point causes streaking of the plate during development.

PRELABORATORY QUESTIONS

1. What are the important considerations in handling TLC plates?
2. Circle the more polar solvent in each pair:
   a. diethyl ether or 1-butanol
   b. chloroform or hexane
   c. ethyl acetate or acetic acid
3. Describe the process whereby iodine visualizes otherwise invisible components of a developed thin-layer plate.
4. Why must the spots at the origin be placed above the level of the solvent?
5. Define the term "develop" as applied to TLC.

REFERENCES

Bobbitt, J.M. *Thin Layer Chromatography*; Van Nostrand Reinhold: New York, 1963.

Fried, B.; Sherma, J. *Thin-Layer Chromatography*, 2nd ed.; Marcel Dekker: New York, 1986.

Kirchner, J.G. Thin Layer Chromatography, in *Techniques of Chemistry*; Perry, E.S.; Weissberger, A., Eds.; Wiley-Interscience: New York, 1978; Vol. XIV.

Lien, V.T. *J. Chem. Educ.* 1971, *48,* 478.

Randerath, K. *Thin-Layer Chromatography*; Academic Press: New York, 1966.

Roswell, D.F.; Zaczek, N.M. *J. Chem. Educ.* 1979, *56,* 834.

Thin Layer Chromatography, A Laboratory Handbook, 2nd ed.; Stahl, E., Ed.; Springer-Verlag: New York, 1969.

Touchstone, J.; Dobbins, M.F. *Practice of Thin Layer Chromatography*; John Wiley and Sons: New York, 1978.

High Performance Liquid Chromatography (HPLC)

We have described solid–liquid chromatography in Chapter 7 (Column Chromatography) and Chapter 8 (TLC). High performance liquid chromatography (HPLC) works in the same way: a sample is applied to one end of a column filled with solid support, and an eluting solvent is then passed through. The sample spends a differential part of its time either in the solvent or adsorbed on the stationary phase, and so is separated into its components.

HPLC differs from column chromatography and TLC in that much more exacting separations can be achieved. To understand how this is done, one must first understand that loss of resolution during chromatographic separation is due primarily to uneven solvent flow over the microscopically rough and irregular surfaces of the silica gel packing. HPLC grade column packings are very even spheres, often 5 or 10 microns in diameter.

Other solid supports can be used with HPLC. Chemically bonding long chain alkyl ("octadeylsilyl") groups to the silica gel particles, for instance, leads to a column packing that is very nonpolar. This is called **reverse phase** silica gel. On this column packing, nonpolar materials are retained, and polar materials come off more quickly. This is the opposite of normal **(straight phase)** silica gel. Often, partially aqueous solvent mixtures are used as the eluate in reverse phase columns. Reverse phase silica gel can also be used for TLC and for open column chromatography. Cationic or anionic exchange resins, size exclusion supports, and many other materials with special properties are also used for HPLC.

HPLC columns operate at several hundred psi, so they are constructed from stainless steel tubing, with pressure fittings on the ends (Fig. 9.1). Special procedures are followed to ensure even packing of the solid support. A solvent pump runs pressurized solvent through the column. The effluent from the column then passes through a detector. Usually the detector measures UV absorbance at a particular wavelength, but other methods of detection such as fluorescence, refractive index, and radioactivity have also been used.

Advances in automation of liquid chromatography have paralleled those in gas chromatography with the result that, in many instances when a nonvolatile mixture is to be separated, automated liquid chromatography is preferred over column chromatography or thin layer chromatography. HPLC is the analytical method of first choice in many industrial settings.

EQUIPMENT

Modern laboratories now use instruments that are completely microprocessor controlled. After the correct parameters have been selected (usually the hardest part of the process), the chemist only has to inject the sample, sit back, and let the computer do its job. The computer proceeds to select preset solvent mixtures, regulate the solvent pumps, and record the data. Many instruments are capable of presenting final reports on hard copy after the experiment is completed.

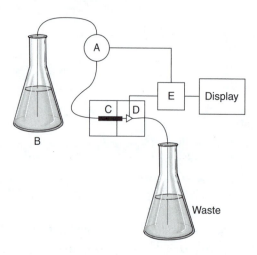

Figure 9.1   The necessary parts of an HPLC system. A high pressure pump (A) forces the solvent (B) through a column of stationary phase (C) and thence to a detector (D), which feeds its information to a computer (E) that manipulates the information and presents it in a usable format.

CAFFEINE IN BEVERAGES

EXPERIMENTS

For this experiment, students may choose from: instant coffee, instant tea, or several colas and soft drinks (some should be caffeine-free) to compare with a known standard.

Preparation of Standard Solutions. Dry a sample of pure caffeine in a 110°C oven for one hour. Prepare a 1000-ppm stock solution by dissolving 100 mg of caffeine in 100 mL of millipure water. Aliquots of this solution should be further diluted to make solutions of 5, 10, 20, and 40 ppm. With these solutions, a standard curve for quantitative analysis can be determined. These solutions can be shared with other members of the class.

Instrumental Conditions. The instrument should be run isocratically at ambient temperature using a 254-nanometer UV absorbance detector. Degassed and filtered HPLC-grade methanol/water (30:70) is used as the mobile phase. A standard C-18–bonded silica column works well.

Determination of Standard Curve for Caffeine. Filter the caffeine solutions through a 1.0-micron filter. Your instructor will show you how to run the HPLC instrument and will monitor your work. Inject 10 microliters of each sample onto a precolumn packed with 37-micron C-18–bonded silica that is connected to a 25 × 4.6-mm (i.d.) 10-micron C-18–bonded column. Plot the area under the caffeine peak vs. known concentration (ppm) for the standard curve.

Sample Selection. Each of the following should be run by at least one group in the class: brewed coffee, instant coffee, instant coffee (decaffeinated), brewed tea, instant tea (decaffeinated), regular cola, decaffeinated regular cola, diet cola, diet cola (decaffeinated), and a fruit cola.

Sample Preparation. Each sample should be filtered through Whatman #42 filter paper (or its equivalent), degassed, diluted 1:10 with deionized water, and, finally, filtered through a 0.5-micron filter prior to injection onto the instrument. Inject 10 microliters of each sample that you have selected (for precise work, several such injections would be necessary). Using the standard curve and data, determine the concentration of caffeine in each of your samples.

QUESTIONS

1. What is the purpose of filtering and degassing your samples prior to HPLC analysis?
2. Which commercial sample that you used has the highest caffeine content?
3. Does brewed tea or brewed coffee contain the higher caffeine content?
4. Do any of the "decaffeinated" drinks contain caffeine?

PRELABORATORY QUESTIONS

1. List three types of stationary phases used in HPLC. Describe when you might want to use each of these.
2. Convert 3000 psi into number of atmospheres.
3. Name two types of detection used in HPLC experiments.
4. Draw a sketch of the essential parts of an HPLC instrument.

REFERENCES

Christie, W.W. *High Performance Liquid Chromatography and Lipids; A Practical Guide;* Pergamon Press: New York, 1987.

DiNunzio, J.E. *J. Chem. Educ.* 1985, *62,* 446.

Englehardt, H. *High Performance Liquid Chromatography*; Springer-Verlag: New York, 1979.

Hancock, W.S.; Sparrow, J.T. *HPLC Analysis of Biological Compounds; A Laboratory Guide*; M. Dekker: New York, 1984.

High Performance Liquid Chromatography; Brown, P.; Hartwick, R.A., Eds.; Wiley-Interscience: New York, 1989.

High Performance Liquid Chromatography; Knox, J.H., et al., Eds.; Edinburg University Press: Edinburg, 1978.

High Performance Liquid Chromatography in Biochemistry; Henschen, A., et al., Eds.; VCH Publishers: Weinheim, 1985.

High Performance Liquid Chromatography in Forensic Chemistry; Lurie, I.A.; Wittwer, J.D., Jr., Eds.; M. Dekker: New York, 1983.

High Performance Liquid Chromatography of Peptides and Proteins; Separation, Analysis, and Conformation; Mant, C.T.; Hodges, R.S., Eds.; CRC Press: Boca Raton, Florida, 1991.

HPLC of Macromolecules; A Practical Approach; Oliver, R.W.A., Ed.; Oxford Press: New York, 1989.

HPLC in Nucleic Acid Research; Methods and Applications; Brown, P., Ed.; M. Dekker: New York, 1984.

HPLC of Small Molecules; A Practical Approach; Lim, C.K., Ed.; Oxford Press: Washington, DC, 1986.

McMaster, M.C. *HPLC, A Practical User's Guide*; VHC Press: New York, 1994.

Steam Distillation of Essential Oils

n Chapter 5 it was seen that distillation of a mixture (more precisely, a solution) of two miscible liquids depends on the vapor pressure and mole fraction of each of the components. There is a different situation in the distillation of a mixture of two compounds that are not mutually soluble. In this case, the vapor pressure above the mixture is the sum of the partial pressures of the components, $P_T = P_A + P_B$; that is, each exerts its own vapor pressure independently of the other (Dalton's Law). Since the vapor pressures are additive, the boiling point of the mixture (i.e., the temperature at which $P_T = 1$ atm) is lower than the boiling point of either of the components (Fig. 10.1).

As long as separate phases are present in the liquid, the mixture will have a constant boiling point. In addition, the distillate will have a constant composition, which is determined by the ratio of the vapor pressures (Eq. 10.1). Since the number of moles of a compound (n) is equal to its weight divided by its molecular weight (mol wt), Equation 10.2 can be derived for the weight ratio of compounds in the distillate.

$$\frac{n_A}{n_B} = \frac{P_A}{P_B} \qquad\qquad [10.1]$$

$$\frac{\text{wt. A}}{\text{wt. B}} = \frac{P_A}{P_B} \times \frac{\text{mw A}}{\text{mw B}} \qquad\qquad [10.2]$$

Thus, if the molecular weights and partial pressures of the two compounds are known, the weight ratio of the components in the distillate can be calculated. Conversely, if the vapor pressure and molecular weight of one component and the total pressure during the distillation are known, the molecular weight of the second component can be calculated from the weight ratio.

For example, assume that one component is water, the other is an unknown organic compound, X, and the distillation occurs at 99.4° at a barometric pressure of 750 mm to give a mixture of water and X in a ratio of 9g/1g. The partial pressure of X can be calculated from the total pressure and the partial pressure of H_2O at 99.4°C, which from handbook tables is 744 mm.

$$P_x = P_T - P_{H_2O}$$

$$= 750 - 744 \text{ mm}$$

$$= 6 \text{ mm}$$

Then

$$\frac{\text{mw X}}{\text{mw H}_2\text{O}} = \frac{\text{wt. X}}{\text{wt. H}_2\text{O}} \times \frac{P_{H_2O}}{P_x}$$

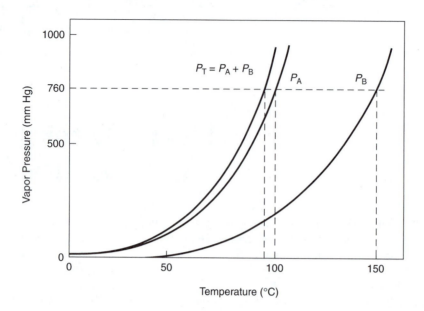

Figure 10.1 Vapor pressure versus temperature for a mixture of two immiscible compounds with boiling points of 100 and 150°C.

$$\frac{mw\ X}{18} = \frac{1}{9} \times \frac{744}{6}$$

$$mw\ X\ = 250$$

When one of the two components is water, this process is termed **steam distillation,** and it provides a means of distilling a slightly volatile compound at a temperature far below its atmospheric boiling point. Steam distillation is useful in the separation of a water-insoluble organic compound that is present in a large amount of nonvolatile material. In this case, simple distillation, even in a vacuum, is ruled out because destructively high temperatures would be needed to distill the relatively small amount of volatile material and because mechanical entrapment will prevent its complete removal. Steam distillation is a useful alternative to extraction for the isolation of volatile organic compounds, such as essential oils from plant material. Extraction with solvents removes gums and fats as well as the volatile oils; the latter are separated selectively by steam distillation.

APPARATUS AND TECHNIQUE

Steam distillation on a macroscale is most conveniently carried out in a two- or three-neck round-bottom flask, or a one-neck flask fitted with a Claisen adapter. A typical setup is illustrated in Figure 10.2. Steam is provided by the addition of water and sufficient heat from a heating mantle or other suitable heat source. Alternatively, steam can be supplied from an external boiling flask or a steam line on the desk; this is necessary to provide agitation when steam distilling a tarry or very bulky mass. When using an external source of steam, a trap must be included between the source of steam and the distilling flask since a drop in steam pressure can cause the contents of the flask to be sucked back through the steam inlet. Steam distillation on a microscale can be carried out similarly, except in a smaller apparatus.

Figure 10.2   Steam distillation setup.

ESSENTIAL OILS

The characteristic aromas of plants are due to volatile or essential oils, which have been used since antiquity as a source of fragrances and flavorings (note that the term "essential" here does not mean necessary but derives from the word "essence"). These oils occur in all living parts of the plant; they are often concentrated in twigs, flowers, and seeds. Essential oils are generally complex mixtures of hydrocarbons, alcohols, and carbonyl compounds, mostly belonging to the broad group of plant products known as terpenes.

Some terpenes, such as the hydrocarbon limonene, are found in many different plant species. Certain other compounds, particularly aromatic aldehydes and phenols, occur as a major constituent in only one or a few plants. Thus, the characteristic aromas of clove and cinnamon oils, for example, are due mainly to one compound in each case. However, the full flavor and bouquet of a spice or flavoring usually depend on a blend of several substances and cannot be reproduced by a single compound.

These essential oils are isolated from plant tissue by steam distillation—sometimes, in remote parts of the world, using a very primitive apparatus. Individual compounds can be separated from the essential oil by fractional distillation at reduced pressure, crystallization, or chromatography. In this experiment, the essential oil will be steam distilled on a macroscale from one of four widely used spices—caraway, clove, cumin, or cinnamon—and the major constituent will be characterized as a derivative. A further experiment describes the isolation of the essential oil of caraway on a microscale.

EXPERIMENTS

A. ISOLATION OF THE ESSENTIAL OILS OF CARAWAY, CLOVE, OR CUMIN

(Macroscale)

SAFETY NOTE

Since steam is being generated, care should be taken to avoid burns. Exposure to methylene chloride fumes should be minimized.

Place 10 g of the freshly ground spice (either caraway, clove, cumin, or cinnamon*) in a 500-mL round-bottom flask, add 150 mL of water and a boiling stone, and mark the water level with a grease pencil (alternately, use a 250-mL flask and 75 mL of water). Assemble the setup similar to that shown in Figure 10.2, with a 125-mL Erlenmeyer flask as the receiver. (Note that for the present experiment the thermometer may be omitted; if so, stopper this opening.) Fill the dropping funnel with water, add a boiling stone, and heat the mixture until a steady distillation begins. Occasionally add water from the funnel to maintain the original level. Collect the distillate until no further droplets of oil can be seen; a minimum of 100 mL should be distilled and collected.

Pour the distillate into a separatory funnel, rinse the receiver with 10 mL of methylene chloride, and add this to the funnel. Shake the funnel, allow the layers to separate completely, and drain the lower organic layer into a 25-mL Erlenmeyer flask. Extract the aqueous layer again, using 5 mL of methylene chloride, and drain the lower layer carefully into the flask. Add a small amount of sodium sulfate, swirl, and transfer the solution with a pipet and bulb through a cotton wad in a funnel into another tared flask. Evaporate the methylene chloride on the steam bath in a hood or on a rotary evaporator until the solution has been concentrated to an oily residue.

Weigh the oil and calculate the percent yield based on the weight of plant material. Continue with either Experiment C, Experiment D, or Experiment E.

B. ISOLATION OF THE ESSENTIAL OIL OF CARAWAY

(Microscale)

Place 2.0 g of ground caraway in a 100-mL round-bottom flask and add 50 mL of water. Assemble the apparatus as shown in Figure 10.2 with a 50-mL Erlenmeyer flask as the receiver. Fill the dropping funnel with water and mark the water level in the round-bottom flask with a grease pencil, as you will want to maintain the level roughly constant. Add a boiling stone and heat the mixture until steady distillation begins, then add water at a rate to maintain the original level. Collect the distillate until the distillate is no longer cloudy; a minimum of 35 to 40 mL should be collected.

Pour the distillate into a separatory funnel, rinse the receiver with 2 mL of methylene chloride, and add this to the funnel. Shake the funnel, allow the layers to separate completely, and

---

* Cinnamon may be used, but to avoid frothing the cinnamon/water mixture must be degassed by repetitive evacuations before distillation (Taber, D.F.; Weiss, A.J. *J. Chem. Educ.* 1998, *75,* 633). Alternatively, frothing can be minimized by the addition of 2 mL of 6N aqueous HCl/10 g of cinnamon (Castrillon, J. *J. Chem. Educ.* 1999, *76,* 318). Cinnamaldehyde gives a semicarbazone of melting point 215°C (see Experiment E).

drain the lower organic layer into a tared 25 × 100-mL test tube or 10-mL Erlenmeyer flask. Extract the aqueous layer again, using 2 mL of methylene chloride, and drain the lower layer carefully into the same test tube or flask. If the layer separations have been made cleanly, drying of the solution is unnecessary. Should there be a significant amount of water present, add a small amount of sodium sulfate, swirl, and filter the solution with a pipet and bulb through another pipet containing a cotton wad into another tared test tube or flask. Carefully, evaporate the methylene chloride on the steam bath in a hood until the solution has been concentrated to an oily residue; then, tilt the warm tube or flask with the open end down and rotate, to spread the oil on the wall and allow the heavy solvent vapor to flow out.

Weigh the oil and calculate the percent yield based on the weight of plant material. Continue with the preparation of the hydrazone, which is described in Experiment C, Microscale.

C. CARAWAY OIL

Caraway seed, which is actually a dried fruit, is obtained from the biennial herb *Carum carvi,* grown mainly in Holland. The essential oil of caraway, varying from 1 to 3% of the seed weight, contains two principal compounds, carvone and limonene. Both compounds contain the 1-methyl-4-isopropylcyclohexane skeleton, which is the most common structural unit among monoterpenes. Carvone is the major constituent in good-quality caraway oil and is the dominant note in the aroma of seed oil. The carvone can be isolated almost quantitatively as the 2,4-dinitrophenylhydrazone, and the carvone content of the essential oil can be estimated from the weight of the derivative.

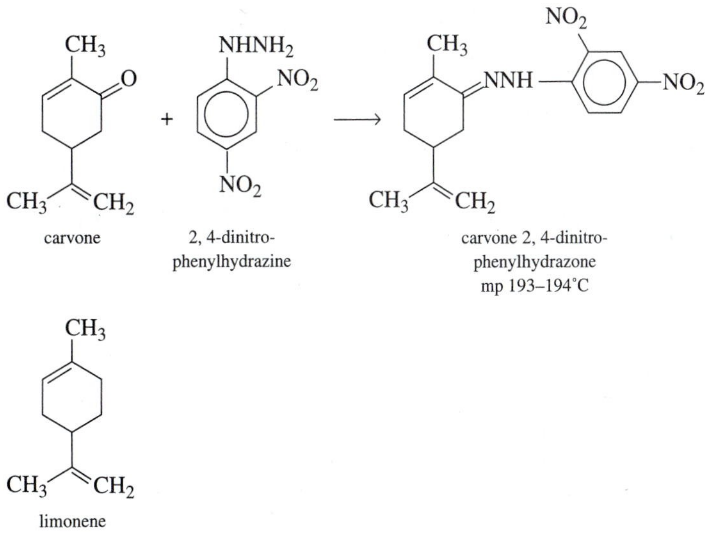

carvone   2, 4-dinitro-
             phenylhydrazine

carvone 2, 4-dinitro-
phenylhydrazone
mp 193–194°C

limonene

Figure 10.3   Derivatization of carvone.

SAFETY NOTE

Care should be taken to avoid contact of the 2,4-dinitrophenylhydrazine reagent with the skin. If skin contact should occur, immediately wash with copious amounts of water.

(Macroscale)

To prepare the hydrazone, dilute the oil with 1 mL of ethanol, transfer it to a test tube, and rinse the flask with another 1 mL of ethanol; then add 1 mL of 2,4-dinitrophenylhydrazine (DNPH) reagent.* Collect the solid on a Hirsch funnel, wash it with a small amount of ethanol, dry the crystals, and determine the melting point. If the yield of oil was insufficient to permit accurate weighing of the oil, weigh the hydrazone and, assuming 100% conversion of carvone to the derivative, calculate the weight and percentage of carvone in the caraway seeds.

(Microscale)

To prepare the hydrazone, dilute the oil with 0.5 mL of ethanol and add 0.25 mL of 2,4-dinitrophenylhydrazine reagent. Collect the solid on a Hirsch funnel, wash it with a small amount of ethanol, dry the product, and determine the melting point. As in the macroscale experiment, if accurate weighing of the oil is not possible, weigh the hydrazone, assume 100% conversion of the carvone to the derivative, and calculate the weight and percentage of carvone in the caraway seeds.

D.　CLOVE OIL

(Macroscale)

Oil of cloves (*Eugenia caryophyllata*) is rich in 4-allyl-2-methoxyphenol (eugenol); other compounds, including eugenol acetate and the sesquiterpene caryophyllene, are present in very small amounts. For characterization, the eugenol can be converted to the benzoate ester.

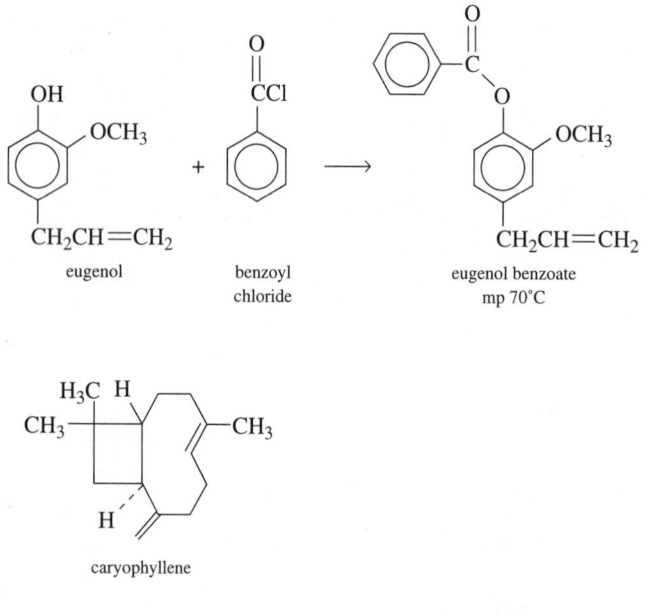

eugenol　　　　　benzoyl
　　　　　　　　　chloride

eugenol benzoate
mp 70°C

caryophyllene

Figure 10.4　Derivatization of eugenol.

---

* 2,4-Dinitrophenylhydrazine reagent is prepared by dissolving 1.0 g of 2,4-dinitrophenylhydrazine in 5 mL of conc H_2SO_4 and adding this solution to a mixture of 8 mL of water and 25 mL of ethanol.

SAFETY NOTE

Avoid breathing benzoyl chloride vapors and avoid skin contact. Should the latter occur, wash with copious amounts of water. The ether used in this extraction is highly flammable.

To prepare the benzoate, use all of the clove oil in your test tube or small Erlenmeyer flask and add 1 mL of water. Add 1 M NaOH solution drop by drop until the oily layer dissolves; a completely clear solution will not result, but there should be no oily droplets (note the phenolic eugenol, ArOH, is converted into the water-soluble ArO⁻). Although clove oil is mostly eugenol, acetyl eugenol and other neutral components will interfere with crystallization of the derivative. To extract these neutral components, add 3 mL of diethyl ether to the vessel, shake to mix, and remove the ether layer with a pipet; dispose of the ether in the waste jug; repeat the extraction with another 3 mL of ether. To the aqueous eugenol solution, add 4 to 5 drops of benzoyl chloride (use a hood); an excess is to be avoided. Stopper and shake the mixture and then warm the unstoppered vessel on a steam bath for approximately 5 minutes. Cool, add 2 to 3 mL of ether, and mix the layers by shaking. With a transfer pipet and bulb, remove the lower aqueous layer from the tube or flask. Add a small amount of anhydrous sodium sulfate to dry the ether solution, and transfer the solution through a pipet containing a cotton plug into another test tube. Add 1 mL of methanol, evaporate the solution to about 0.5 mL volume and scratch to crystallize the ester. If the ester does not crystallize, add about 1 mL of petroleum ether to the mixture and cool and scratch again. Collect and dry the crystals and determine the melting point.

E. CUMIN OIL

(Macroscale)

The dried fruit of cumin (*Cuminum cynimum*), a small shrub cultivated in eastern Europe and in India, is an important seasoning in curries, goulashes, and sausages. It is also a major component of chili powder used in Mexican food. The major volatile constituent of cumin is 4-isopropylbenzaldehyde (cuminaldehyde). Cumin oil also contains limonene and other compounds that contribute to the aroma of the condiment, as can be observed by comparing the odor of the oil before and after isolating the cuminaldehyde as the semicarbazone.

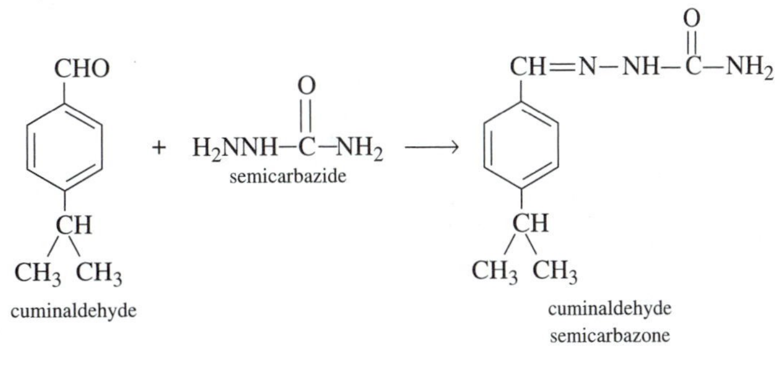

Figure 10.5 Derivatization of cuminaldehyde.

To prepare the semicarbazone, dissolve your oil in 3 mL of ethanol and transfer it to a test tube. Add to this a solution of 0.20 g of semicarbazide hydrochloride and 0.30 g of sodium acetate in 2 mL of water. Warm this mixture briefly on a steam bath, cool, and allow the derivative to crystallize. Collect the crystals in a Hirsch funnel and recrystallize from methanol; add enough methanol so that the solid just dissolves when heated on a steam bath, then cool in an ice bath to get crystals. Determine the melting point (literature value 216°C); recrystallize again if necessary.

QUESTIONS

1. A mixture of water and nitrobenzene distills at 99.3°C when the barometric pressure is 760 mm Hg. The vapor pressure of water at 99.3°C is 740 mm. Calculate the weight ratio of water to nitrobenzene that will distill.
2. The vapor pressure of aniline, $C_6H_5NH_2$, near 100°C is approximately 46 mm Hg. Estimate the amount of aniline that would steam distill if 500 g of water (steam) is collected at a barometric pressure of 760 mm Hg.
3. When an unknown liquid is steam distilled at 93°C (vapor pressure of water at this temperature is 589 mm) and 760 mm, the distillate contains 6.0 g of water and 11.0 g of oil. Calculate the molecular weight of the oil.

PRELABORATORY QUESTIONS

1. What is meant by "steam distillation"?
2. When one has a mixture of non–mutually soluble components, what is the relationship of the total pressure (P_T) in relation to the partial pressures (P_A and P_B) of the components?
3. When is it desirable to separate components by steam distillation rather than normal distillation?
4. Why is the boiling point of a steam distillation always less than 100°C?
5. Define the term "essential oil."

REFERENCE

Guenther, E. *The Essential Oils;* Van Nostrand: New York, 1949–1952; 6 vols.

Infrared Absorption Spectroscopy

nfrared absorption spectroscopy is the measure of the amount of radiation absorbed by compounds within the infrared region of the electromagnetic spectrum. The infrared region consists of radiation with wavelengths between 2.5 and 15 μm (1 μm = 1 micrometer or micron = 10^{-6} m), which corresponds to frequencies between 4000 and 650 cm^{-1} (cm^{-1} = "reciprocal centimeters" or wave numbers). An infrared spectrometer measures the amount of radiation absorbed as a function of the frequency of the radiation and provides absorption spectra such as those shown in Figures 11.2 through 11.8. Spectrometers can be constructed with a chart drive that is linear in either frequency or wavelength; the latter is in common use. The wavelength and frequency scales have a reciprocal relationship, and conversion from one to the other is very simple:

$$\bar{\nu}(\text{cm}^{-1}) = \frac{1}{\lambda(\text{cm})} = \frac{10,000}{\lambda(\mu\text{m})}$$

$$\lambda(\mu\text{m}) = \frac{10,000}{\bar{\nu}(\text{cm}^{-1})}$$

When electromagnetic radiation in the infrared region is absorbed by molecules, the molecules are excited to a higher energy state. Infrared radiation absorption causes energy changes on the order of 2 to 10 kcal/mole, and the excited state is one involving a greater amplitude of molecular vibration. Each absorption band or peak in the spectrum corresponds to the excitation of a different type (mode) of vibration of the atoms, and the positions of these peaks (measured in μm or cm^{-1}) provide useful information about the structure of the molecule. The intensity of the absorbance is proportional to the change in dipole moment for the motion, so a $C=O$ stretch will, for instance, be much more intense than a $C=C$ stretch.

For simple molecules containing few atoms, the number and position of peaks in the infrared spectrum can be calculated, and the spectrum can be completely analyzed. The major use of infrared spectra in organic chemistry, however, depends on empirical correlations of band positions with structural units. These correlations have been derived from spectra of a large number of compounds of known structure. By the use of this information, the presence or absence of certain functional groups or other structural features of a new compound can be determined.

Two types of vibrations, stretching and bending, are responsible for most of the peaks of importance in the identification of organic compounds. A few of the types of vibrations observed are shown in Figure 11.1, using CH_2 as a typical group. Also shown are the approximate frequencies of the radiation absorbed in exciting each type of vibration. When expressed in cycles per second (Hz), these are also the vibrational frequencies of the indicated motions. The vibrational frequency, and thus the frequency of radiation absorbed, is determined by the force constant for the deformation (i.e., the rigidity or strength of the bond) and the masses of the atoms that are involved in the vibration; specifically, the stronger the bond and the lighter the atoms,

93

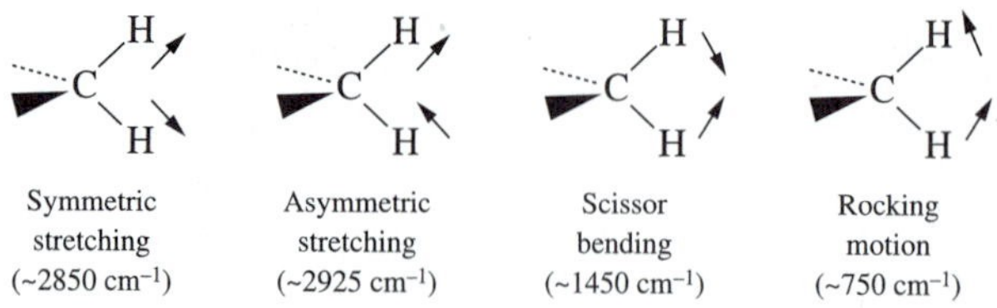

Symmetric stretching (~2850 cm^{-1}) Asymmetric stretching (~2925 cm^{-1}) Scissor bending (~1450 cm^{-1}) Rocking motion (~750 cm^{-1})

Figure 11.1 Normal modes of CH$_2$ vibration.

the higher is the vibration frequency. Bending motions are easier than stretching motions, so the former absorb at lower frequencies. These trends are illustrated in Table 11.1, which lists the types of vibrations that appear in various regions of the infrared spectrum.

Peaks in the first four regions listed in Table 11.1 are largely due to vibrations of specific types of bonds, and these are by far the most useful in compound classification. A few of the peaks appearing below 1500 cm^{-1} are characteristic of certain functional groups, but most of the absorption bands in this region are associated with vibrations of larger groups or the molecule as a whole. An infrared spectrum can be divided into two portions for examination. The region 4000 to 1500 cm^{-1} is useful for the identification of various functional groups. The region from 1500 to 600 cm^{-1}, sometimes called the "fingerprint region," is quite complex yet represents a unique pattern for each organic compound and, so, is useful for comparing two compounds for identification.

Table 11.1 Absorption Frequencies and Types of Vibrations

FREQUENCY (CM^{-1})	VIBRATION
3600–2700	X—H single bond stretching: O—H, N—H, C—H
3300–2500	Hydrogen-bonded O—H--X stretching
	Ammonium ion $\overset{+}{—N—}$H stretching
2400–2000	Triple bond and cumulated double bond stretching: C≡C, C≡N, N=C=O, C=C=O, N=C=N
1850–1550	Double bond stretching: C=O, C=N, C=C
1600–650	Single bond bending: NH$_2$, CH$_3$, C—C—C
	Single bond stretching: C—C, C—O, C—N

Fourier transform infrared spectroscopy (FTIR) is now being widely used. In this method, the electromagnetic radiation is split into two beams and one is made to travel a longer path than the other. Recombination of the two beams creates an interference pattern or **interferogram.** Mathematical manipulation (Fourier transformation) of this interference pattern by a computer converts it into the usual infrared (IR) spectrum. The advantages of FTIR are that the whole spectrum is measured in a few seconds, very small samples can be analyzed, and high resolution is obtained.

STRUCTURAL GROUP ANALYSIS

Extensive correlations exist between absorption peak positions and structural units of organic molecules; the most useful of these are summarized in Table 11.2. In spectra to be presented later, the vertical scale is percent transmittance, 100% being the top of the spectrum and 0% the bottom. Thus, absorption of radiation at a certain frequency results in a decrease in transmittance and appears as a dip in the curve (Figs. 11.2 through 11.8).

Table 11.2 Structural Units and Absorption Frequencies

BOND		TYPE OF COMPOUND	FREQUENCY (CM^{-1})
$-\overset{\mid}{\underset{\mid}{C}}-H$	(stretch)	Alkane	2800–3000
$=\overset{\mid}{C}-H$	(stretch)	Alkenes, aromatics	3000–3100
$\equiv C-H$	(stretch)	Alkynes	3300
$-O-H$	(stretch)	Alcohols, phenols	3600–3650 (free)
			3200–3500 (H-bonded)
			(broad)
$-OH$	(stretch)	Carboxylic acids	2500–3300
$-\overset{\mid}{N}-H$	(stretch)	Amines	3300–3500 (doublet for NH$_2$)
$-\overset{O}{\overset{\parallel}{C}}-H$	(stretch)	Aldehyde	2720 and 2820
$-\overset{\mid}{C}=\overset{\mid}{C}-$	(stretch)	Alkenes	1600–1680
$-\overset{\mid}{C}=\overset{\mid}{C}-$	(stretch)	Aromatic	1500 and 1600
$-C\equiv C-$	(stretch)	Alkynes	2100–2270
$-\overset{O}{\overset{\parallel}{C}}-$	(stretch)	Aldehydes, ketones	1680–1740
$-C\equiv N$	(stretch)	Nitriles	2220–2260
$C-N$	(stretch)	Amines	1180–1360
$-C-H$	(bending)	Alkane	1375 (methyl)
$-C-H$	(bending)	Alkane	1460 (methyl and methylene)
$-C-H$	(bending)	Alkane	1370 and 1385 (isopropyl split)
$-C-H$	(bending)	R—CH$=$CH$_2$	1000–960 and 940–900
$-C-H$	(bending)	R$_2$C$=$CH$_2$	915–870
$-C-H$	(bending)	*cis* RCH$=$CHR	790–650
$-C-H$	(bending)	*trans* RCH$=$CHR	990–940
$-C-H$	(out-of-plane bending)	*mono* subst. benzene	770–730 and 710–690
$-C-H$	(out-of-plane bending)	*o*-subst. benzene	770–735
$-C-H$	(out-of-plane bending)	*m*-subst. benzene	810–750 and 710–690
$-C-H$	(out-of-plane bending)	*p*-subst. benzene	860–800
$-C-O$	(stretch)	Primary alcohol	1050–1085
$-C-O$	(stretch)	Secondary alcohol	1085–1125
$-C-O$	(stretch)	Tertiary alcohol	1125–1200
$-C-O$	(stretch)	Phenol	1180–1260

For the more common structural units shown in Table 11.2, more specific assignments and deductions may often be made. Some of these are discussed in the following sections. For even more specific correlations and for similar data on the other functional groups, any of several books on this subject can be consulted (see References).

ALKANES

The most prominent peaks in infrared spectra of saturated hydrocarbons and from saturated portions of more complicated organic compounds are those due to C—H stretching and bending (see Fig. 11.2). The stretching frequencies are in the regions 3000 to 2800 cm^{-1} and are usually rather strong. Since most organic compounds contain several CH$_3$, CH$_2$, and/or CH groupings, the presence or absence of peaks in this region is simply taken to indicate the presence or absence of aliphatic C—H bonds in the molecule. The major bending modes of CH$_2$ and CH$_3$ groups appear in the 1470 to 1420 cm^{-1} and 1340 to 1380 cm^{-1} regions, but interpretation is usually complicated by the presence of several bands of this type or additional bands from other sources. The usually large number of different C—C bonds in an organic molecule makes the C—C stretching vibrations in the 1300 to 1100 cm^{-1} region uninterpretable in most cases. The "breathalyzer" sometimes used to check motorists for inebriation is an infrared device that measures "C—H" stretch on the breath.

ALKENES

Olefinic C—H stretching peaks generally appear in the region 3100 to 3000 cm^{-1}, thus differentiating between saturated and unsaturated hydrocarbons. Of the C—H bending modes, the out-of-plane vibrations in the 1000 to 650 cm^{-1} region were, before the advent of ^{1}H and ^{13}C-NMR, often used to deduce the substitution pattern of a double bond. This is illustrated in Table 11.2. The C=C stretching frequency in the 1600 to 1675 cm^{-1} region also varies with substitution, but to a lesser degree. The out-of-plane bending and C=C stretching absorptions of an alkene are illustrated in Figure 11.2.

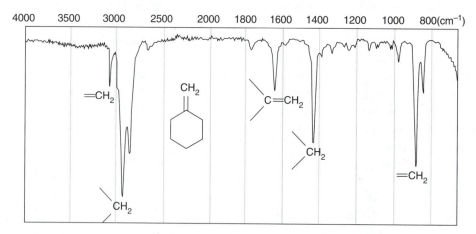

Figure 11.2   Infrared spectrum of an alkene.

ALKYNES

The C—H stretching vibration of terminal acetylenes generally appears at about 3300 cm^{-1} as a strong sharp band. The C≡C stretching band is found in the region 2150 to 2100 cm^{-1} if the alkyne is monosubstituted (Fig. 11.3) and at 2270 to 2150 cm^{-1} if disubstituted. Because the

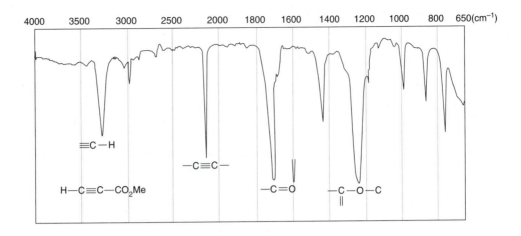

Figure 11.3   Infrared spectrum of an alkyne.

change in dipole moment is small, the latter are usually quite weak absorptions, unless the alkyne has a polarizing substituent such as an alcohol nearby (the partial peak at 1601 cm^{-1} in the figure is a frequency marker, from a polystyrene standard).

AROMATIC RINGS

Aromatic C—H stretching absorption appears in the region 3100 to 3000 cm^{-1}. This fact, taken with the corresponding frequencies of aliphatic and olefinic C—H stretching bands, allows a reliable determination of the types of carbon-bound hydrogen in the molecule. Aromatic C—H out-of-plane bending bands in the 900 to 690 cm^{-1} region are reasonably well determined by the substitution pattern of the benzene ring, as indicated in Table 11.2. In the absence of other interfering absorptions, such as those from nitro groups, these strong, usually sharp bands can be used in distinguishing positional isomers of substituted benzenes. Sharp peaks at approximately 1600 and approximately 1500 cm^{-1} are very characteristic of all benzenoid compounds; a band at 1580 cm^{-1} appears when the ring is conjugated with a substituent (Fig. 11.4).

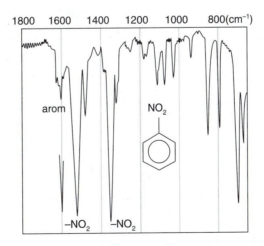

Figure 11.4   Infrared spectrum of an aromatic compound.

ALCOHOLS, PHENOLS, AND ENOLS

The very characteristic infrared band due to O—H stretching appears at 3650 to 3600 cm^{-1} in dilute solutions. In spectra of neat (undiluted) liquids or solids, intermolecular hydrogen bonding broadens the band and shifts its position to lower frequency (3500 to 3200 cm^{-1}) (Fig. 11.5). Intramolecular hydrogen bonding (to C=O, —NO$_2$ groups), as in enols, lowers the frequency and broadens the absorption even more. Strong bands due to O—H bending and C—O stretching are observed at 1500 to 1300 cm^{-1} and 1220 to 1000 cm^{-1}, respectively. In simple alcohols and phenols, the exact position of the latter is useful in classification of the hydroxyl group (see Table 11.2). Within the range given, typical frequencies observed are phenols > tertiary > secondary > primary. Further branching and unsaturation also affect the frequency somewhat.

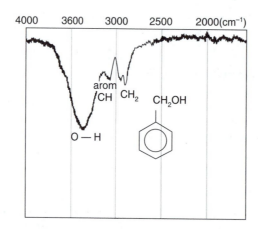

Figure 11.5   Infrared spectrum of an alcohol.

ALDEHYDES AND KETONES

The C=O stretching frequencies of saturated aldehydes and acyclic ketones are observed at 1735 to 1710 cm^{-1} and 1725 to 1700 cm^{-1}, respectively. Adjacent unsaturation lowers the frequency by 25 to 50 cm^{-1}. Thus, aryl aldehydes generally absorb at 1700 to 1690 cm^{-1} and diaryl ketones at 1670 to 1660 cm^{-1}. Intramolecular hydrogen bonding to the carbonyl oxygen also lowers the frequency by 25 to 50 cm^{-1}. Aldehydes are also recognizable by the C—H stretching vibration, which appears as two peaks in the 2850 to 2700 cm^{-1} region (Fig. 11.6). The former may be obscured by aliphatic C—H stretching, but the latter is usually quite prominent. Cyclic ketones (four- or five-membered rings) absorb at higher frequencies (in the area of 1780 and 1745 cm^{-1}, respectively).

CARBOXYLIC ACIDS

The most characteristic absorption of carboxylic acids is a broad peak extending from 3300 to 2500 cm^{-1} due largely to hydrogen-bonded O—H stretching (Fig. 11.7). The C—H stretching vibrations appear as small peaks on top of this band. The carbonyl group of aliphatic acids appears at 1730 to 1700 cm^{-1} and is shifted to 1720 to 1680 cm^{-1} by adjacent unsaturation.

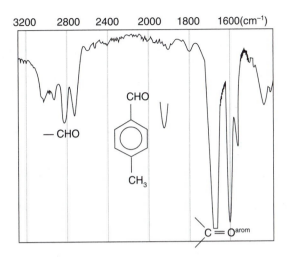

Figure 11.6   Infrared spectrum of an aldehyde.

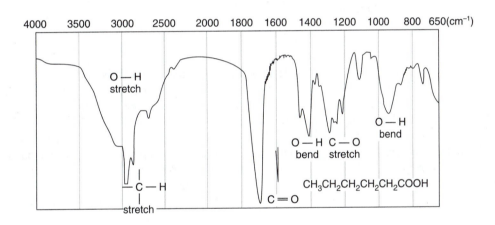

Figure 11.7   Infrared spectrum of a carboxylic acid.

CARBOXYLIC ESTERS AND LACTONES

Saturated ester carbonyl stretching is observed at 1750 to 1730 cm^{-1}. Unsaturation adjacent to the carbonyl group lowers the frequency by 10 to 15 cm^{-1} (see Fig. 11.3), whereas unsaturation adjacent to the oxygen (enol and phenol esters) increases the frequency by 20 to 30 cm^{-1}. The C—O—C stretching of esters appears as two bands in the 1280 to 1050 cm^{-1} region. The asymmetric stretching peak at 1280 to 1150 cm^{-1} is usually strong and varies with substitution, much as does the corresponding ether band. Cyclic esters (lactones), like cyclic ketones, absorb at higher frequencies as the ring size decreases: five-membered rings (γ-lactones) absorb around 1770 cm^{-1}, and four-membered rings (β-lactones) around 1820 cm^{-1}. **The most important use of infrared spectroscopy in deducing the structure of an organic unknown is the assignment of the ring size of a cyclic carbonyl compound.**

ANHYDRIDES

Acid anhydrides are readily recognized by the presence of two high-frequency (1830 to 1800 cm^{-1} and 1775 to 1740 cm^{-1}) carbonyl absorptions. As with other carbonyl stretching vibrations, the frequency is increased by incorporating the group in a ring and decreased by adjacent unsaturation. Cyclic anhydrides differ from acyclic anhydrides also in that the lower frequency band is stronger in the former, while the reverse is true of the latter.

AMIDES AND LACTAMS

Amide carbonyl stretching is observed in the 1700 to 1630 cm^{-1} region. In contrast to other carbonyl groups, both adjacent unsaturation and ring formation (lactams) cause the absorption to shift to higher frequencies. Primary and secondary amides also show N—H stretching at 3500 to 3100 cm^{-1} (see discussion of amines), and N—H bending at 1640 to 1550 cm^{-1}.

ETHERS

The asymmetric C—O stretching absorption of ethers appears in the region 1280 to 1050 cm^{-1}. As in alcohols, the exact position of this strong peak is dependent on the nature of the attached groups. Phenol and enol ethers generally absorb at 1275 to 1200 cm^{-1} and dialkyl ethers at 1150 to 1050 cm^{-1}. Epoxides have three characteristic absorptions in the 1270 to 1240, 950 to 810, and 850 to 750 cm^{-1} regions of the spectrum.

AMINES

Primary and secondary amines show N—H stretching vibrations in the 3500 to 3300 cm^{-1} region (Fig. 11.8). Primary amines generally have two bands approximately 70 cm^{-1} apart due to asymmetric and symmetric stretching modes. Secondary amines show only one band. Intermolecular or intramolecular hydrogen bonding broadens the absorptions and lowers the frequency. In general, the intensities of N—H bands are less than of O—H bands. The N—H bending and C—N stretching absorptions are not as strong as the corresponding alcohol bands and occur at approximately 100 cm^{-1} higher frequencies. In addition, NH$_2$ groups give an additional broad band at 900 to 700 cm^{-1} caused by out-of-plane bending.

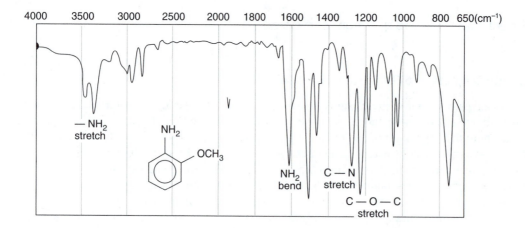

Figure 11.8   Infrared spectrum of an amine.

NITRO GROUPS

Due to the high polarity of the N—O bonds in nitro compounds, the absorptions at the N—O stretching frequencies are very strong. In nitroaromatics, the asymmetric stretching band appears at 1560 to 1520 cm^{-1}, and the symmetric stretching at 1360 to 1320 cm^{-1} (see Fig. 11.4). Since peaks due to several groups other than NO_2 appear in these regions, the presence or absence of a nitro group often cannot be determined with certainty from infrared data alone.

NITRILES

A sharp, usually strong absorption at 2260 to 2220 cm^{-1} due to C≡N stretching is very characteristic of nitriles.

SAMPLE PREPARATION

FILMS

Infrared spectroscopy is quite convenient for the identification of films such as cellophane, Saran Wrap, and so forth. To do this, cut a hole (approximately 2 × 3 cm) in a 3 × 5-inch file card and tape the film over the hole in the card, taking care that none of the tape overlaps the hole. The sample is then ready for insertion in the sample beam.

LIQUIDS

The simplest method for mounting a liquid sample consists of placing a thin film of the liquid between two transparent windows. The most common material used for the windows is NaCl, which is transparent throughout the normally used region of the infrared spectrum (10,000 to 650 cm^{-1}). Large polished plates ground from single crystals are used, *and it is important to remember when handling them that NaCl is water soluble.* They should be picked up *only* by the edges, preferably with gloves, to avoid marring the polished surface with moisture from your fingers.

For mounting the salt plates with the liquid sample between them, the holder illustrated in Figure 11.9 may be used. The bottom metal plate is placed on a flat surface, and one of the rubber gaskets is placed around the opening in the plate. This serves to cushion the relatively fragile salt plate, which is placed on top of it. A drop of the liquid compound is placed in the center of the lower window, and the second salt plate is placed carefully on top, spreading the drop into a thin film. The other rubber gasket and the face plate are then added to the top of the "sandwich" and the entire stack is held together by the thumb nuts on the threaded studs as shown. All four nuts should be tightened firmly *but not excessively.* The assembly is placed in the holder provided on the instrument.

After obtaining the spectrum, disassemble the cell and rinse the windows well with a dry, volatile solvent (CH_2Cl_2, hexane, or the like). Let the solvent evaporate, and store them in a desiccator to protect them from atmospheric moisture.

As alternatives to the cell illustrated in Figure 11.9, smaller circular salt plates may be used and held together in threaded holders that simply screw together or, if the liquid is somewhat viscous, the two plates will stick together by themselves and can be set in the sample beam without mounting brackets.

SOLIDS

Solutions

Sealed cells are available for volatile samples or solutions of compounds in volatile solvents. Solution spectra in nonpolar liquids are useful for minimizing intermolecular association of polar groups in the molecule.

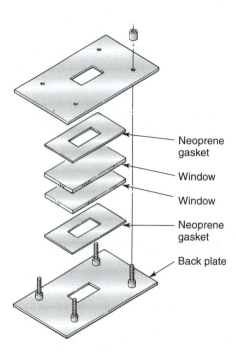

Figure 11.9   Infrared salt plate assembly.

An approximately 10% solution of the compound is prepared using a solvent such as carbon tetrachloride, and the solution is placed in the sealed cell through an injection port. Another (carefully matched) cell is filled with pure solvent and placed in the reference beam. In this manner the absorption of the solvent in the reference beam and in the sample will cancel each other.

KBr Pellet

As an alternative to measuring the spectrum of a liquid solution of a solid compound, a solid solution or dispersion in KBr can be prepared. A 1- to 2-mg sample of the compound is ground together with 50 to 100 mg of dry KBr powder. The mixing must be complete and vigorous enough to reduce the particle size below which it will diffract or scatter the light (less than 2 microns). This can be accomplished by grinding the mixture thoroughly with an agate mortar and pestle or by using the pounding action of a miniature ball mill. The latter technique is preferable because it minimizes the possibility of contamination with moisture. It will be described here.

Weigh the specified quantities of material into the dry plastic vial provided and add a glass ball that fits loosely into the vial. Place the stoppered vial in the shaker and mix for 30 seconds at full speed. Remove the ball from the vial and loosen the powder from the sides of the vial by tapping or by scraping with a spatula.

The powder is formed into a transparent pellet by pressure in a small die. One type of die that may be used consists of a threaded metal block and two bolts with polished end surfaces (Fig. 11.10). To use the die, screw one of the bolts in five or six turns, so that one or two threads remain showing. Pour the powder into the open end and distribute it evenly over the end of the bolt by tapping gently. Screw the other bolt down on top of the sample and tighten the bolts securely with the wrenches provided (use of a torque wrench is recommended). Let the die stand for about 1 minute, during which time the KBr flows to fill the empty spaces between the original particles.

Figure 11.10   Pellet press.

Gently loosen the bolts with the wrenches and remove them, leaving the KBr as a window in the middle of the die. If the pellet is very cloudy, either the compound was not ground well or the bolts were not tightened enough, and a poor spectrum will result due to light scattering. Slight scattering due to a translucent pellet can be partially compensated for with an attenuating device in the reference beam of the instrument. In the absence of a commercial attenuator, a piece of wire screen works adequately to balance the lowered transmittance of the pellet.

Nujol Mull

The infrared spectra of solid compounds can also be measured in a Nujol (a trade name for a purified grade of mineral oil) mull, prepared by grinding 5 mg of the solid to a fine dispersion in a drop of Nujol. The mixture is then placed between salt plates and the spectrum of the thin film is measured. Since Nujol consists of a mixture of saturated hydrocarbons, absorption bands will be present at 2850 to 3000, 1460, and 1375 cm^{-1} in all Nujol mull spectra.

Use of IR in Deducing Unknown Organic Structures

While a typical IR spectrum will cover 600–4500 cm^{-1}, there are two areas of particular interest: 1600–2800 cm^{-1} and 3000–3600 cm^{-1}. The area 2800–3000 cm^{-1} is dominated by the C—H stretch. Pertinent absorptions are summarized in the table in this chapter. IR is diagnostic. If there is a terminal alkyne in the molecule, for instance, then there should be a band between 2100 and 2150 cm^{-1}. The corollary is particularly important. If a structure that could otherwise fit the ^{13}C and ^{1}H NMR data includes a four-membered ring ketone, there should be a band around 1780 cm^{-1}. If there is not, then that structure is *not* acceptable.

While IR is diagnostic for a variety of functional groups, it is especially valuable for establishing the ring size of cyclic esters ("lactones"), and cyclic ketones. Note that α,β-unsaturated carbonyl compounds in general absorb 20–40 cm^{-1} lower frequency than the saturated analogues.

To work a problem with IR data, note which of the (often many!) reported absorptions fall in the key diagnostic areas outlined above, and try to assign *those absorptions*. Note also that any functional groups thought to be present must be consistent with the molecular formula, including the Index of Hydrogen Deficiency (see Chapter 12). At the end, check prospective solutions against the IR. It may be possible that several of the otherwise acceptable alternative ways of assembling the unknown molecule will be seen to be not compatible with the IR spectrum.

E X P E R I M E N T S

Precautions

1.   No aqueous solution or other material that may dissolve the sodium chloride optics should be brought near the instrument.

2. The sample holders (except for film holders) are sodium chloride crystals and should be treated with extreme care! Avoid using water or hydroxylic solvents because these dissolve sodium chloride, and avoid mechanical shock because the salt crystals break easily.

3. Do not operate the instrument without prior detailed instructions. You may inadvertently cause considerable damage.

4. Keep the instrument clean.

SAFETY NOTE

Nothing in this experiment is particularly hazardous; however, usual precautions should be taken to avoid excessive exposure to vapors of solvents and/or samples. Chloroform and carbon tetrachloride, sometimes used as solvents, are suspected carcinogens.

PROCEDURE

A demonstration of the use of IR spectrophotometers will be given. Students are reminded to sign the log book when using instruments and to promptly label each spectrum (date, sample identity, instrument number, etc.) for future use.

A. Determine the infrared spectrum of polystyrene film. The 1601 cm^{-1} band is often used as a frequency marker (e.g., see Figs. 11.3 and 11.7); note particularly its location on your spectrum (it is important to correctly position the chart paper before recording a spectrum). Label this spectrum and secure it in your laboratory notebook.

B. Determine the infrared spectrum of mineral oil (Nujol). Note the simplicity of the spectrum and the location of the absorption bands near 3000 cm^{-1} (above or below 3000 cm^{-1}?). Use this spectrum for future reference should you have occasion to measure a spectrum in a Nujol mull.

C. Determine the infrared spectrum of cyclohexanol. Examine this spectrum and identify the absorption bands due to the O—H stretch, the C—O stretch, and the C—H stretching.

D. Determine the spectrum of ethyl benzoate:

Examine this spectrum and note the location of the absorption band due to the

stretch. Does absorption occur above and below 3000 cm^{-1}? Note the absence of absorption above 3100 cm^{-1} (O—H stretch). Attempt to identify the C—O—C stretching bands and as many of the aromatic absorption bands as you can. Label the spectrum and secure it in your laboratory notebook.

E. You will be given an unknown to characterize. It will be either an alkane, aromatic hydrocarbon, alcohol, or ketone. Assign the major peaks in your spectrum and conclude from your data the nature of your sample.

QUESTIONS

1. Show the following conversions to the units requested.
   2.5 microns = _____ cm^{-1}
   1700 cm^{-1} = _____ μm
2. Give absorption frequency ranges (in cm^{-1}) for each of the following.

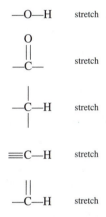

3. How could you distinguish between cyclohexane and cyclohexene using IR?
4. In general, how could you identify a compound as an alkane, alkene, alkyne, or arene using IR?
5. An unknown has the following physical data:
   Elemental analysis: 60.0% C, 13.3% H, 26.7% O
   Molecular weight = 60
   IR: 3400, 2950, 1460, 1385, 1365, 1100 cm^{-1}.
   Draw three structures consistent with the analysis and molecular weight data. Which one is consistent with the IR data? Assign absorptions to support your answer.
6. Figure 11.11 A(C$_{10}$H$_{12}$O) is the infrared spectrum of one of the essential oils in Chapter 10. Interpret the spectrum for functional groups, and if possible, suggest the structure of this compound.

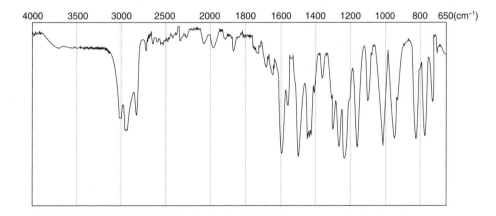

Figure 11.11   Compound A.

PRELABORATORY QUESTIONS

1. Why should aqueous solutions not be used in infrared cells?
2. Describe the procedure to prepare a liquid sample for infrared examination.
3. What is Nujol? What is a Nujol mull?
4. What is a "KBr pellet"? Why not just dissolve a solid sample in a suitable solvent to take its IR spectrum?
5. How could you tell if an unknown is an alcohol by examining its IR spectrum?

REFERENCES

Anderson, D.H.; Woodall, N.B.; West, W. *Techniques of Organic Chemistry,* 3rd ed.; Weissberger, A., Ed.; Interscience: New York, 1959; Vol. I, pt. III.

Cooper, J.W. *Spectroscopic Techniques for Organic Chemists;* Wiley: New York, 1980.

Guzler, H. *Infrared Spectroscopy;* John Wiley & Sons: New York, 1999.

Nicolet IR Spectral Search and Retrieval Program for Aldrich Library of FT-IR Spectra; Nicolet Instrument Corp.: Madison, WI, 1985.

Pouchert, C.J.; *The Aldrich Library of FT-IR Spectra,* 1st ed.; Aldrich Chemical Co., Inc.: Milwaukee, WI, 1989.

Shriner, R.L.; Hermann, C.K.F.; Morrill, T.C.; Fuson, R.C.; Silverstein, R.M.; Webster, F.X. *The Systematic Identification of Organic Compounds Seventh Edition and Spectrometric Identification of Organic Compounds;* John Wiley and Sons: New York, 1998.

Silverstein, R.M.; Bassler, G.C.; Morrill, T.C. *Spectrometric Identification of Organic Compounds,* 5th ed.; John Wiley and Sons, Inc.: New York, 1991.

Williams, D.H.; Fleming, I. *Spectroscopic Methods in Organic Chemistry,* 4th ed.; McGraw-Hill: London, 1987.

Spectroscopic Structure Determination: A Lifelong Quest

INTRODUCTION

Why are there wiggles?
And why are there squiggles?
It's all in the quantum mechanics...

Assume that you have a pure organic compound. How can you tell what it is? You can't tell just by looking at it—it usually looks like a blob of vegetable oil, or like crystals of sugar. Instead, you apply several spectroscopic techniques, and deduce from the data—the squiggles that the machine made on a piece of chart paper—what the structure is. This is immediately enjoyable, as within a week you will learn to decipher simple structures—and it can become a lifelong quest, figuring out more and more complicated structures.

THE QUEST, LEVEL ONE: FINDING THE PIECES

This I know, that whereas I was blind, now I see.

The pieces that make up an organic molecule are the **organic functional groups** and the **hydrocarbon framework**. You will be able to find both the organic functional groups and the pieces of the hydrocarbon framework from the ^{13}C NMR (nuclear magnetic resonance), the theory and practice of which are explained in detail in your lecture text.

As it comes from the spectrometer, a ^{13}C NMR spectrum might look like Figure 12.1.

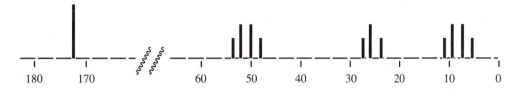

Figure 12.1 ^{13}C NMR spectrum of methyl propanoate.

Altogether, there are twelve lines in this spectrum. These can be divided into four sets: a single line at 173, four lines centered at 51, three lines centered at 27, and four lines centered at 9. The structure that gave rise to this spectrum is shown in Figure 12.2.

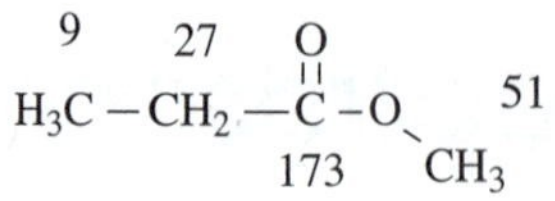

$$\overset{9}{H_3C} - \overset{27}{CH_2} - \overset{\overset{\displaystyle O}{||}}{\underset{173}{C}} - O \overset{51}{\diagdown CH_3}$$

Figure 12.2 ^{13}C NMR Chemical shifts of methyl propanoate.

There are four different carbons in this structure, labelled respectively 9, 27, 173 and 51. It can be seen that 173 gave a single line (a singlet, s) in the ^{13}C NMR spectrum, 51 gave four lines (quartet, q), 27 gave three lines (triplet, t), and 9 also gave four lines (q). If there had been a carbon atom with one H on it, it would have given two lines (doublet, d).

This spectrum introduces two new terms, **chemical shift** and **multiplicity**. Chemical shift is the distance removed from zero at which the center of the pattern of lines appears (in this case 9, 27, 51, and 173). Multiplicity is the number of lines in each pattern. We will summarize ^{13}C NMR spectra as shown in Figure 12.3.

9, q
27, t
51, q
173, s

Figure 12.3 ^{13}C spectrum the way it will be reported.

So . . . how do we use this information to find the **organic functional groups** in the molecule? Actually, there are two kinds of organic functional groups, those that have sp^2- or sp-hybridized carbons and those that have only sp^3-hybridized carbons. We will call the former **unsaturated functional groups** and the latter **saturated functional groups**. Representative samples of each of these kinds of organic functional groups are depicted in Figure 12.4.

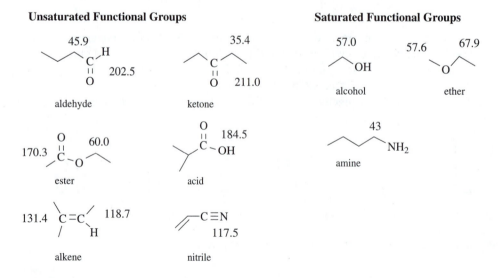

Figure 12.4   Unsaturated and saturated organic functional groups.

It is easy to tell if a molecule has any **unsaturated functional groups** by considering the **molecular formula**. In this example, the molecular formula is $C_4H_8O_2$. A saturated hydrocarbon will have the formula C_nH_{2n+2}. If the ends of the hydrocarbon meet to form a ring, or if two H's are removed to make an alkene, the formula becomes C_nH_{2n}. This count is not affected by oxygen. In our example, one pair of H's is missing, so there is one ring or double bond in the molecule—we can be sure that we have one unsaturated functional group. The IHD (Index of Hydrogen Deficiency) is the number of double bonds plus rings in the molecule. For $C_{13}H_{16}O_2$ for instance, the IHD = 6. There are other considerations. While oxygen does not affect the IHD, nitrogen does. Because N is trivalent, each nitrogen brings with it an extra hydrogen. For instance, the IHD of $C_9H_{14}N_2O_2$ is four. The H_{14} includes two extra H's, one for each of the N's. Halogens (F, Cl, Br, I) in a formula count as H's.

A double bond can be either a C=C bond or a C=O bond. There is O in the formula of our example, so there could be a C=O bond. From the examples above (there are more examples in Tables 12.8 and 12.9), we know that the C=O carbon of a ketone or aldehyde will come around 210, while the C=O carbon of an ester or acid will come around 170. Since we observe a singlet at 173, and we have two O's in the formula, we know that we have either an ester or an acid (Fig. 12.5)—but, which is it?

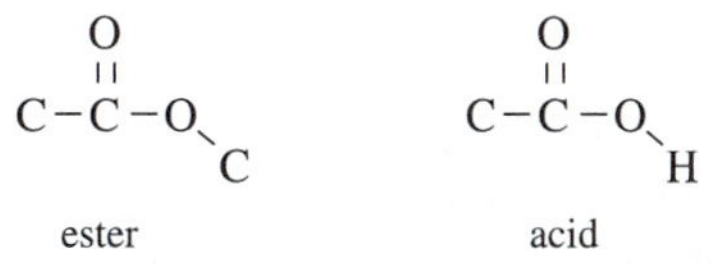

$$
\begin{array}{cc}
\overset{\displaystyle O}{\overset{\displaystyle \|}{C-C}}-O_{\diagdown C} & \overset{\displaystyle O}{\overset{\displaystyle \|}{C-C}}-O_{\diagdown H} \\
\text{ester} & \text{acid}
\end{array}
$$

Figure 12.5   Is it an ester or an acid?

A key difference between an ester and an acid is that with an ester, all of the H's are attached to C. With an acid, one of the H's is attached to an O. We can count in the ^{13}C NMR spectrum how many H's are attached to each C by the multiplicity: s = 0 H, d = 1 H, t = 2 H, q = 3 H. In this example, when we add up all the H's attached to C the total comes to 8. That is the same number of H's as in the formula, so there are no H's attached to O. Therefore, the **unsaturated functional group** in this example must be an ester.

Next, we **explore around** the unsaturated functional group. An ester has two carbons attached to it, one connected to the sp^2-hybridized C and one connected to the sp^3-hybridized O. We would like to know how many H's are attached to each of these two C's. From the pictures in Table 12.9, we find that the C that is attached to the sp^3-hybridized O comes around 50 or 60. That must be the signal in our data at 51. That signal is a quartet (q), so there must be three H's attached to the C that is attached to the sp^3-hybridized O. From the same pictures in Table 12.9, we find that the C connected to the sp^2-hybridized C comes in the range 20–35. We can conclude that this must be the signal from our data at 27. That signal is a triplet (t), so there must be two H's attached to the C that is connected to the sp^2-hybridized C of the carbonyl. Putting all of this together gives us the **partial structure** shown in Figure 12.6.

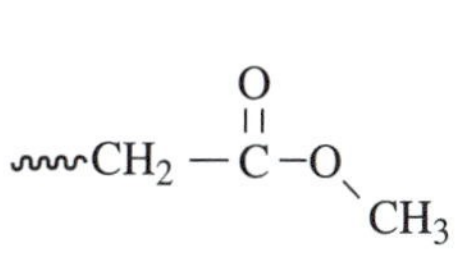

$$
\text{\textasciitilde CH}_2-\overset{\displaystyle O}{\overset{\displaystyle \|}{C}}-O_{\diagdown CH_3}
$$

Figure 12.6   The partial structure of the ester.

SATURATED FUNCTIONAL GROUPS: After you are finished with all of the unsaturated functional groups, are there heteroatoms (O, N, S) in the formula that you have not yet dealt with? If so, now would be the time to find the saturated functional groups. In this example, we had 2 O's in the formula, and we have already accounted for both of them, so in this case there are no saturated functional groups.

HYDROCARBON FRAMEWORK: Which signals in the carbon spectrum have you not yet accounted for? In this case, all that is left is "9, q"—so there is one more -CH$_3$ to attach to the structure.

PUTTING IT ALL TOGETHER: In this case, there is only one way to connect the pieces, so the structure must be as shown in Figure 12.7.

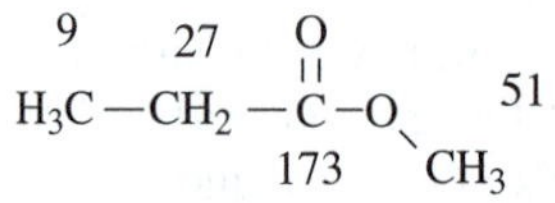

$$\underset{173}{\overset{9}{H_3C}}-\overset{27}{CH_2}-\overset{\overset{\displaystyle O}{\|}}{C}-O\underset{CH_3}{\overset{51}{\diagdown}}$$

Figure 12.7   The full structure of the ester.

Note that there is a pattern to ^{13}C chemical shifts—the fewer the H atoms on a carbon, the larger the chemical shift. Consider ethyl methyl ether. (Figure 12.8) The methyl carbon directly attached to the oxygen comes at 57.6, whereas the methylene carbon directly attached to oxygen comes at 67.9. If you study the examples in Table 12.7, you will see that this is a general pattern.

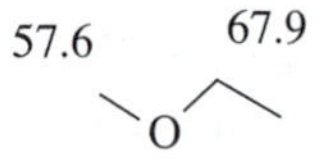

57.6   67.9

$\diagdown O \diagup$

ether

Figure 12.8   Chemical shifts of the carbons attached to the oxygen of an ether.

To solve a given structure, it is necessary to assemble a significant amount of data in an orderly way. To aid in this process, we will use a worksheet. A sample of the worksheet is reproduced on the following page. Especially in the early going, you may find it useful to make several copies of this worksheet and write directly on them in working the problems.

You should now be ready to work through Examples A–C. It would probably work best to do Example A on your own, then study the way it is worked out in the book before going on to Example B. After you have worked Example B on your own, study the way it is worked out in the book before going on to Example C.

Example A		**Example B**		**Example C**	
23, q (2)	C$_7$H$_{14}$O	11, q (2)	C$_6$H$_{14}$O	16, t (2)	C$_5$H$_6$N$_2$
28, d		23, t (2)		22, t	
29, q		44, d		119, s (2)	
33, t		65, t			
42, t					
206, s					

WORKSHEET

Step 1.   Molecular formula and IHD. Any OH or N-H?

Step 2.   Exploring around the unsaturated functional groups

C=O (160–220):

C=C (100–160):

R-CN (105–120, s):

Alkyne (65–90):

Number of rings:

_____

IHD =

Step 3.   Other heteroatoms—Exploring around the saturated functional groups

Step 4.   Other pieces

   a.   Methyl groups:

   b. ^{1}H NMR greater than 2.0 δ:

   c.   Other carbons:

Step 5.   Putting it all together

Solution for A

Step 1.   (The steps here are the same as the steps on the worksheet.)

From the molecular formula, $C_7H_{14}O$, the IHD is one, so we have one ring or multiple bond. All 14 H's are attached to C.

Step 2.

We know from the signal at 206 that we have either an aldehyde or a ketone (see Table 12.8). Because it is a singlet, it must be a ketone. That means that there is a carbon on either side of the carbonyl. The question is, how many H's are on each of the two carbons? Again from Table 12.8, we can observe that a methyl attached to a ketone carbonyl comes at about 30 (this would be a quartet) and a methylene comes at about 40 (this would be a triplet). Methines (doublet) and carbons with no H's (quaternary carbons, singlets) would be further downfield (larger numbers) than that.

   Looking at the data, we see that we do indeed have a quartet at 29 and a triplet at 42, so we can draw the following partial structure as shown in Figure 12.9.

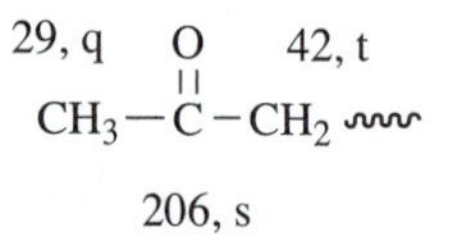

$$29, q \quad O \quad 42, t$$
$$\text{CH}_3 - \overset{\text{||}}{\text{C}} - \text{CH}_2 \sim$$
$$206, s$$

Figure 12.9   Partial structure of the unknown ketone.

   We draw the wiggly line to remind ourselves that there is more to attach—we just don't know what it is yet.

Step 3.

There is only one heteroatom in the molecular formula, and we have already found it.

Step 4.

Since we are not yet using ^{1}H NMR, only 4c is relevant. From the ^{13}C NMR, there are four carbons we have not yet accounted for, two methyls, a methylene, and a methine. Note, however, that the two methyl groups have the same chemical shift. This makes it likely that there is a plane of symmetry in the molecule. This gives us the pictures in Figure 12.10:

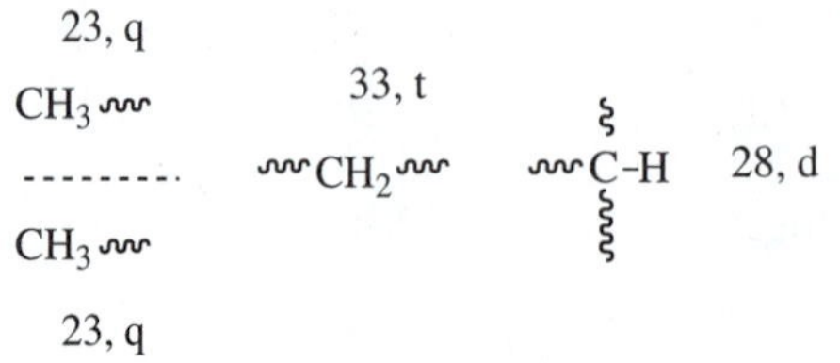

$$23, q$$
$$\text{CH}_3 \sim \qquad 33, t$$
$$\text{-------} \qquad \sim \text{CH}_2 \sim \qquad \sim \text{C-H} \quad 28, d$$
$$\text{CH}_3 \sim$$
$$23, q$$

Figure 12.10   The pieces of the hydrocarbon skeleton of the unknown ketone.

Step 5.

In this step we put the pieces together, like assembling a jigsaw puzzle. Because the methyl groups must be symmetrical, and nothing else in the molecule is symmetrical (none of the other carbon signals are doubled), there is only one way to assemble the final structure. (See Fig. 12.11.)

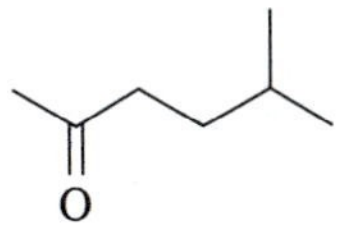

Figure 12.11   The final structure of the unknown ketone.

Solution for B

Step 1.   From the molecular formula, $C_6H_{14}O$, the IHD is zero—there are no rings or double bonds in this molecule. From the hydrogen count, there are only 13 H's attached to carbon, so one must be attached to some other atom. The only other atom in this example is the O, so we must have an alcohol.

Step 2.

As the IHD is zero, there are no unsaturated functional groups, and there is no need to look for rings.

Step 3.

Following the examples in Table 12.6, the carbon directly attached to -OH comes around 60 or 70. Looking at the data, we have 65, t, which gives us the partial structure in Figure 12.12.

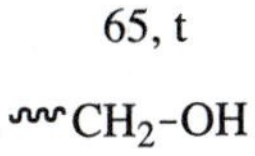

65, t

∿CH$_2$-OH

Figure 12.12   Partial structure of the unknown alcohol.

Step 4.

Both the methyl (11, q) and the methylene (23, t) carbon signals are doubled, so we must have symmetrical ethyl groups. We also have a methine. (See Fig. 12.13.)

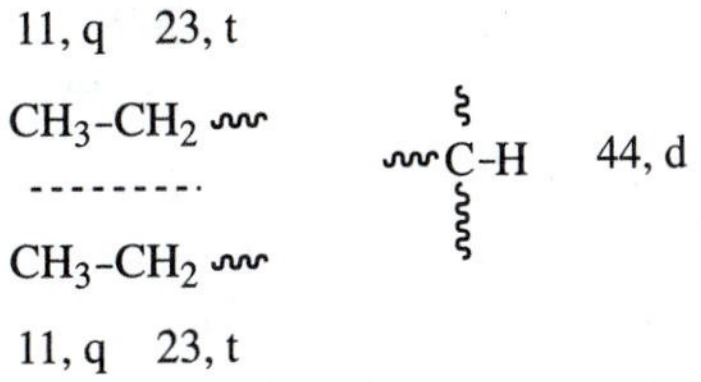

11, q   23, t

CH$_3$-CH$_2$ ∿   ∿C-H   44, d

CH$_3$-CH$_2$ ∿

11, q   23, t

Figure 12.13   The pieces of the hydrocarbon skeleton of the unknown alcohol.

Step 5.

There is only one way to assemble the molecule. (See Fig. 12.14.)

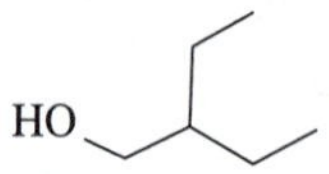

Figure 12.14   The final structure of the unknown alcohol.

Solution for C

Step 1.

From the molecular formula, $C_5H_6N_2$, the IHD is 4. Remember, each N brings an extra H with it, so to calculate the IHD, this is "C_5H_4". From the ^{13}C NMR, all 6 H's are attached to carbon.

Step 2.

The presence of N in the molecule, and the carbon singlet at 119, together suggest a nitrile. There are two of them, and they are symmetrical. This accounts for all of the IHD.(See Fig. 12.15.)

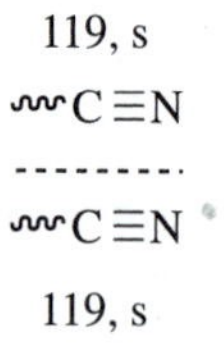

119, s

〰C≡N

- - - - - - - -

〰C≡N

119, s

Figure 12.15   The unknown is a symmetrical nitrile.

Step 3.

All heteroatoms are already accounted for.

Step 4.

There are three methylenes. Two of them are symmetrical. (See Fig. 12.16.)

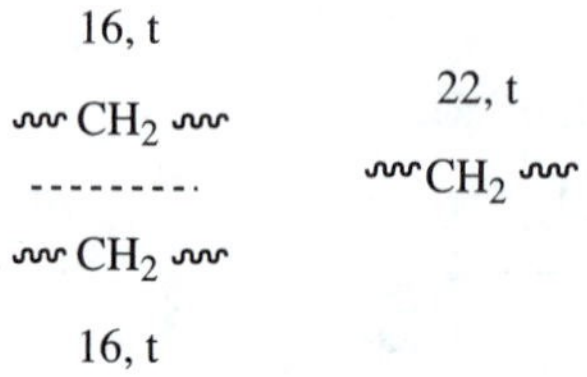

16, t

〰 CH₂ 〰

- - - - - - - -   22, t

〰 CH₂ 〰   〰CH₂ 〰

16, t

Figure 12.16   The pieces of the hydrocarbon skeleton of the unknown nitrile.

Step 5.

There is only one way to combine these fragments. (See Fig. 12.17.)

$$N \equiv C \diagup\diagdown\diagup C \equiv N$$

Figure 12.17   The final structure of the unknown nitrile.

PROBLEMS

Using this same approach, you should be able to figure out the structures of unknowns 1–20 using the molecular formula and the ^{13}C NMR.

1. $C_3H_8O_2$

    ^{13}C NMR

    57.5, q
    60.1, t
    73.1, t

2. $C_5H_{10}O$

    ^{13}C NMR

    29.6, q (2)
    70.9, s
    110.6, t
    146.8, d

3. $C_5H_{10}O_2$

    ^{13}C NMR

    19.3, q (2)
    34.4, d
    51.4, q
    176.9, s

4. $C_5H_{10}O$

    ^{13}C NMR

    212.5, s
    41.6, d
    27.4, q
    18.1, q (2)

5. $C_6H_{12}O$

    ^{13}C NMR

    80.0, s
    41.3, t
    28.2, q
    24.1, t

6. $C_5H_{11}N$

    ^{13}C NMR

    56.3, t (2)
    42.2, q
    24.1, t (2)

7. C_4H_9NO

    ^{13}C NMR

    170.4, s
    34.4, t
    23.1, q
    14.7, q

8. $C_6H_{12}O$

    ^{13}C NMR

    211.5, s
    44.3, t
    35.9, t
    17.4, t
    13.8, q
    7.8, q

9. $C_9H_{16}O$

    ^{13}C NMR

    204.7, d
    50.7, d
    26.7, t (2)
    26.2, t (2)
    25.6, t
    25.3, t (2)

10. $C_5H_{12}O$

    ^{13}C NMR

    73.4, t
    32.6, s
    26.0, q (3)

11. $C_5H_{10}O$

    ^{13}C NMR

    135.5, d
    125.5, d
    68.8, d
    23.3, q
    17.5, q

12. $C_6H_{14}O$

    13C NMR

    70.5, t
    66.1, t
    32.0, t
    19.5, t
    15.3, q
    14.0, q

13. $C_5H_{10}O$

    13C NMR

    205.2, d
    47.7, d
    23.6, t
    12.9, q
    11.4, q

14. C_3H_8O

    13C NMR

    68.3, d
    67.8, t
    18.7, q

15. $C_5H_{10}O$

    13C NMR

    67.2, t
    37.4, d
    24.5, t (2)
    18.4, t

16. C_4H_8O

    13C NMR

    145.7, s
    110.5, t
    67.4, t
    19.9, q

17. $C_5H_{12}O$

    13C NMR

    72.7, t
    58.5, q
    31.8, t
    19.4, t
    13.9, q

18. $C_5H_{12}O$

    ¹³C NMR

    67.0, d
    41.6, t
    23.3, q
    19.1, t
    14.0, q

19. C_4H_9NO

    ¹³C NMR

    170.5, s
    38.0, q
    35.1, q
    21.5, q

20. $C_{11}H_{20}O_2$

    ¹³C NMR

    173.8, s
    64.5, t
    34.4, t
    26.1, t
    24.8, t
    24.5, t
    24.0, t
    23.9, t
    23.5, t
    23.3, t
    22.1, t

REFERENCES

For further information about organic spectroscopy, relevant to Chapters 12–14, see:

Bates, R.B.; Beavers, W.A. *Carbon-13 NMR Spectral Problems*; Humana Press: Clifton, NJ, 1981.

Davis, R.; Wells, C.H.J. *Spectral Problems in Organic Chemistry*; Chapman and Hall: New York, 1984.

Fuchs, P.L.; Bunnell, C.A. *Carbon-13 NMR Based Organic Spectral Problems*; John Wiley and Sons, Inc.: New York, 1979.

Graybeal, J.D. *Molecular Spectroscopy*; McGraw-Hill: New York, 1988.

Kemp, W. *Organic Spectroscopy*, 3rd ed.; W.H. Freeman: New York, 1991.

Lambert, J.B.; Shurvell, H.F.; Lightner, D.A.; Cooks, G. *Introduction to Organic Spectroscopy*; Macmillan: New York, 1987.

Macomber, R.S. *NMR Spectroscopy, Basic Principles and Applications;* Harcourt Brace Jovanovich: New York, 1987.

Pavia, D.L.; Lampman, G.M.; Kriz, G.S., Jr. *Introduction to Spectroscopy, a Guide for Students of Organic Chemistry,* 2nd ed.; Saunders: Philadelphia, 1996.

Pretsch, E.; Clerc, T.; Seibl, J.; Simon, W. *Tables of Spectral Data for Structure Determination of Organic Compounds*, 2nd ed.; Springer-Verlag: Berlin, 1989.

Richards, S.A. *Laboratory Guide to Proton NMR Spectroscopy*; Blackwell Scientific Publications: Oxford, 1988.

Silverstein, R.M.; Bassler, G.C.; Morrill, T.C. *Spectrometric Identification of Organic Compounds,* 5th ed.; John Wiley and Sons, Inc.: New York, 1991.

Sorrell, T.N. *Interpreting Spectra of Organic Molecules;* University Science Books: Mill Valley, CA, 1988.

Sternhall, S.; Kalman, J.R. *Organic Structures from Spectra;* John Wiley & Sons: New York, 1986.

Williams, D.H.; Fleming, I. *Spectroscopic Methods in Organic Chemistry*, 4th ed.; McGraw-Hill: New York, 1989.

Yoder, C.H.; Schaeffer, C.D., Jr. *Introduction to Multinuclear NMR*; Benjamin/Cummings: Menlo Park, CA, 1987.

Table 12.1 Chemical Shifts of Cycloalkanes (ppm from TMS)

C_3H_6	−2.9	C_7H_{14}	28.4
C_4H_8	22.4	C_8H_{16}	26.9
C_5H_{10}	25.6	C_9H_{18}	26.1
C_6H_{12}	26.9	$C_{10}H_{20}$	25.3

From R.M. Silverstein, G.C. Bassler, and T.C. Morrill, *Spectrometric Identification of Organic Compounds,* 5th ed., New York: John Wiley and Sons, Inc., 1991. Reprinted with permission of John Wiley and Sons, Inc. Table 5.4, p. 237.

Table 12.2 Chemical Shifts for Saturated Heterocyclics (ppm from TMS, neat)

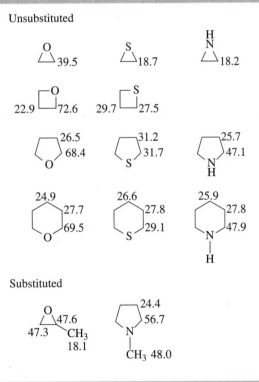

From R.M. Silverstein, G.C. Bassler, and T.C. Morrill, *Spectrometric Identification of Organic Compounds,* 5th ed., New York: John Wiley and Sons, Inc., 1991. Reprinted with permission of John Wiley and Sons, Inc. Table 5.5, p. 237.

Table 12.3   Alkene and Cycloalkene Chemical Shifts (ppm from TMS)

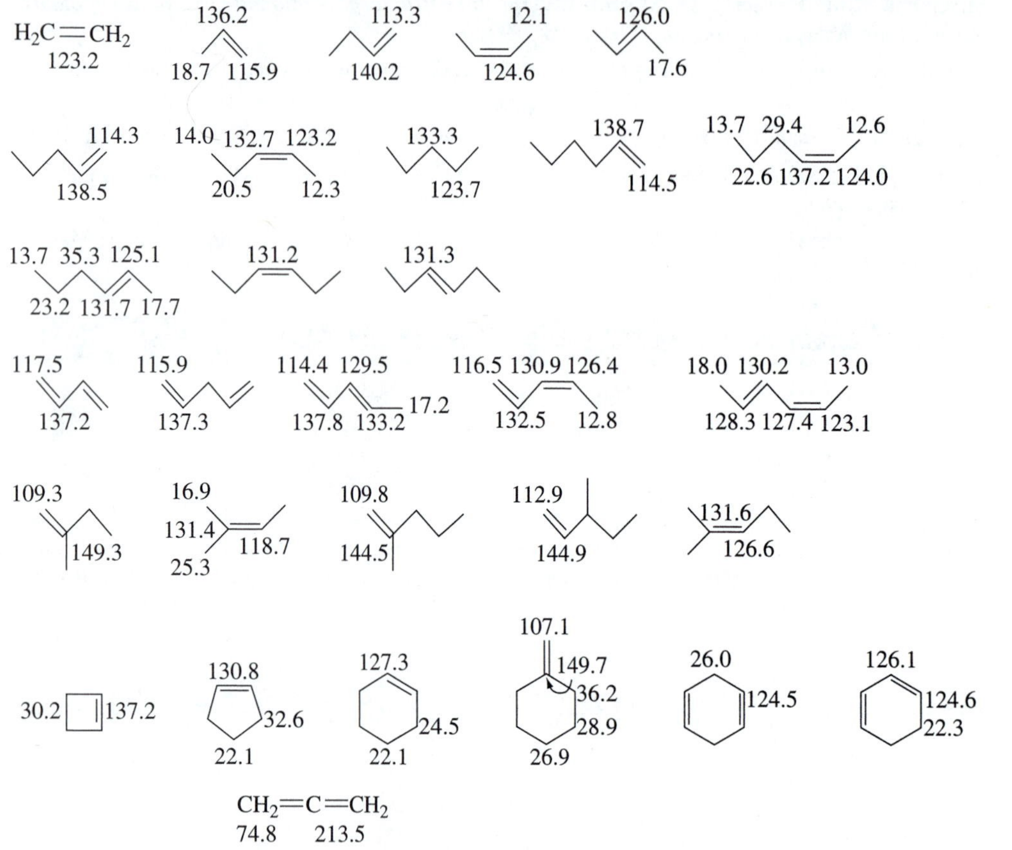

$H_2C = CH_2$
123.2

136.2
18.7   115.9

113.3
140.2

12.1
124.6

126.0
17.6

114.3
138.5

14.0 132.7 123.2
20.5   12.3

133.3
123.7

138.7
114.5

13.7 29.4   12.6
22.6 137.2 124.0

13.7 35.3 125.1
23.2 131.7 17.7

131.2

131.3

117.5
137.2

115.9
137.3

114.4 129.5
137.8 133.2   17.2

116.5 130.9 126.4
132.5   12.8

18.0 130.2   13.0
128.3 127.4 123.1

109.3
149.3

16.9
131.4   118.7
25.3

109.8
144.5

112.9
144.9

131.6
126.6

30.2 ☐ 137.2

130.8
32.6
22.1

127.3
24.5
22.1

107.1
149.7
36.2
28.9
26.9

26.0
124.5

126.1
124.6
22.3

$CH_2 = C = CH_2$
74.8   213.5

From R.M. Silverstein, G.C. Bassler, and T.C. Morrill, *Spectrometric Identification of Organic Compounds,* 5th ed., New York: John Wiley and Sons, Inc., 1991. Reprinted with permission of John Wiley and Sons, Inc. Table 5.6, p. 238.

Table 12.4 Chemical Shifts of Substituted Alkenes (ppm from TMS)

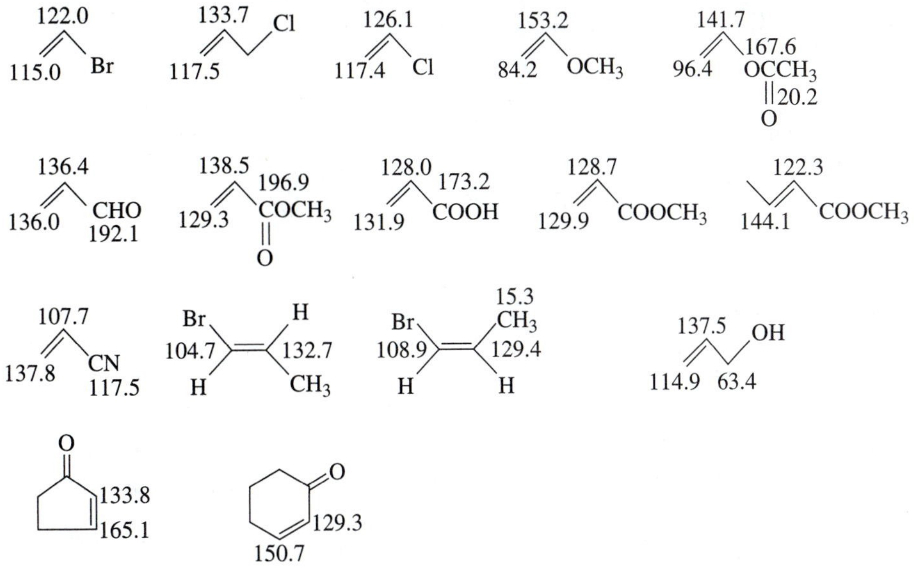

122.0
115.0 Br

133.7 Cl
117.5

126.1
117.4 Cl

153.2
84.2 OCH₃

141.7
167.6
96.4 OCCH₃
‖ 20.2
O

136.4
136.0 CHO
192.1

138.5 196.9
129.3 COCH₃
‖
O

128.0 173.2
131.9 COOH

128.7
129.9 COOCH₃

122.3
144.1 COOCH₃

107.7
137.8 CN
117.5

Br   H
104.7   132.7
H   CH₃

Br   15.3 CH₃
108.9   129.4
H   H

137.5 OH
114.9 63.4

O
133.8
165.1

O
129.3
150.7

From R.M. Silverstein, G.C. Bassler, and T.C. Morrill, *Spectrometric Identification of Organic Compounds,* 5th ed., New York: John Wiley and Sons, Inc., 1991. Reprinted with permission of John Wiley and Sons, Inc. Table 5.7, p. 239.

Table 12.5 Alkyne Chemical Shifts (ppm)

COMPOUND	C-1	C-2	C-3	C-4	C-5	C-6
1-Butyne	67.0	84.7				
2-Butyne		73.6				
1-Hexyne	67.4	82.8	17.4	29.9	21.2	12.9
2-Hexyne	1.7	73.7	76.9	19.6	21.6	12.1
3-Hexyne	14.4	12.0	79.9			

From R.M. Silverstein, G.C. Bassler, and T.C. Morrill, *Spectrometric Identification of Organic Compounds,* 5th ed., New York: John Wiley and Sons, Inc., 1991. Reprinted with permission of John Wiley and Sons, Inc. Table 5.8, p. 239.

Table 12.6   Chemical Shifts of Alcohols (neat, ppm from TMS)

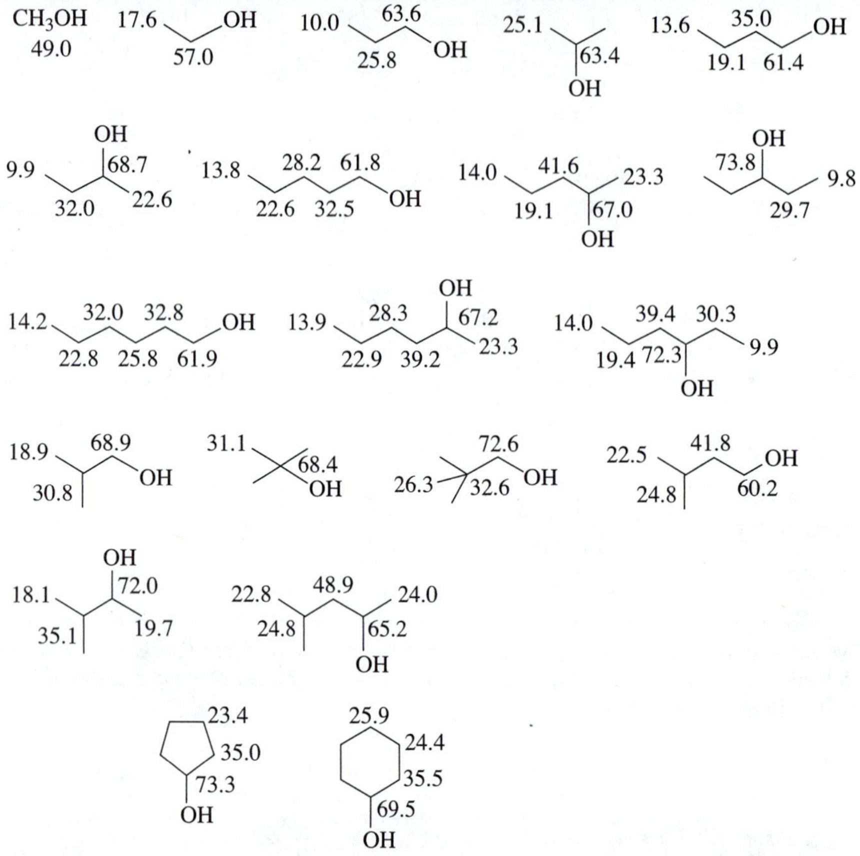

CH_3OH 49.0

17.6 / 57.0 — OH

10.0 / 25.8 / 63.6 — OH

25.1 / 63.4 / OH

13.6 / 19.1 / 35.0 / 61.4 — OH

9.9 / 32.0 / 68.7 / 22.6 — OH

13.8 / 22.6 / 28.2 / 32.5 / 61.8 — OH

14.0 / 19.1 / 41.6 / 67.0 / 23.3 — OH

73.8 / 29.7 / 9.8 — OH

14.2 / 22.8 / 32.0 / 25.8 / 32.8 / 61.9 — OH

13.9 / 22.9 / 28.3 / 39.2 / 67.2 / 23.3 — OH

14.0 / 19.4 / 39.4 / 72.3 / 30.3 / 9.9 — OH

18.9 / 30.8 / 68.9 — OH

31.1 / 68.4 — OH

26.3 / 32.6 / 72.6 — OH

22.5 / 24.8 / 41.8 / 60.2 — OH

18.1 / 35.1 / 72.0 / 19.7 — OH

22.8 / 24.8 / 48.9 / 65.2 / 24.0 — OH

cyclopentanol: 23.4 / 35.0 / 73.3 — OH

cyclohexanol: 25.9 / 24.4 / 35.5 / 69.5 — OH

Table 12.7   Chemical Shifts of Ethers, Acetals, and Epoxides (ppm from TMS)

From R.M. Silverstein, G.C. Bassler, and T.C. Morrill, *Spectrometric Identification of Organic Compounds,* 5th ed., New York: John Wiley and Sons, Inc., 1991. Reprinted with permission of John Wiley and Sons, Inc. Table 5.12, p. 243.

Table 12.8   Shift Positions of the C=O Group and Other Carbon Atoms of Ketones and Aldehydes (ppm from TMS)

From R.M. Silverstein, G.C. Bassler, and T.C. Morrill, *Spectrometric Identification of Organic Compounds,* 5th ed., New York: John Wiley and Sons, Inc., 1991. Reprinted with permission of John Wiley and Sons, Inc. Table 5.15, p. 245.

Table 12.9 Shift Positions for the C=O Group and Other Carbon Atoms of Carboxylic Acids, Esters, Lactones, Chlorides, Anhydrides, Amides, Carbamates, and Nitriles (ppm from TMS)

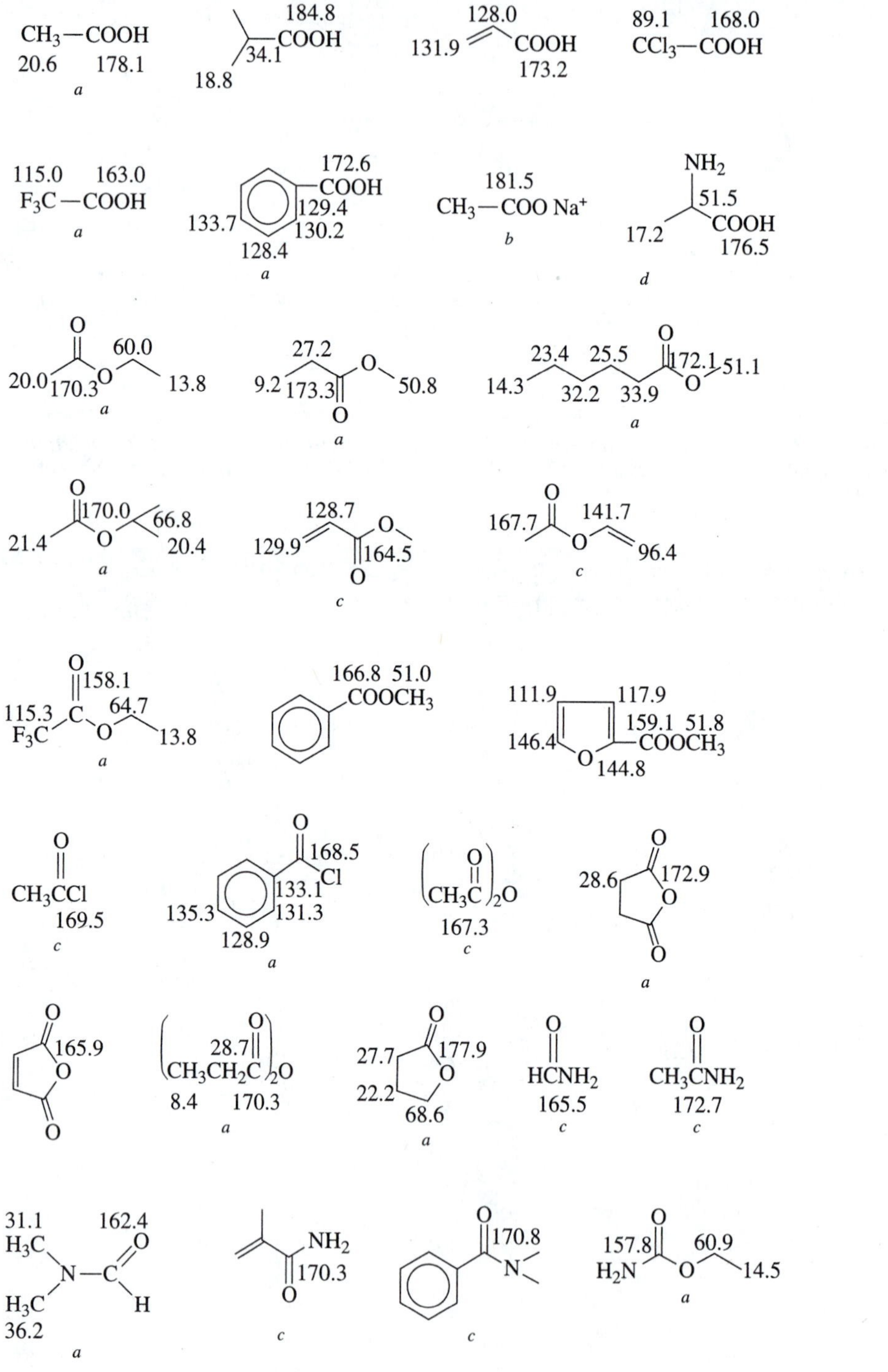

CH_3—COOH
20.6   178.1
a

184.8
—COOH
34.1
18.8

128.0
131.9   COOH
173.2

89.1   168.0
CCl_3—COOH

115.0   163.0
F_3C—COOH
a

172.6
—COOH
133.7   129.4
130.2
128.4
a

181.5
CH_3—COO Na⁺
b

NH_2
51.5
17.2   COOH
176.5
d

O
60.0
20.0 170.3 O   13.8
a

27.2
9.2 173.3 O   50.8
O
a

23.4 25.5
O
172.1 51.1
14.3   32.2 33.9 O
a

O
170.0 66.8
21.4   O   20.4
a

128.7
O
129.9   164.5
O
c

O
167.7   141.7
O   96.4
c

O
158.1
115.3 C   64.7
F_3C O   13.8
a

166.8 51.0
COOCH₃

111.9   117.9
146.4   159.1 51.8
O   COOCH₃
144.8

O
CH_3CCl
169.5
c

O
168.5
Cl
135.3   133.1
131.3
128.9
a

O
(CH_3C)₂O
167.3
c

O
28.6   172.9
O
O
a

165.9
O
O

O
28.7
(CH_3CH_2C)₂O
8.4   170.3
a

27.7   177.9
O
22.2
68.6
a

O
HCNH₂
165.5
c

O
CH_3CNH_2
172.7
c

31.1   162.4
H₃C   O
N—C
H₃C   H
36.2
a

NH₂
170.3
O
c

O
170.8
N
c

O
157.8   60.9
H₂N   O   14.5
a

From R.M. Silverstein, G.C. Bassler, and T.C. Morrill, *Spectrometric Identification of Organic Compounds,* 5th ed., New York: John Wiley and Sons, Inc., 1991. Reprinted with permission of John Wiley and Sons, Inc. Table 5.16, p. 246.

¹H Nuclear Magnetic Resonance

THE QUEST, LEVEL TWO: CONNECTING THE PIECES TOGETHER

Hip bone connected to the thigh bone ...

So far, you have learned how to find the functional groups in an unknown molecule. You have also learned to find the pieces of the molecule, and in some cases you have been able to find the carbons directly attached to the functional groups. For the compounds you have done so far, once you found the pieces, it was easy to put them together to get the final structure. With more advanced compounds, you will also need information from the ¹H NMR spectrum to settle on the structure. With ¹H NMR, you will be able to see how one piece of the unknown structure is connected to the other pieces. The theory of ¹H NMR (proton magnetic resonance) is discussed in detail in your organic text.

A typical ¹H NMR spectrum can be seen in Figure 13.1.

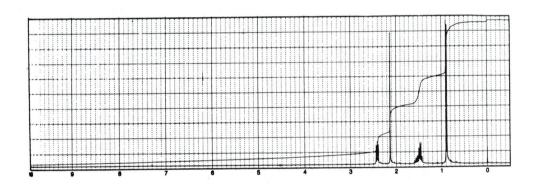

Figure 13.1   A typical ¹H nuclear magnetic resonance spectrum.

In this text, a ¹H NMR spectrum will be summarized as shown in Figure 13.2.

¹H NMR
0.90, d, 6H, J = 6.5 Hz
1.2–1.6, m, 3H
2.13, s, 3H
2.40, t, 2H, J = 6.8 Hz

Figure 13.2   Summary of a typical ¹H nuclear magnetic resonance spectrum.

ANALYZING THE ¹H SPECTRUM

Taking the first entry of this spectrum as an example, with each entry you will see the chemical shift (0.90), the multiplicity (d), the number of protons at that chemical shift (6), and the coupling constants (J values, in this case = 6.5 Hz).

THE CHEMICAL SHIFT

The chemical shift δ (parts per million downfield from the internal standard, tetramethylsilane) of the protons attached to a carbon is a function of the environment of those protons. These effects are summarized in Table 13.1 on page 138. In this case, 0.90 is typical for a methyl group that is not shifted by some other functional group.

THE MULTIPLICITY

The multiplicity (d, in this case) is the number of lines in the signal. As with Chapter 12, s = singlet, one line; d = doublet, two lines; t = triplet, three lines; q = quartet, four lines. In addition, you will see examples such as dd = doublet of doublets, and m = multiplet. In the ¹H NMR spectrum, multiple lines tell you the number of **H's** on neighboring carbons. In this case, there would be one H on a neighboring carbon, which would suggest the partial structure shown in Figure 13.3. **Multiplicity** is explained in more detail under "Coupling Constant (J)" below.

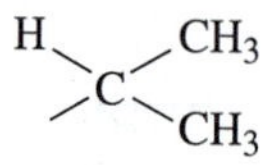

Figure 13.3   A fragment of the unknown structure.

THE NUMBER OF PROTONS—INTEGRATION

In the ¹H spectrum, it is typical to have two, three, or even more protons with the same chemical shift. The vertical distance on the integral of the signal at a given chemical shift is proportional to the peak area, and thus also to the number of protons having that chemical shift. If the total number of hydrogens is known, one can divide the total vertical integration by that number to get the vertical distance per hydrogen. In the summaries, you will be given the actual number of protons in the signal. In this first entry, there are 6 H's, so you might guess (correctly!) that this represents two methyl groups (see Fig. 13.3).

COUPLING CONSTANT (J)

The effective magnetic field at a given nucleus is the sum of the imposed external magnetic field, H_0, added to all the smaller magnetic fields from surrounding nuclei. Consider the case (Fig. 13.4) of H_a, attached to carbon A, with one proton, H_b, on the adjacent carbon. The actual magnetic field experienced by H_a will be the sum of H_0 plus the field due to spin of H_b. On average, half the H_a's will see an H_b having a spin aligned with the external field, and half the H_a's will see an H_b with the spin opposed to the magnetic field. Thus, there will be two populations of H_a, and so two resonances. We say that the signal due to H_a is **split,** because of coupling to H_b. The magnitude of the coupling, the **coupling constant (J),** is measured in **Hertz** (= one cycle per second). For values of J in different situations, see Table 13.2 on page 140–141.

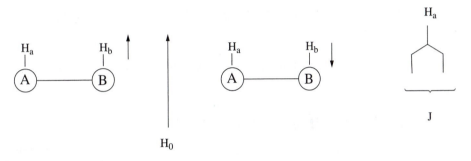

Figure 13.4 Illustration of ¹H NMR coupling.

Protons with the same chemical shift will not show coupling. Protons on the same carbon ("geminal") **will** show coupling if they have different chemical shifts, as, for instance, if the carbon is part of a ring (Fig. 13.5).

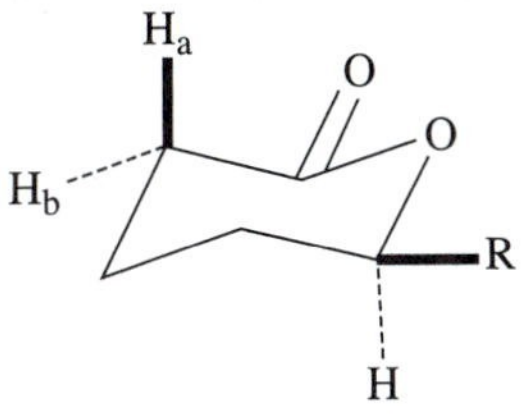

Figure 13.5 H_a and H_b are not equivalent. They will have different chemical shifts, and will couple to each other.

If protons are coupled to each other, they will show the same numerical value for J ($J_{AB} = J_{BA}$). This is very useful for following connectivity in an unknown structure.

MULTIPLICITY—THE NUMBER OF NEIGHBORS

If a proton is coupled to more than one other proton, the J values may or may not be equal. If they are equal, then the multiplicity of the signal for that first proton will be the number of hydrogens to which it is coupled with that J value, plus one (Fig. 13.6).

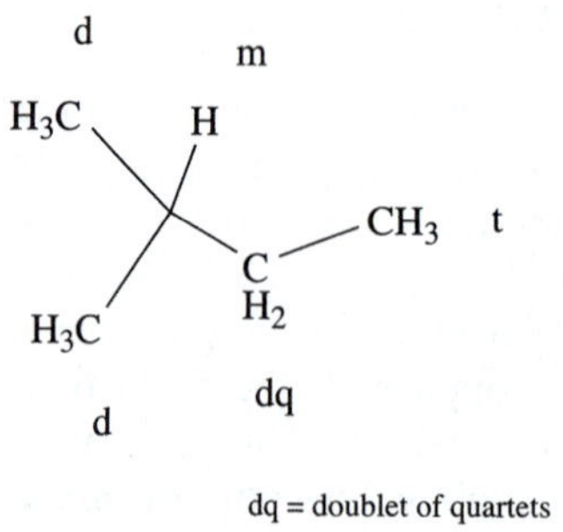

dq = doublet of quartets

Figure 13.6 The multiplicity of a ¹H NMR signal is the number of neighbors, plus one.

If there are no protons attached to the neighboring carbons, the ¹H signal will appear as a singlet.

OTHER INFORMATION IN THE ¹H SPECTRUM

In general, you should look for signals that stand out from the others (Step 4). Methyl groups often stand out, as do protons shifted downfield by adjacent functional groups. It is often possible to make a reasonable guess as to what that functional group is, first based on the ¹³C data, and then using the ¹H chemical shift data in Table 13.1. It should be remembered that a given proton could be adjacent to two or even three functional groups. As you will see, the proton shifts due to the nearness of two or more functional groups are very nearly additive.

SUMMARY

The proton spectrum can be used in two ways: (a) confirming deductions from the ¹³C spectrum about the presence and substitution pattern of organic functional groups in the molecule, and (b) establishing carbon–carbon connectivity.

EXAMPLE A

¹³C NMR	$C_7H_{14}O$
22.3, q (2)	
27.8, d	
29.7, q	
32.8, t	
41.8, t	
206.8, s	

¹H NMR
0.90, d, 6H, J = 6.5 Hz
1.2–1.6, m, 3H
2.13, s, 3H
2.40, t, 2H, J = 6.8 Hz

The spectrum shown in Figure 13.1 is, in fact, for the same substance as Example A in Chapter 12. From the ^{13}C spectrum, we had deduced that a ketone was present. Using approximate carbon shifts, we had concluded that the carbons flanking the ketone were a methyl and a methylene. From the data in Table 13.1, we can read that a methyl group directly adjacent to a ketone should have a chemical shift of 2.1 δ. The signal should integrate for three protons. Because there are no protons attached to adjacent carbons, it should appear as a singlet. In fact, there is just such a signal in the ¹H NMR spectrum listed above: 2.13, s, 3H

The other carbon flanking the carbonyl carbon was a methylene. Referring to the data in Table 13.1, that methylene should have a chemical shift of about 2.3 δ. In fact, we do have a signal at 2.40 δ. It integrates for two hydrogens, as we would expect. The critical information is that this signal appears as a **triplet.** That means that there are **two** protons on the adjacent carbon. In other words, the adjacent carbon is a methylene. That gives us the partial structures illustrated in Figure 13.7 (We know that we have two methyls attached to a C-H from 0.90, d, 6H, as discussed above).

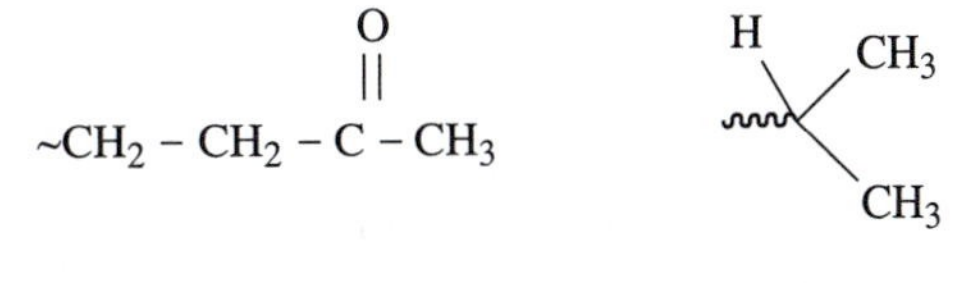

Figure 13.7 Partial structures of Example A.

It is helpful to use a wiggly line on partial structures, to remind yourself of where an additional fragment will eventually be attached. In this case, there is only one way to put the molecule together (Fig. 13.8). Note that undifferentiated H's, often with more than one coupling, appear as a **multiplet, "M".**

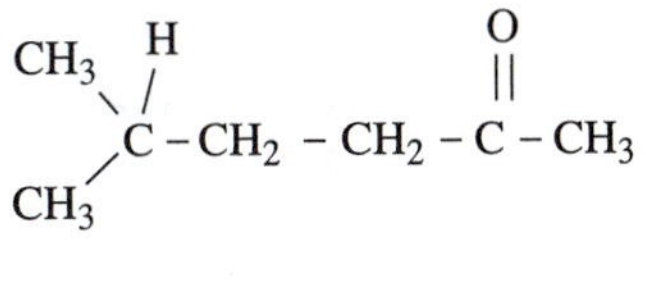

Figure 13.8 Complete Structure of Example A.

EXAMPLE B

^{13}C NMR	$C_6H_{14}O$
11.1, q (2)	
22.9, t (2)	
43.6, d	
64.8, t	

¹H NMR
0.90, t, 6H, J = 7.0 Hz
1.2–1.6, m, 5H
1.95, 1H, bs (exchanges)
3.52, d, 2H, J = 6.5 Hz

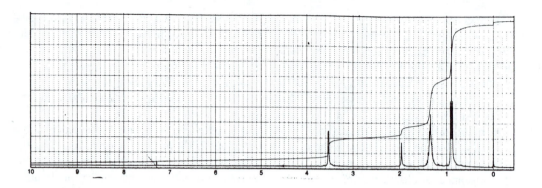

Figure 13.9   ¹H NMR spectrum for Example B.

This is the same substance as Example B in Chapter 12. By counting that there were 13 protons bonded to carbon in the ^{13}C data, we deduced that the other proton must be bonded to oxygen, and that we therefore have an alcohol functional group. That we have one proton attached to a heteroatom is also apparent from the proton spectrum. When a $CDCl_3$ solution of a sample having such a proton is shaken with D_2O, the signal for the O-H proton disappears, having exchanged with the D_2O. Further, because such protons are rapidly exchanging with each other, broadening the signal somewhat, one does not usually observe coupling of O-H or N-H to protons attached to adjacent carbon atoms. Thus, protons attached to heteroatoms appear, as in this problem, "bs, 1H, exchanges," where "bs" stands for "broadened singlet" and "exchanges" means "exchanges with D_2O."

Using approximate carbon shifts, we had concluded that the carbon to which the alcohol was attached was a $-CH_2-$. From the data in Table 13.1, this methylene should have a chemical shift of about 3.4 δ. It should integrate for two protons. In fact, we do have a two-proton signal at 3.52 δ. Now we look at the multiplicity of that signal. It is a doublet. That means that the group of two protons at 3.4 δ has exactly one proton on a neighboring carbon. We can thus discern the partial structure shown in Figure 13.10.

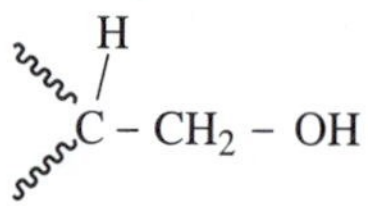

$$\overset{\displaystyle H}{\underset{\displaystyle}{\overset{\displaystyle |}{C}}} - CH_2 - OH$$

Figure 13.10   Partial structure for Example B.

Methyl groups also stand out in the proton spectrum, because they are intense signals (three hydrogens each) that can have at most only two protons on neighboring carbons. We know from the ^{13}C spectrum that there are two methyl groups in this molecule, with the same apparent chemical shift. Further, we know that these methyl groups are not near any heteroatoms, so they should have a normal hydrocarbon shift. Consulting the data in Table 13.1, we would expect the signal for these methyl groups at about 0.9 δ. In fact, we do observe a six-proton signal at 0.90 δ. This signal has a multiplicity of t, a triplet. We can thus discern that there must be a methylene adjacent to each methyl. From the ^{13}C spectrum, the methylenes must also be symmetrical. This analysis gives us the partial structures illustrated in Figure 13.11.

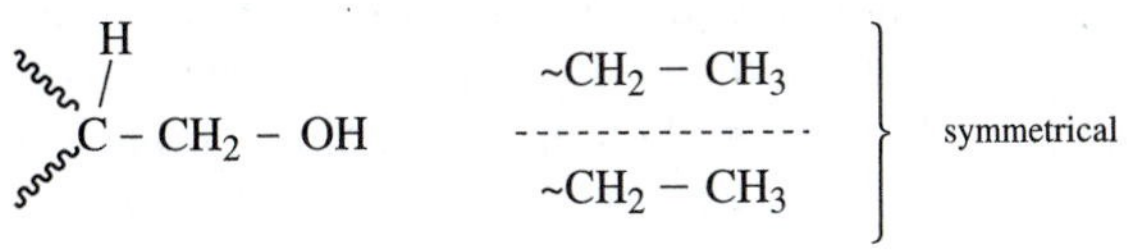

$$\overset{\displaystyle H}{\underset{\displaystyle\sim}{\overset{|}{C}}} - CH_2 - OH \qquad\qquad \left.\begin{array}{c} \sim CH_2 - CH_3 \\ \text{-----------------} \\ \sim CH_2 - CH_3 \end{array}\right\} \text{symmetrical}$$

Figure 13.11   All the partial structures for Example B.

There is only one way to assemble these partial structures, so the final structure is that illustrated in Figure 13.12.

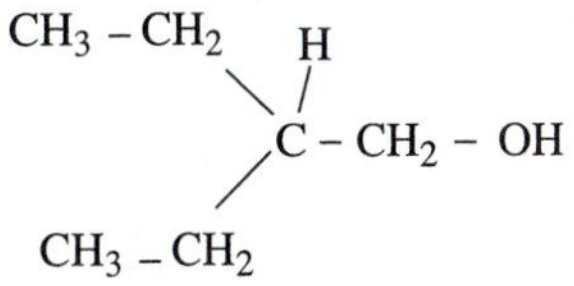

$$\begin{array}{c} CH_3 - CH_2 \quad H \\ \diagdown \quad / \\ C - CH_2 - OH \\ \diagup \\ CH_3 - CH_2 \end{array}$$

Figure 13.12   Complete structure for Example B.

THE QUEST, LEVEL THREE: RINGS

Around and around and around we go . . .

You have assigned all of the unsaturated functional groups, but you still have IHD for which you have not accounted—you must have a ring! So . . . how do we approach rings?

Branch Points and End Groups: Imagine three cartoon rings.   These cartoons are drawn as eight-membered rings, but are meant to be generic, "any rings". The rings (Fig. 13.13) could be unbranched (A), singly branched (B), or multiply branched (C). How are you going to be able to tell?

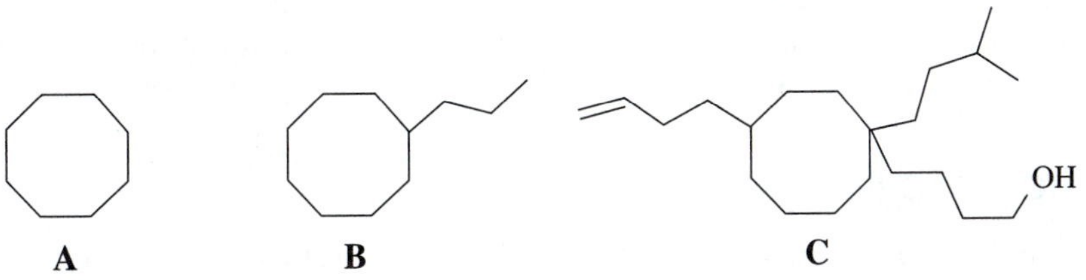

A   **B**   **C**

Figure 13.13   Representative rings.

We will use branch points and end groups to narrow down the possibilities for our unknown ring. An end group is a group that ends a chain, such as a methyl group or a primary alcohol. A branch point is a methine, such as that in B. Example C also includes a methine, and it also includes an sp³-hybridized quaternary carbon, a double branch point. Ring A has no methine or quaternary carbons, and it has no end groups, so there could not be any branches off the ring. Ring B has one branch point and one end group, so it must have one branch off the ring. Ring C has multiple branch points and end groups, including a branch point on a chain away from the ring. By considering how many branch points and end groups a molecule has, it is possible to more quickly come to a beginning idea of how to assemble the structure.

Problems 13.1-13.10, beginning on page 133, will give further practice in deducing structures from spectra.

EXPERIMENTS

SAFETY NOTE

All samples should be prepared in a well-ventilated area to avoid unnecessary breathing of solvent vapors. Although the capped NMR tubes are reasonably resilient, care must be taken not to crush the thin walls inadvertently.

For all of the experiments listed below, the instructor should demonstrate the proper use of the spectrometer and explain the precautions to be taken in determining a spectrum. A Varian model 360EM or any more powerful spectrometer would be appropriate.

The amount of sample used will depend on the spectrometer. With a modern Fourier transform instrument, 30 mg of sample is ample for both ^{1}H and ^{13}C spectra. In general, spectra are acquired using a 5-mm NMR tube. Make sure that the tube and its plastic cap are clean and dry. Take up the sample in 0.3 mL of deuterated NMR solvent (usually CDCl$_3$ containing 0.1% tetramethylsilane as an internal standard) and add the solution to the tube. If the solution appears turbid, filter it through a small plug of glass wool and then into the NMR tube. Cap the NMR tube securely.

A.   PROTON SPECTRUM OF ETHYLBENZENE

In a 5-mm NMR tube, prepare a solution of ethylbenzene in deuterated chloroform containing TMS. Place the cap on the tube, and mix the contents well by inverting the tube several times. Wipe the sample tube clean with a disposable tissue. Place the sample tube in the spectrometer and allow it to come to the temperature of the instrument. Determine and record the spectrum and integration curve over the range of 10 ppm. Record the instrument parameters on the spectrum and identify all of the peaks. Secure the spectra in your laboratory notebook using staples or tape.

B.   PROTON SPECTRA OF ALCOHOLS

The purpose of this experiment is to illustrate the effect of proton exchange on the splitting patterns of nearest neighbor hydrogens by the hydroxyl hydrogens of alcohols. A suitable deuterated nonhydroxylic solvent such as dimethylsulfoxide (DMSO-d$_6$), dimethylformamide (DMF-d$_7$), or chloroform (CDCl$_3$) is necessary. When sufficiently rapid hydrogen exchange involving the hydroxyl hydrogen occurs, the nearest-neighbor splitting patterns due to this hydrogen disappear and the hydroxyl hydrogen itself appears as a broad singlet. Other compounds with exchangeable hydrogens will also demonstrate this effect.

Determine the spectra of the following samples:

1. Neat absolute ethanol and 3 drops of TMS.
2. A 25% solution of absolute ethanol in DMSO-d$_6$.
3. Sample 2 plus 10 drops of water.

Identify each peak and comment on the utility of this effect in the interpretation of NMR spectra. Repeat the experiment using 2-propanol or 2-methyl-2-propanol.

C. CARBON-13 SPECTRUM OF ETHYLBENZENE

This experiment is designed to illustrate the effect of proton decoupling on a carbon-13 spectrum. Decoupled spectra are less complex than "nondecoupled" spectra because the splitting of the carbon signal by hydrogen is absent.

1. Determine the ¹³C spectrum of the ethylbenzene sample used in Experiment A using the following conditions:

   a. Normal, decoupled power applied.
   b. Decoupling power turned off.

   Overlay the two spectra, line up the TMS peaks, and record the identity and splitting patterns for each peak of the sample.

D. ACQUIRE THE SPECTRA OF AN UNKNOWN COMPOUND SUPPLIED BY YOUR INSTRUCTOR, AND DETERMINE THE STRUCTURE

PROBLEM 13.1

$C_7H_{12}O$

¹³C NMR	¹H NMR
206.7, d	1.08, s, 6H
133.1, d	2.21, d, J = 7.2 Hz, 2H
118.4, t	5.08, d, J = 11.8 Hz, 1H
45.7, s	5.11, d, J = 15.5 Hz, 1H
41.5, t	5.75, ddt, J = 11.8, 15.5, 7.2 Hz, 1H
21.2, q (2)	9.49, s, 1H

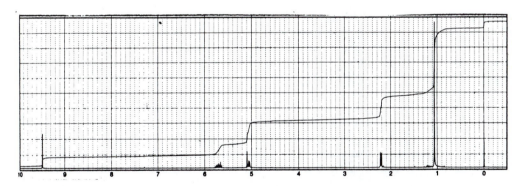

PROBLEM 13.2

$C_{10}H_{20}O_2$

¹³C NMR

177.0, s
64.4, t
34.1, d
31.6, t
28.8, t
25.7, t
22.6, t
19.1, q (2)
14.0, q

¹H NMR

0.89, t, J = 7.3 Hz, 3H
1.26, d, J = 6.5 Hz, 6H
1.4, m, 6H
1.64, m, 2H
2.52, m, 1H
4.05, t, J = 7.1 Hz, 2H

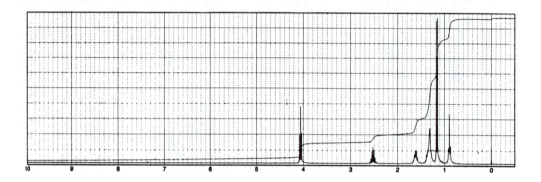

PROBLEM 13.3

$C_9H_{16}O_2$

¹³C NMR

173.6, s
51.3, q
42.0, t
34.9, d
33.1, t (2)
26.2, t (2)
26.1, t

¹H NMR

3.67, s, 3H
2.19, d, J = 6.4 Hz, 2H
1.70, m, 6H
0.9–1.3, m, 5H

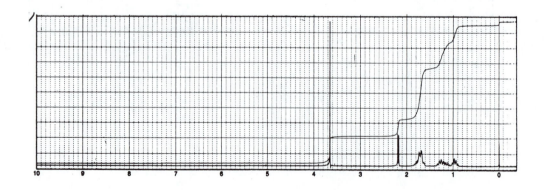

PROBLEM 13.4

C₈H₁₁N

¹³C NMR	**¹H NMR**
127.1, s	5.79, t, J = 6.2 Hz, 1H
126.3, d	2.97, s, 2H
117.7, s	2.02, m, 4H
28.0, t	1.70, m, 4H
25.8, t	
25.1, t	
22.5, t	
21.8, t	

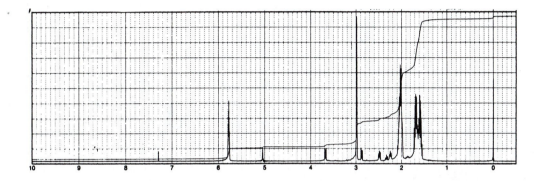

PROBLEM 13.5

C₈H₁₄O₂

¹³C NMR	**¹H NMR**
170.5, s	4.75, m, 1H
72.6, d	2.03, s, 3H
31.7, t (2)	1.2–1.7, m, 10H
25.4, t	
23.8, t (2)	
21.4, q	

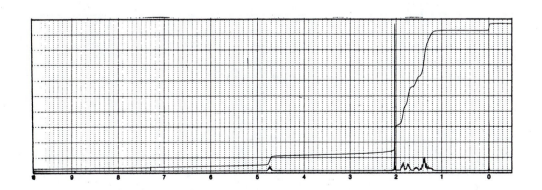

PROBLEM 13.6

$C_8H_{17}NO$

¹³C NMR	¹H NMR
208.2, s	1.02, t, J = 7.1, 6H
47.5, t	2.16, s, 3H
46.9, t	2.52, t, J = 7.1 Hz, 4H
41.6, t (2)	2.61, t, J = 7.6 Hz, 2H
30.2, q	2.75, t, J = 7.6 Hz, 2H
11.8, q (2)	

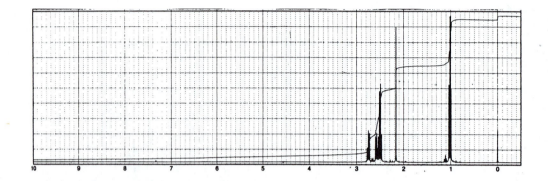

PROBLEM 13.7

$C_8H_{12}O_3$

¹³C NMR	¹H NMR
212.3, s	1.28, t, J = 7.2 Hz, 3H
169.4, s	1.9–2.1, m, 2H
61.3, t	2.3, m, 4H
54.8, d	3.15, t, J = 8.4 Hz, 1H
38.0, t	2.39, q, J = 7.2 Hz, 2H
27.4, t	
21.0, t	
14.2, q	

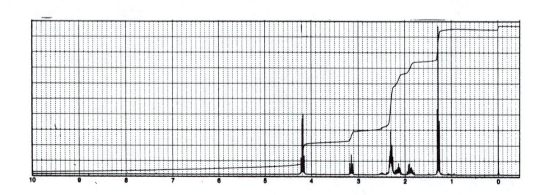

PROBLEM 13.8

$C_6H_{11}NO_2$

¹³C NMR	**¹H NMR**
119.4, s	4.46, t, J = 6.2 Hz, 1H
102.7, d	3.36, s, 6H
53.8, q (2)	2.41, t, J = 7.4 Hz, 2H
28.5, t	1.92, dt, J = 6.2, 7.4 Hz, 2H
12.4, t	

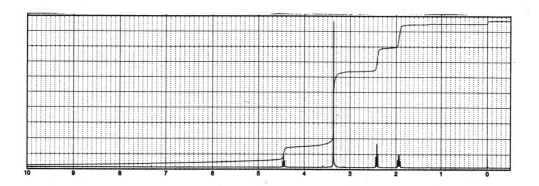

PROBLEM 13.9

$C_8H_{12}O_3$

¹³C NMR	**¹H NMR**
210.1, s	4.04, s, 4H
107.1, s	2.50, t, J = 6.6 Hz, 4H
64.6, t (2)	2.02, t, J = 6.6 Hz, 4H
38.2, t (2)	
33.9, t (2)	

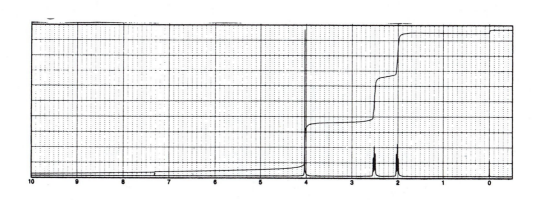

PROBLEM 13.10

$C_9H_{17}NO$

¹³C NMR	**¹H NMR**
171.1, s	0.96, d, J = 7.8 Hz, 6H
46.8, t (2)	1.9, m, 4H
43.7, t	2.18, d, J = 6.8 Hz, 2H
26.1, t (2)	2.40, m, 1H
25.5, d	3.44, m, 4H
22.7, q (2)	

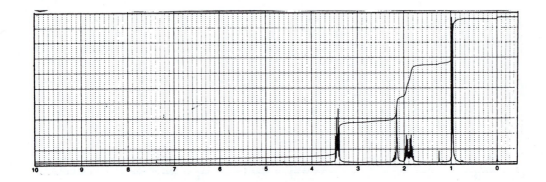

Table 13.1   Chemical Shifts of Protons on a Carbon Atom Adjacent (α Position) to a Functional Group in Aliphatic Compounds (M—Y)

❚ M = methyl

8 M = methylene

❚ M = methine

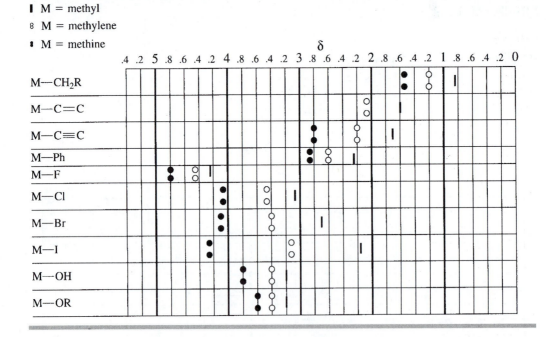

Table 13.1 (Continued)

▮ M = methyl
8 M = methylene
▮ M = methine

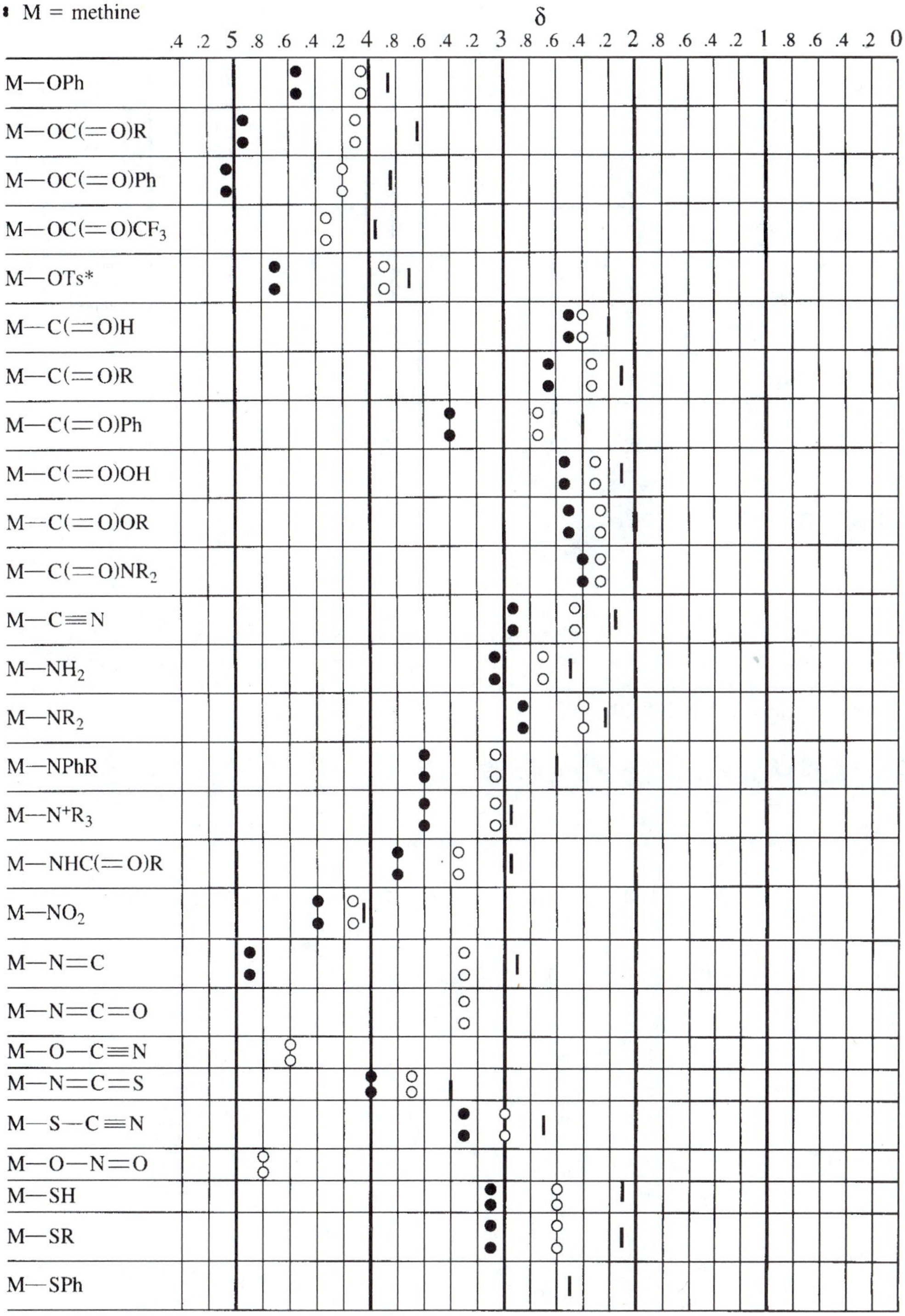

δ

Scale (left to right): .4 .2 5 .8 .6 .4 .2 4 .8 .6 .4 .2 3 .8 .6 .4 .2 2 .8 .6 .4 .2 1 .8 .6 .4 .2 0

M—OPh
M—OC(=O)R
M—OC(=O)Ph
M—OC(=O)CF₃
M—OTs*
M—C(=O)H
M—C(=O)R
M—C(=O)Ph
M—C(=O)OH
M—C(=O)OR
M—C(=O)NR₂
M—C≡N
M—NH₂
M—NR₂
M—NPhR
M—N⁺R₃
M—NHC(=O)R
M—NO₂
M—N=C
M—N=C=O
M—O—C≡N
M—N=C=S
M—S—C≡N
M—O—N=O
M—SH
M—SR
M—SPh

Table 13.1 (Continued)

❙ M = methyl
8 M = methylene
❙ M = methine

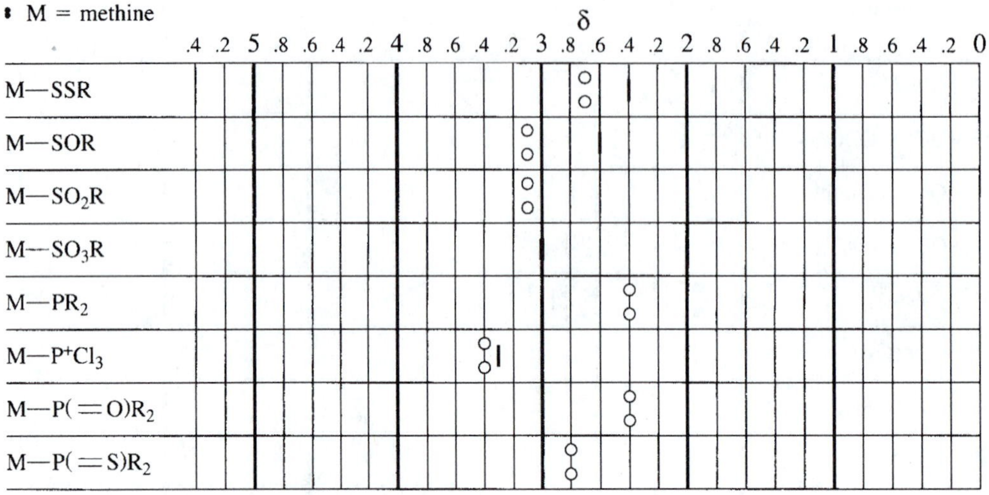

δ

.4 .2 5 .8 .6 .4 .2 4 .8 .6 .4 .2 3 .8 .6 .4 .2 2 .8 .6 .4 .2 1 .8 .6 .4 .2 0

Group	Chemical shift range (δ)
M—SSR	~3.1 (○○)
M—SOR	~3.7–3.8 (○○)
M—SO$_2$R	~3.7–3.8 (○○)
M—SO$_3$R	~3.6
M—PR$_2$	~2.5 (○○)
M—P$^+$Cl$_3$	~4.1 (○○)
M—P(=O)R$_2$	~2.5 (○○)
M—P(=S)R$_2$	~3.0 (○○)

*OTs is

$$-O-\overset{\displaystyle O}{\underset{\displaystyle O}{\overset{\|}{\underset{\|}{S}}}}-\bigcirc-CH_3$$

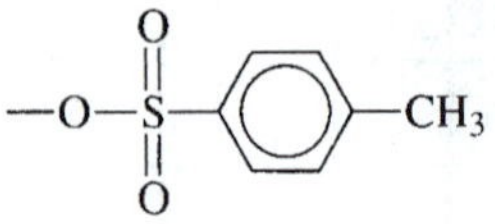

From R. M. Silverstein, G. C. Bassler, and T. C. Morrill, *Spectrometric Identification of Organic Compounds,* 5th ed., New York: John Wiley and Sons, Inc., 1991. Reprinted with permission of John Wiley and Sons, Inc. Chart A. 1, p. 208/209.

Table 13.2 Proton Spin-Coupling Constants

TYPE	J_{ab} (Hz)	J_{ab} TYPICAL	TYPE	J_{ab} (Hz)	J_{ab} TYPICAL
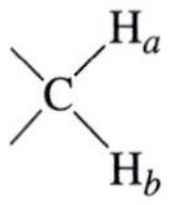 $\overset{\displaystyle}{\underset{\displaystyle}{C}}\overset{H_a}{\underset{H_b}{}}$	0–30	12–15	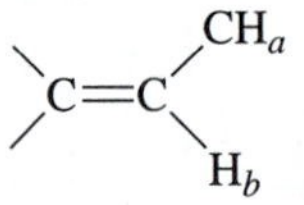 $C=C\overset{CH_a}{\underset{H_b}{}}$	4–10	7
CH$_a$—CH$_b$ (free rotaion)	6–8	7	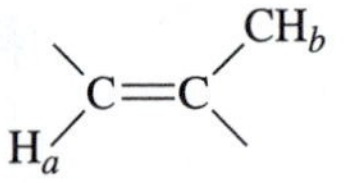 $\overset{}{\underset{H_a}{}}C=C\overset{CH_b}{\underset{}{}}$	0–3	1.5
CH$_a$—C—CH$_b$	0–1	0	$\overset{H_a}{\underset{}{}}C=C\overset{CH_b}{\underset{}{}}$	0–3	2
H$_a$ ... H$_b$ (ring)			C=CH$_a$—CH$_b$=C	9–13	10
ax–ax	6–14	8–10	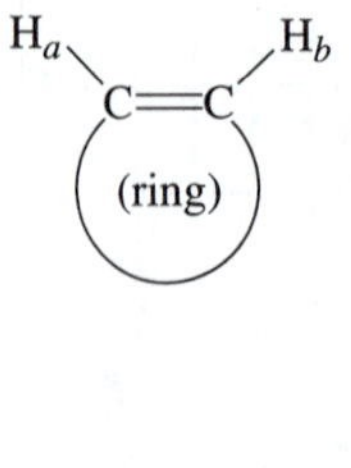 $\overset{H_a}{\underset{}{}}C=C\overset{H_b}{\underset{}{}}$ (ring)	3 member	0.5–2.0
ax–eq	0–5	2–3		4 member	2.5–4.0
eq–eq	0–5	2–3		5 member	5.1–7.0
H$_a$ ... H$_b$ *cis*	5–10			6 member	8.8–11.0
(cis or trans) *trans*	5–10			7 member	9–13
				8 member	10–13

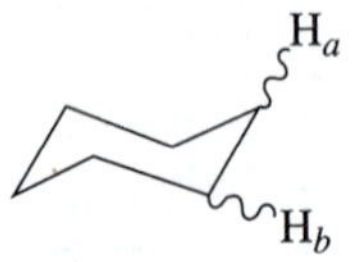

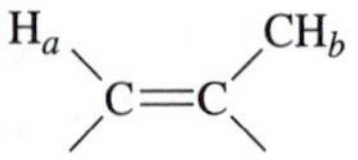

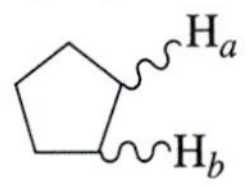

Table 13.2 (Continued)

TYPE	J_{ab} (Hz)	J_{ab} TYPICAL	TYPE	J_{ab} (Hz)	J_{ab} TYPICAL
			CH_a—C≡CH_b	2–3	
			—CH_a—C≡C—CH_b—	2–3	

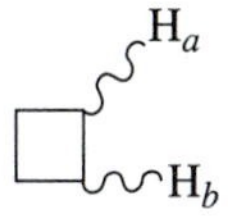

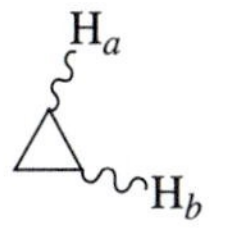

cis 4–12
trans 2–10
(cis or trans)

6

cis 7–13
trans 4–9
(cis or trans)

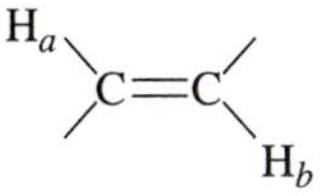

4

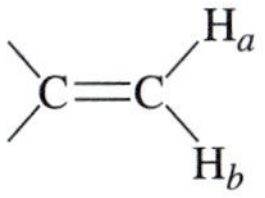

2.5

CH_a—OH_b (no exchange) 4–10 5

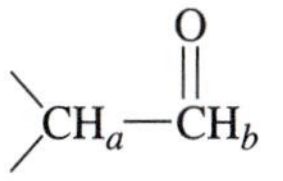

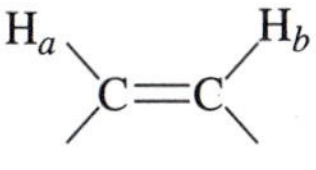

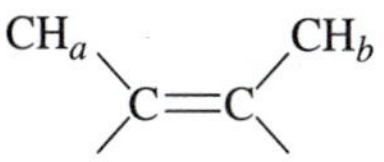

1–3 2–3

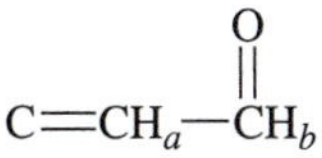

			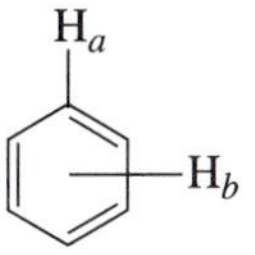 J (ortho)	6–10	9
			J (meta)	1–3	3
			J (para)	0–1	~0

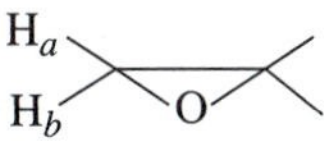

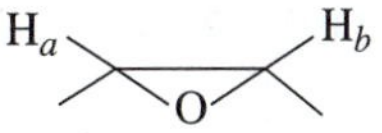

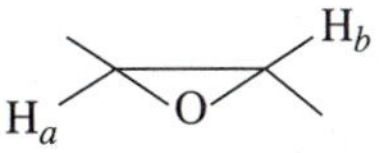

C=CH_a—CH_b 5–8 6

			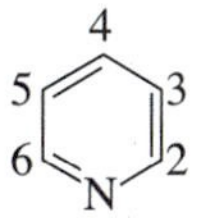 J (2–3)	(5–6)	5
			J (3–4)	(7–9)	8
			J (2–4)	(1–2)	1.5
			J (3–5)	(1–2)	1.5
			J (2–5)	(0–1)	1
			J (2–6)	(0–1)	~0

12–18 17

0–3 0–2

			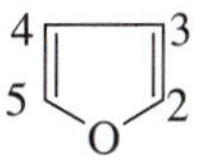 J (2–3)	1.3–2.0	1..8
			J (3–4)	3.1–3.8	3.6
			J (2–4)	0–0	~0
			J (2–5)	1–2	1.5

6–12 10

			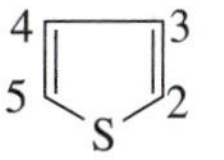 J (2–3)	4.9–6.2	5.4
			J (3–4)	3.4–5.0	4.0
			J (2–4)	1.2–1.7	1.5
			J (2–5)	3.2–3.7	3.4

0–3 1–2

From R. M. Silverstein, G. C. Bassler, and T. C. Morrill, *Spectrometric Identification of Organic Compounds,* 5th ed., New York: John Wiley and Sons, Inc., 1991. Reprinted with permission of John Wiley and Sons, Inc. Chart Appendix F, p. 221, and compiled by Varian Associates.

Table 13.3   Chemical Shifts in Alicyclic Rings

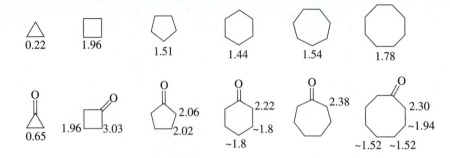

From R. M. Silverstein, G. C. Bassler, and T. C. Morrill, *Spectrometric Identification of Organic Compounds,* 5th ed., New York: John Wiley and Sons, Inc., 1991. Reprinted with permission of John Wiley and Sons, Inc. Table C.1, p. 214.

14

Advanced Structure Determination

... down the rabbit hole, into Wonderland ...

In this chapter, you will learn about more advanced concepts in structural elucidation, including the use of spectroscopic tables to predict ^{13}C and ^{1}H chemical shifts. These concepts, however, are just the beginning. Every day, the detailed three-dimensional structures of complex natural products are elucidated by professionals in the field. Leading references to this work, including detailed instruction on the several useful variations of NMR pulse sequences, are listed at the end of this chapter and at the end of Chapter 12.

THE QUEST, LEVEL FOUR: CALCULATING ^{13}C CHEMICAL SHIFTS

... hear the calculators clicking ...

Consider the unknown listed in Figure 14.1. Using the approach outlined in Chapter 12, it is easy to arrive at the fragments listed in Figure 14.2. The question is, how to assemble those fragments? Two structures are possible, A and B, illustrated in Figure 14.3. How can we tell which it is?

14.1 $C_6H_{14}O$

^{13}C NMR
72.3, d
39.4, t
30.3, t
19.4, t
14.0, q
9.9, q

Figure 14.1 The data for unknown 14.1.

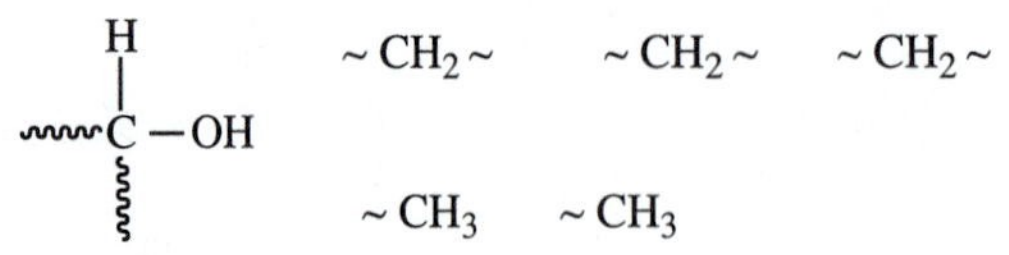

Figure 14.2   Structural fragments for unknown 14.1.

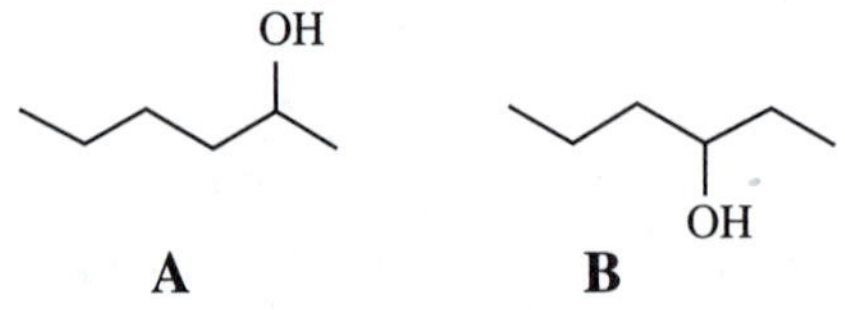

Figure 14.3   Possible structures for unknown 14.1.

To calculate ^{13}C chemical shifts, you will use the data in Tables 14.1 and 14.2. Table 14.1 lists chemical shifts of representative hydrocarbons. Table 14.2 lists the changes in chemical shifts induced by the attachment of functional groups.

To illustrate, both structure A and structure B (Fig. 14.3) are derived from *n*-hexane. The chemical shifts for *n*-hexane are listed in Figure 14.4.

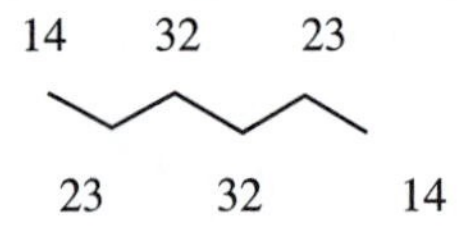

Figure 14.4   The chemical shifts for *n*-hexane.

Using the data in Table 14.2, we can calculate the chemical shifts expected for structure A (Fig. 14.5). Note that we use the internal shifts for the alcohol functional group, since we know that it is attached in the middle of the chain. The α shift refers to the carbon where the functional group is attached. The β carbon is next to the α carbon, and the γ carbon is next to the β carbon.

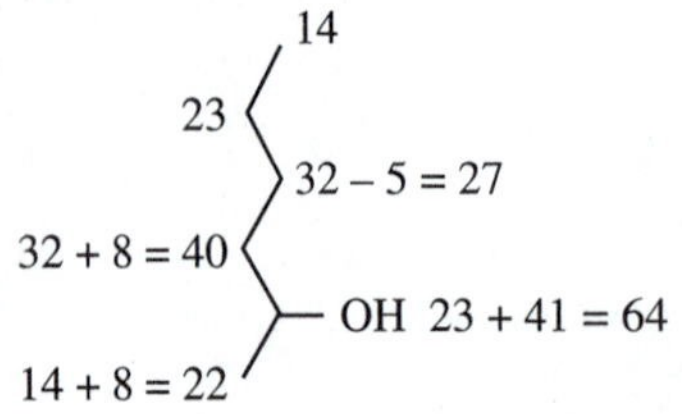

Figure 14.5   Calculated ^{13}C chemical shifts for 2-hexanol (A).

Using this same approach, we can calculate the chemical shifts expected for structure B (Fig. 14.6).

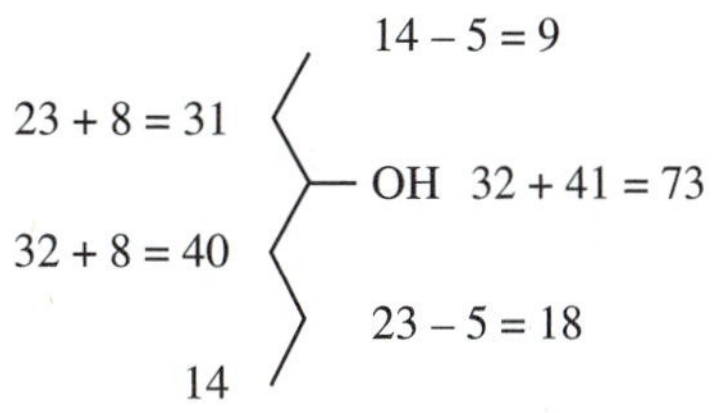

$$14 - 5 = 9$$

$$23 + 8 = 31$$

$$OH \quad 32 + 41 = 73$$

$$32 + 8 = 40$$

$$23 - 5 = 18$$

$$14$$

Figure 14.6   Calculated ^{13}C chemical shifts for 3-hexanol (B).

With both sets of data calculated, it is easy to list each set in order and compare each list to the unknown (Fig. 14.6). It is apparent that B fits much more closely, so unknown 14.1 is 3-hexanol.

A	B	unknown 14.1
14, q	9, q	9.9, q
22, q	14, q	14.0, q
23, t	18, t	19.4, t
27, t	31, t	30.3, t
40, t	40, t	39.4, t
64, d	73, d	72.3, d

Figure 14.7   Comparison of calculated ^{13}C chemical shifts for A and B with unknown 14.1.

It would be a useful exercise at this point to practice by calculating the ^{13}C chemical shifts of the unknowns in Chapters 12 and 13. In addition to the data in Table 14.1, the structures in Tables 12.1–12.9 can be used as "starting hydrocarbons."

THE QUEST, LEVEL FIVE: RING SIZE

... around and around and around again ...

While IR is diagnostic for a variety of functional groups, it is especially valuable for ring size of cyclic esters (known as "lactones") and cyclic ketones. Note that α,β-unsaturated carbonyl compounds in general absorb at 20 to 40 cm^{-1} lower frequency than the saturated analogues.

Many of the problems from here on will have IR data. In Step 5, check prospective solutions against the IR. It may be possible that several otherwise acceptable alternative ways of assembling the unknown molecule will not be compatible with the IR spectrum.

We will practice this approach with unknown 14.2, the data for which is in Figure 14.8. The ^{13}C signal at 176.6, s tells us that we have an ester or a carboxylic acid. Since all the H's are attached to carbon, it must be an ester. That accounts for one IHD, but the calculated IHD = 2. There are no other sp^2- or sp-hybridized atoms, so the molecule must also have a ring. The only branch point (see Chapter 12) is the methine, so the ester must be included in the ring, making a lactone. There is one branch off the ring, terminating in a methyl group that must be adjacent to a -CH$_2$-(^{13}C δ 13.1, q; ^{1}H δ 0.96, t, J = 6.8, 3H). The unknown, then, is either **C** or **D** (Fig. 14.9).

Unknown 14.2

$C_7H_{12}O_2$

IR: 2980, 2890, 1775, 1470, 1370, 1350, 1190, 1020, 980, 925 cm^{-1}

^{13}C NMR	^{1}H NMR
13.1, q	0.96, t, J = 6.8, 3H
17.9, t	1.4–1.9, m, 6H
27.3, t	2.4-2.6, m, 2H
28.1, t	4.5, m, 1H
36.9, t	
80.1, d	
176.6, s	

Figure 14.8   The data for unknown 14.2.

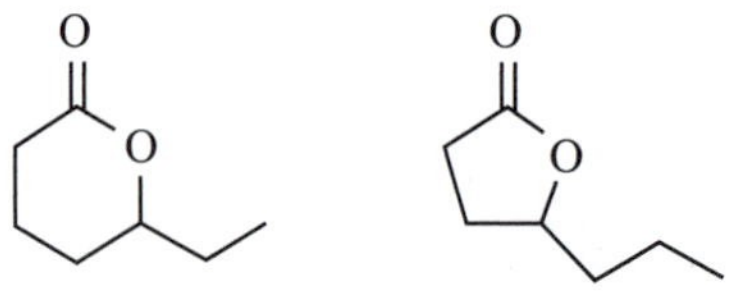

C   D

Figure 14.9   Possible structures for unknown 14.2.

Referring to "Carboxylic Esters and Lactones" in Chapter 11, we learn that **C** would be expected to have a carbonyl stretch in the range 1730 to 1750 cm^{-1}, whereas **D** would be expected to have a carbonyl stretch around 1770 cm^{-1}. We can therefore conclude that the structure of unknown 14.2 is **D**.

THE QUEST, LEVEL SIX: ALKENES

How to find the number of alkenes?
Count the number of alkene carbons, and divide by two!

The ^{13}C chemical shifts of representative alkenes are summarized in Tables 12.3 and 12.4. A careful perusal of Table 12.3 reveals a consistent pattern: alkene carbons bearing two hydrogens resonate in the range 105 to 120, those bearing one hydrogen resonate in the range 120 to 140, and those bearing no hydrogen resonate in the range 130 to 150.

In Table 12.6, one observes some carbons that fall in the ranges specified, but many that do not. This is because these alkenes have either heteroatoms or electron-withdrawing groups directly attached to ("conjugated with") them. The chemical shift of a carbon is a function of the electron density observed at that carbon. An increase in electron density will cause a carbon to resonate at higher field (smaller numbers), while a decrease in electron density will cause the carbon to resonate at lower field (larger numbers). A carbonyl conjugated with an alkene shifts the β-carbon downfield. It does not affect the α-carbon. This can be rationalized by considering the polarization of the bonding electrons. The β-carbon has lost electron density, so it will resonate at lower field (larger numbers). The α-carbon is unchanged (Figure 14.10).

A heteroatom conjugated with an alkene shifts *both* carbons. The α-carbon is shifted downfield (larger chemical shift) and the β-carbon is shifted upfield (smaller chemical shift). This can be rationalized by considering that the alkene electron cloud is repelled by the nonbonding electrons of the heteroatom, while at the same time the electronegative heteroatom withdraws electron density from the α-carbon.

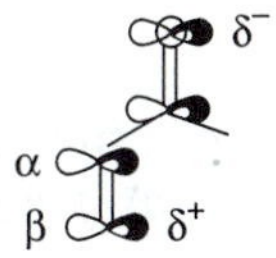

Figure 14.10 Polarization of alkenes by electron-withdrawing and electron-donating groups.

These effects are summarized by Figure 14.11. Remember that "normal" chemical shift for an alkene carbon depends on how many alkyl groups are attached. From Table 12.5, we can see that a "normal" alkene carbon with two protons attached would come at 105–120 δ, a "normal" alkene with one proton attached would come at 120–140 δ, and a "normal" alkene with no protons attached would come at 130–150 δ.

Note that each alkene has two carbons. If you think a signal upfield (e.g., 100 δ) might be half of a heteroatom-polarized alkene, look for the other alkene carbon shifted downfield. If it is not there, then that signal at 100 δ is not an alkene carbon.

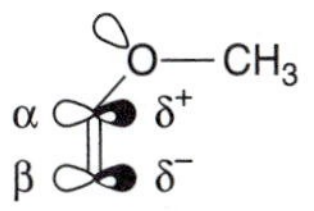

Figure 14.11 Expected chemical shift ranges for polarized and nonpolarized alkenes.

CALCULATING THE ^{1}H NMR CHEMICAL SHIFTS OF ALKENES

^{1}H NMR shifts of alkenes are easily calculated using Table 14.4. You will note that for polarized alkenes, the protons are shifted the same way as the carbons. For carbonyl-polarized alkenes, the β-protons are shifted downfield and the α-protons are normal. For heteroatom-polarized alkenes, the α-protons are shifted downfield and the β-protons are shifted upfield. Note (Table 13.2) that protons *trans* on an alkene share a J value of about 17 Hz, whereas protons *cis* on an alkene usually share a J value of about 10 Hz. Figure 14.12 is the diagram on which Table 14.4 is based, and Figure 14.13 is an example that illustrates the use of the data in Table 14.4.

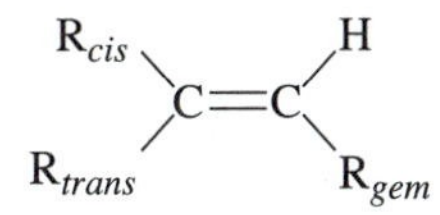

$$\delta_H = 5.25 + Z_{gem} + Z_{cis} + Z_{trans}$$

Figure 14.12 Diagram on which Table 14.4 is based.

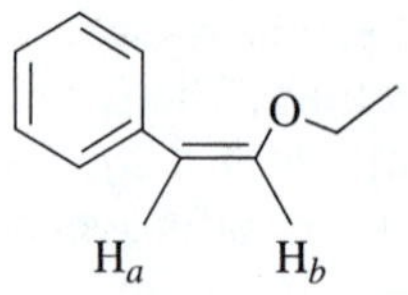

Figure 14.13 Using Table 14.4 to calculate the ^1H NMR chemical shifts of an alkene.

Practice using Table 14.4 with the alkene illustrated (Fig. 14.13). The starting value is always 5.25 δ. H substituents neither add to nor subtract from this value. For H_a and H_b, then, we need only to consider the shifts due to the geminal substituent and to the trans substituent. The predicted chemical shift for H_a is then $5.25 + 1.04 - 1.28 = 5.32$ δ, and for H_b the predicted chemical shift is $5.25 + 1.18 - 0.10 = 6.33$ δ.

THE QUEST, LEVEL SEVEN: BENZENE DERIVATIVES

^{13}C **of Benzene Derivatives:** The ^{13}C chemical shift of a carbon in a benzene derivative can be calculated using the data in Table 14.5. This can best be illustrated with an example (Fig. 14.14).

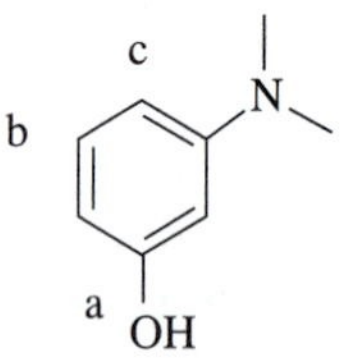

Figure 14.14 Using Table 14.5 to calculate the ^{13}C NMR chemical shifts of a benzene derivative.

Carbon a: The starting value for any benzene carbon is 128.5 δ, from the data in Table 14.5. We then add the incremental contribution from each substituent. The increment shift from H is zero, so we only need to account for effects due to substituents on the ring. For carbon a there are two substituents, a C-1 OH and a C-3 -NMe_2. The C-1 OH contributes +26.6. The C-3 -NMe_2 contributes +0.8. The calculated chemical shift for carbon a is therefore $128.5 + 26.6 + 0.8 = 155.9$ δ.

Carbon b: The starting value for carbon b is 128.5 δ. With regard to b, the -OH is at C-3, and so contributes +1.6. The -NMe_2 substituent is at the other C-3, and so contributes an increment of +0.8. The calculated chemical shift for carbon b is therefore $128.5 + 1.6 + 0.8 = 130.9$ δ.

Carbon c: The starting value for carbon c is 128.5 δ. With regard to c, the -OH is at C-4, and so contributes -7.3. The -NMe_2 substituent is at C-2, and so contributes an increment of -15.7. The calculated chemical shift for carbon c is therefore $128.5 - 7.3 - 15.7 = 105.5$ δ.

Often, the only way to decipher the substitution pattern on a highly substituted benzene derivative is to calculate the chemical shifts of the aromatic carbons for each possible substitution pattern, and compare the calculated values to the data given. Remember to consider multiplicity as well as chemical shift in making these comparisons.

¹H NMR OF BENZENE DERIVATIVES

The ¹H NMR chemical shifts of a benzene derivative can be calculated using the data in Table 14.6. These are approximately additive. Again, the use of this table can best be illustrated with an example (Fig. 14.15).

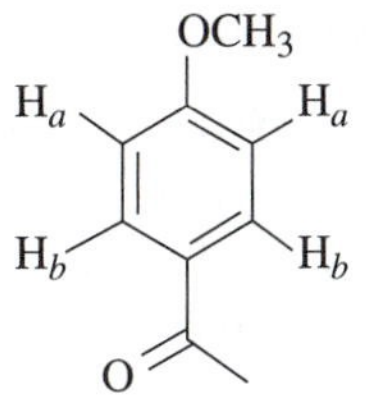

Figure 14.15   Using Table 14.6 to calculate the ¹H NMR chemical shifts of a benzene derivative.

The protons on an unsubstituted or alkyl-substituted benzene (first entries in Table 14.6) come at about 7.26 δ. A proton *ortho* ("1,2-") to an -OCH₃ would, according to this table, come at about 7.0 δ, a shift upfield of 0.26 δ. A proton *meta* ("1,3") to a ketone carbonyl would come at 7.5 δ, a downfield shift of 0.24 δ. It follows that H_a should resonate at about 7.26 − 0.26 + 0.24 = 7.24 δ. For H_b, a proton *meta* to an -OCH₃ would, according to Table 14.6, come at about 7.35 δ, a shift downfield of 0.09 δ. A proton *ortho* to a ketone carbonyl would come at 7.9 δ, a downfield shift from 7.26 of 0.64 δ. It follows that H_b should resonate at about 7.26 + 0.09 + 0.64 = 7.99 δ.

Protons *ortho* to each other (Table 13.2) have a 9-Hz coupling constant, protons *meta* to each other have a 3-Hz coupling constant, and protons *para* to each other do not couple. Remember that protons with the same chemical shift (even if not chemically the same) do *not* couple to each other.

For the example illustrated, H_a and H_b would share a 9-Hz coupling constant. There would be no *meta* coupling, since protons with the same chemical shift do not couple to each other. The aromatic portion of the ¹H spectrum would then be summarized: 7.24, 2H, d, J = 9.0 Hz; 7.99, 2H, d, J = 9.0 Hz.

SOLVING PROBLEMS WITH ALKENES AND ARENES

The key to solving problems with alkenes and arenes is in Step 2 on the worksheet, "assigning and exploring around the unsaturated functional groups." First establish the alkene carbon count, and, where possible, try to pair the alkene carbons. The first clue that an unknown might contain a benzene ring will come when the unknown has at least one ring and at least three double bonds. Confirmation comes if there are also arene-type protons (6.5–8.0 δ).

It is important at this point to figure out the substitution pattern of the arene, especially if there are more than three double bonds in the molecule. Count the arene-type hydrogens. There are six positions on the benzene ring. If there are only three aromatic protons, then there must also be three substituents on the ring.

Once you know how many substituents there are, look for symmetry. For instance, a disubstituted aromatic that has only two bands of protons (two doublets, each 2H, J = 9 Hz) must be *para*-disubstituted. Then, decide what the substituents might be. This can often be deduced from the chemical shifts of the arene protons, and other information about functional groups in the molecule. Once the substituents on the benzene ring are known and you have deduced the pattern of attachment, it should be possible to calculate the approximate chemical shifts (and multiplicity) of the arene carbons. Once these carbon signals are subtracted from the spectrum, assignment of the residual alkenes should be more straightforward.

THE QUEST, LEVEL EIGHT: MASS SPECTROMETRY

A typical mass spectrometer is illustrated in Figure 14.16. A sample introduced into the source is vaporized, then ionized (an electron is removed). The resultant group of ions (some of which are falling apart!) is accelerated from the source. This group of ions then encounters a magnetic field. In that field, the moving ions are deflected by an amount inversely proportional to the mass/charge (m/z) ratio. Since most ions carry a single charge, they are thus separated, when they reach the detector, according to mass. An ion current is then measured at each value of m/z, and the result tabulated. That tabulated result is the **mass spectrum.**

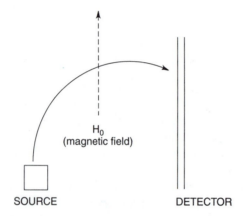

Figure 14.16   Diagram of a mass spectrometer.

Some portion of that ion current will be due to the molecular ion, the ion that is the entire molecule. Other "fragment" ions are also generated. In this section, we will learn how to use these fragment ions to assist in deducing the structure of an unknown. There are four processes that are important to understand: α-cleavage, β-cleavage, McLafferty rearrangement, and decarbonylation.

α-**Cleavage:** In the ionization process, an electron is removed from the molecule. If there is a heteroatom in the molecule, one of the nonbonding electrons on the heteroatom is most easily removed. This gives a radical cation, the molecular ion. Radical cleavage (Fig. 14.17) of the adjacent carbon-carbon bond then gives a stable carbocation, a fragment ion.

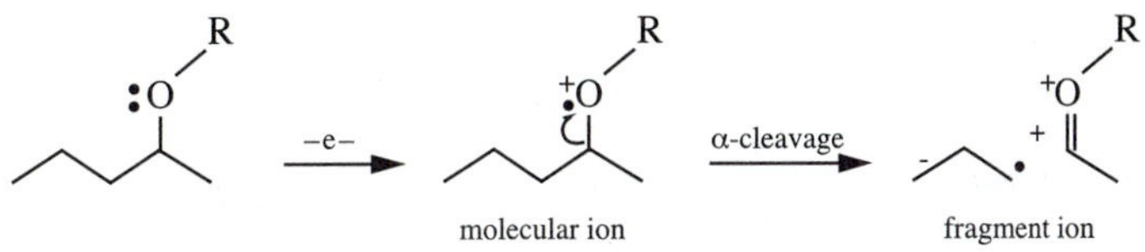

Figure 14.17   Illustration of α-cleavage.

β-Cleavage: Another process that is important, especially in cyclic systems, is β-cleavage. For example, α-cleavage of the molecular ion in a cyclic ketone, as shown in Figure 14.18, gives a new species that has the same molecular weight as the molecular ion. Radical cleavage of the carbon-carbon bond β to the radical center then gives a fragment ion. β-cleavage can proceed whenever there is a carbon-carbon bond β to a radical center.

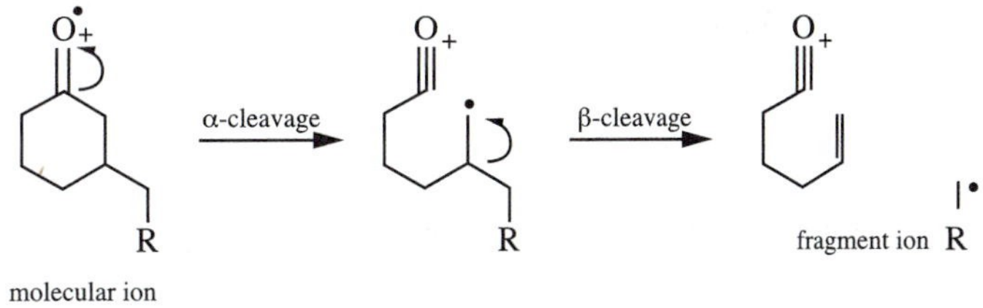

Figure 14.18   Illustration of β-cleavage.

McLafferty Rearrangement: When a carbonyl derivative has a proton on a carbon three carbons from the carbonyl, McLafferty rearrangement can proceed (Fig. 14.19). The product fragment ion rearrangement from McLafferty rearrangement is itself the molecular ion of a new ketone, which can then proceed with α-cleavage, β-cleavage, or McLafferty rearrangement. The fragments resulting from McLafferty rearrangement stand out in the mass spectrum, because (if the starting molecular weight is even) they are of **even mass.** Ions resulting from α-cleavage and β-cleavage, while more abundant, are all **odd mass.**

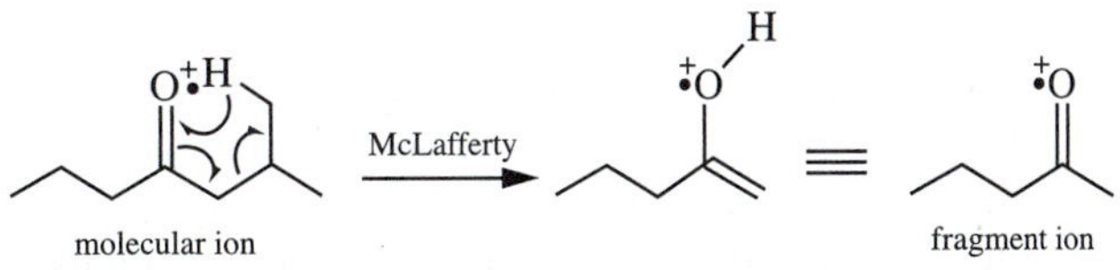

Figure 14.19   Illustration of McLafferty rearrangement.

Decarbonylation: The product of α-cleavage of a ketone, as shown in Figure 14.20, is an acylium ion. Such an ion will often go on to lose carbon monoxide ("decarbonylation"), to give a new fragment ion. This can then further lose methylene units, as illustrated.

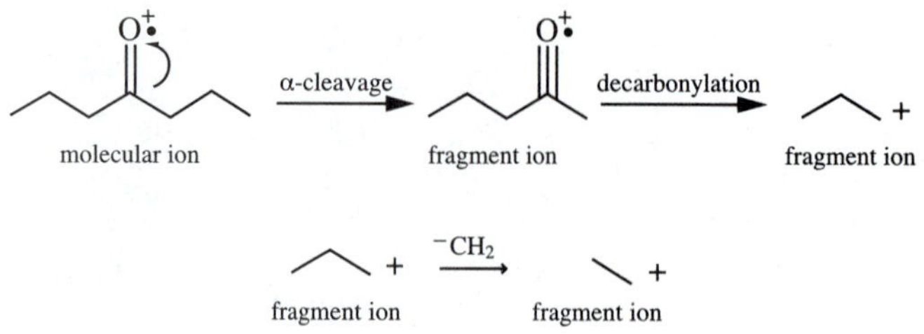

Figure 14.20   Illustration of decarbonylation.

Analysis of the Mass Spectrum: While a variety of other fragmentation and rearrangement processes can and do proceed in the mass spectrometer, the four processes outlined above will often be sufficient to rationalize most of the prominent fragment ions. Use the mass spectrum in Step 5, focusing on fragmentations that differentiate between alternative structures. This is illustrated by unknown 14.3 (Fig. 14.21).

Unknown 14.3

$C_{19}H_{39}N$

MS: 280 ($M^+ - H$, 4), 266 (10), 124 (3), 111 (5), 98 (100)

13C NMR		**1H NMR**
50.7, d	26.3, t	3.07, 1H, m
45.7, d	22.6, t	2.88, 1H, m
33.9, t	21.0, q	2.05, 1H, bs (exchanges)
32.8, t	19.4, t	1.6, 8H, m
31.8, t	14.0, q	1.3, 22H, m
30.6, t		1.07, 3H, d, J = 7.1 Hz
30.4, t (2)		0.88, 3H, t, J = 6.6 Hz
29.7, t (2)		
29.6, t (2)		
29.2, t (2)		

Figure 14.21   The data for unknown 14.3.

From the data, this is a secondary amine, with two methines (C-H) attached to the N. The methyl group that is a doublet must also be attached to one of those same methines. Unknown 14.3 has a ring, and a side chain that ends in a methyl that is attached to a -CH₂-. This approximate structure is summarized in Figure 14.22. The question is, how many methylenes are in the ring, and how many are in the side chain?

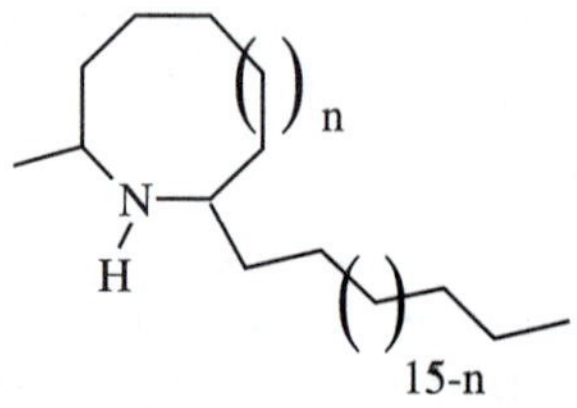

Figure 14.22   The approximate structure of unknown 14.3.

The final structure can be found by analyzing the mass spectrum. α-cleavage of the radical cation derived from the amine would proceed on either side. When the methyl group is lost, an ion of m/z = 266 will be generated. This is observed, but that does not tell us anything new—we already knew that the methyl group was attached to one of the methines.

Loss of the other side chain would mean loss of the terminal methyl on that side chain, plus the accompanying methylenes. If the side chain were ethyl, for instance, α-cleavage would generate a fragment having m/z = 252. Loss of a propyl group would give m/z = 238. Working down, we eventually come to m/z = 98. This is loss of a C_{12} sidechain, so the structure of unknown 14.3 must be as depicted in Figure 14.23.

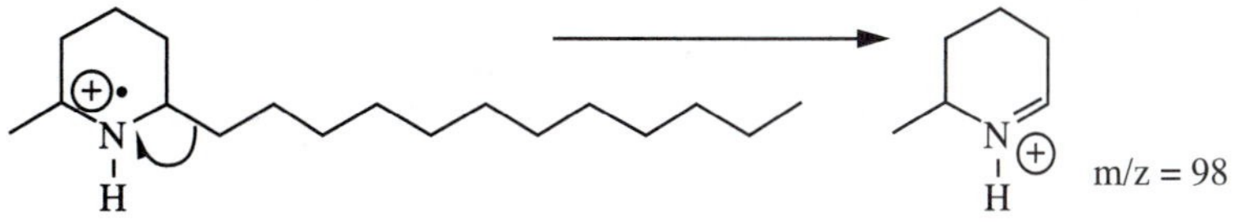

Figure 14.23   The structure of unknown 14.3, and its α-cleavage.

PROBLEMS

Problems 14.1–14.10 will give further practice in deducing structures using spectroscopic data.

REFERENCE

Crews, P.; Rodriguez, J.; Jaspars, M. *Organic Structure Analysis;* Oxford University Press: New York, 1998.

PROBLEM 14.1

$C_{12}H_{22}O_2$
IR: 2926, 2856, 1728, 1645, 1436, 1197, 1175, 819 cm^{-1}
MS: 198 (M$^+$, 2), 167 (7), 124 (9), 113 (100), 100 (33), 87 (47), 74 (43)

^{13}C NMR

166.5, s	29.0, t
150.7, d	28.9, t
119.0, d	28.8, t
50.6, q	22.5, t
31.7, t	13.9, q
29.3, t	
29.1, t	

^1H NMR

6.14, dt, 1H, J = 7.5, 11.5 Hz
5.68, d, 1H, J = 11.5 Hz
3.61, s, 3H
2.57, dq, 2H, J = 1.4, 7.4 Hz
1.35, m, 2H
1.18, m, 10H
0.80, m, 3H

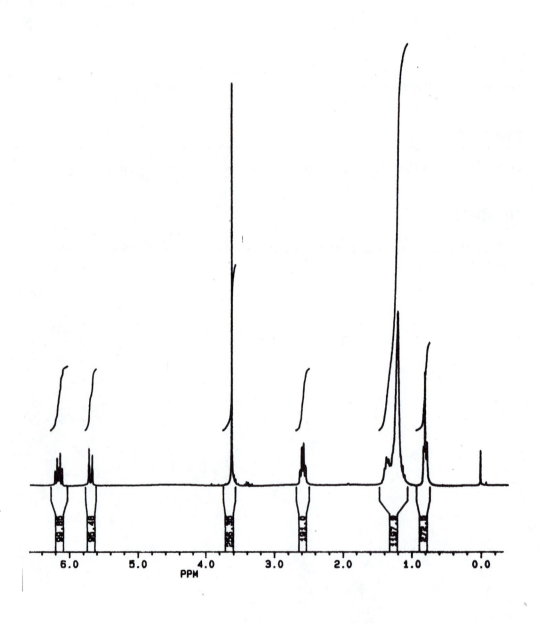

PROBLEM 14.2

$C_{11}H_{16}O_3$

IR (film) 2982, 1732, 1377, 1183, 1023 cm^{-1}
MS: 196 (M$^+$, 2), 150 (20), 122 (35), 108 (36), 107 (28), 104 (36), 81 (100)

^{13}C NMR	^{1}H NMR
193.8, d	9.61, d, 1H, J = 7.9 Hz
173.0, s	7.44, ddd, 1H, J = 0.9, 11.1, and 15.2 Hz
146.3, d	6.32, t, 1H, J = 11.1 Hz
142.1, d	6.16, dd, 1H, J = 7.9 and 15.2 Hz
132.0, d	5.99, dt, 1H, J = 7.9 and 11.1 Hz)
127.5, d	4.14, q, 2H, J = 7.1 Hz
60.3, t	2.41, q, 2H, J = 7.9 Hz
33.3, t	2.36, t, 2H, J = 7.3 Hz
27.5, t	1.81, quint, 2H, J = 7.3 Hz
24.2, t	1.26, t, 3H, J = 7.1 Hz
14.1, q	

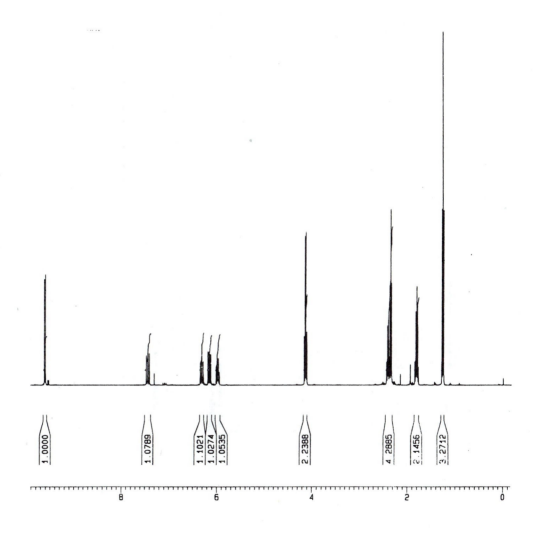

PROBLEM 14.3

$C_{14}H_{16}O_3$

IR: 2926, 2855, 1743, 1718, 1644, 1466, 1358, 1242, 1151, 1120 cm^{-1}

MS: 232 (M$^+$, 34), 190 (26), 189 (51), 168 (33), 156 (100), 131 (54), 130 (39), 119 (69), 116 (56), 104 (18), 91 (88), 77 (18)

^{13}C NMR

202.1, s	57.3, q
169.6, s	52.3, d
136.9, s	31.5, t
132.7, d	29.2, q
128.4 , d (2)	
127.3, d	
126.1, d (2)	
125.6, d	

^{1}H NMR

7.31, m, 5H
6.46, d, 1H, J = 15.8 Hz
6.12, dt, 1H, J = 7.1, 15.8 Hz
3.74, s, 3H
3.62, t, 1H, J = 7.4 Hz
2.76, t, 2H, J = 7.3 Hz
2.26, s, 3H

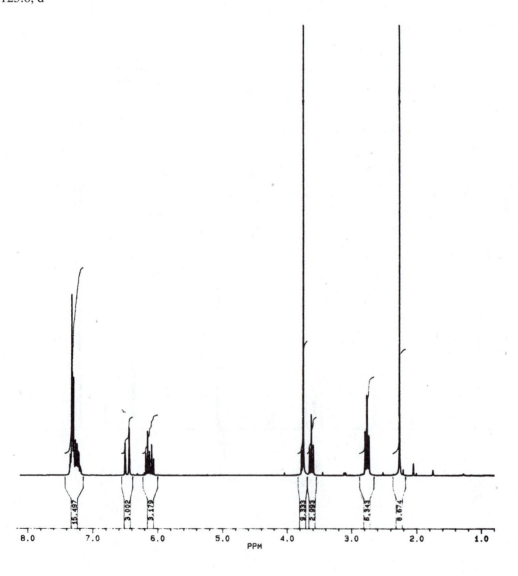

PROBLEM 14.4

$C_{15}H_{22}O_2$

IR: 2931, 1733, 1692, 1452, 1373, 1155, 1036, 757 cm^{-1}
MS: 234 (M$^+$, 10), 206 (63), 188 (30), 156 (26), 131 (21), 117 (100)

^{13}C NMR		^{1}H NMR
173.6, s	35.8, t	7.2, m, 5H
145.3, s	34.4, t	4.09, 2H, q, J = 7.1 Hz
128.2, d (2)	29.6, t	2.40, 1H, m
127.7, d	23.1, t	2.22, 2H, t, J = 7.5 Hz
125.9, d (2)	14.2, q	1.60, 6H, m
60.1, t	12.1, q	1.21, 3H, t, J = 7.1 Hz
47.6, d		0.76, 3H, t, J = 7.4 Hz

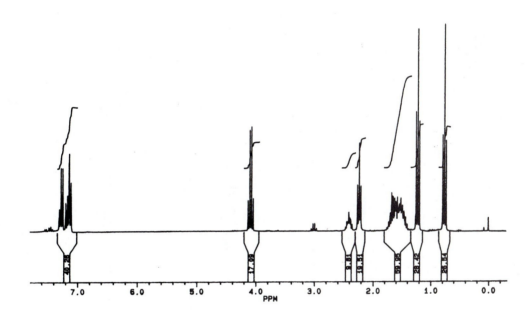

PROBLEM 14.5

$C_{15}H_{22}O_2$

IR : 2931, 2872, 1713, 1455, 1414, 1367, 1102, 738, 697 cm^{-1}
MS: 234 (0.28), 143 (28), 128 (24), 127 (26), 107 (24), 99 (23), 97 (12), 91 (100),

^{13}C NMR

212.1, s	36.1, t
138.5, s	32.1, d
129.2, d (2)	27.5, t
128.2, d	15.5, q
127.5, d (2)	9.5, q
75.7, t	
72.8, t	
39.0, t	

^{1}H NMR

7.30, 5H, m
4.49, 2H, s
3.29, 2H, dd, J = 1.8, 6.0 Hz
2.38, 4H, m
1.74, 2H, m
1.42, 1H, m
0.98, 3H, t, J = 7.3 Hz
0.93, 3H, d, J = 6.6 Hz

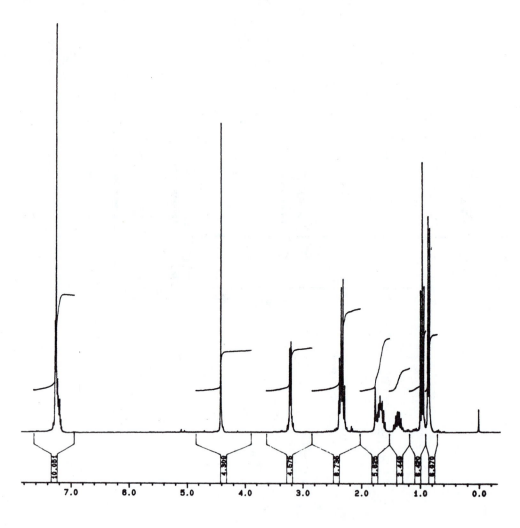

PROBLEM 14.6

$C_{16}H_{20}O_2$

IR: 2922, 1673, 1245, 755, 692 cm^{-1}
MS: 244 (M$^+$, 43), 151 (42), 121 (16), 109 (100)

^{13}C NMR		**^{1}H NMR**
199.8, s	32.3, t	7.24–7.29, 2H, m
158.7, s	32.1, t	6.86–6.94, 3H, m
144.6, d	26.2, t	6.70, 1H, t, J = 4.5 Hz
135.4, s	15.6, q	3.93, 2H, t, J = 6.2 Hz
129.3, d (2)		2.55, 1H, dd, J = 11.4, 8.2 Hz
120.5, d		2.44, 1H, m
114.2, d (2)		2.10, 3H, m
67.4, t		1.80, 2H, m
44.4, t		1.77, 3H, s
35.3, d		1.52, 2H, m

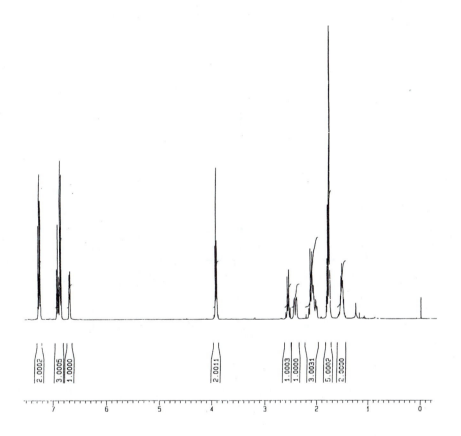

PROBLEM 14.7

$C_{13}H_{18}O$

IR: 3075, 2963, 1640, 1600 cm^{-1}
MS: 190 (M$^+$, 24), 94 (100), 81 (24)

^{13}C NMR	^{1}H NMR
159.0, s	7.27, 2H, m
136.6, d	6.91, 3H, m
129.4, d (2)	5.80, 1H, ddt, J = 16.2, 10.4, 6.8 Hz
120.4, d	5.10, 1H, d, J = 16.2 Hz
116.4, t	4.98, 1H, d, J = 10.4 Hz
114.5, d (2)	3.84, 2H, d, J = 5.8 Hz
69.7, t	2.23, 2H, m
39.4, d	1.82, 1H, m
35.2, t	1.48, 2H, m
23.5, t	0.95, 3H, t, J = 7.4 Hz
11.2, q	

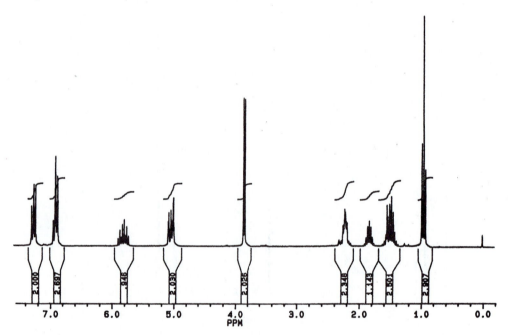

PROBLEM 14.8

$C_{17}H_{22}O_4$

IR: 2938, 1731, 1681, 1596, 1448, 1359, 1182, 1096, 1031, 972, 773, 695 cm^{-1}

MS: 291 (M$^+$ +1, 100), 275 (33), 249 (39), 204 (11), 187 (26), 163 (15), 157 (11), 141 (12), 105 (41)

^{13}C NMR

204.1, s	33.8, t
196.2, s	28.5, t
173.3, s	27.8, q
136.3, s	27.0, t
133.7, d (2)	24.6, t
128.8, d	14.1, q
128.6, d (2)	
63.1, d	
60.2, t	

^{1}H NMR

8.00, 2H, d, J = 8.5 Hz
7.61, 1H, m
7.45, 2H, m
4.44, 1H, t, J = 7.0 Hz
4.10, 2H, q, J = 7.1 Hz
2.29, 2H, t, J = 7.4 Hz
2.13, 3H, s
2.05, 2H, m
1.78, 2H, m
1.40, 2H, m
1.23, 3H, t, J = 7.1 Hz

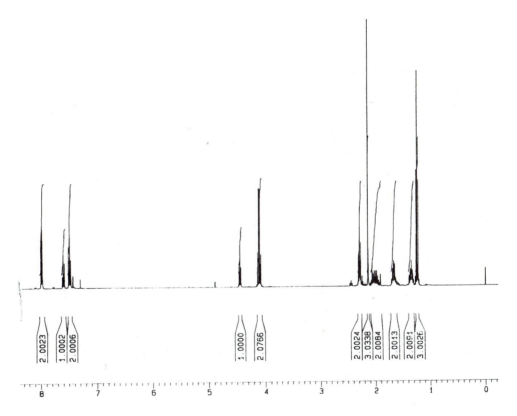

PROBLEM 14.9

$C_{17}H_{18}O$

IR: 1665 cm^{-1}
MS: 258 (M$^+$), 143, 115

^{13}C NMR
199.7, s
137.6, s (2)
133.5, s (2)
133.4, s (2)
130.3, d (2)
129.4, d (2)
129.1, d (2)
19.7, q (2)
18.5, q (2)

^{1}H NMR
7.2, 6H, m
2.37, 6H, s
2.29, 6H, s

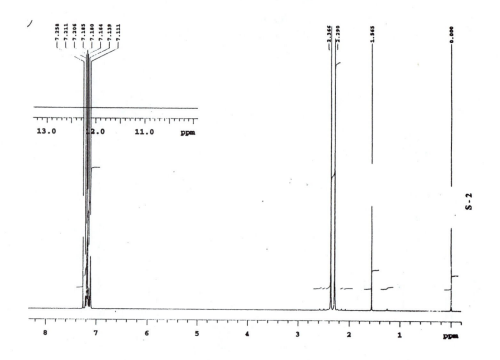

PROBLEM 14.10

$C_{15}H_{13}BrO_2$

IR: 1666 cm^{-1}
MS: 306 (M$^+$), 304 (M$^+$), 185, 183, 157, 155, 149, 121

^{13}C NMR

194.1, s	128.5, d
153.8, s	126.5, s
135.3, s	126.4, s
131.2, d	110.0, d
130.0, d (2)	54.2, q
129.7, d (2)	18.8, q
128.6, s	

^{1}H NMR

7.67, 2H, d, J = 8.6 Hz
7.57, 2H, d, J = 8.6 Hz
7.25, 1H, dd, J = 2.06, 8.44 Hz
7.17, 1H, d, J = 2.06 Hz
6.89, 1H, d, J = 8.44 Hz
3.68, 3H, s
2.33, 3H, s

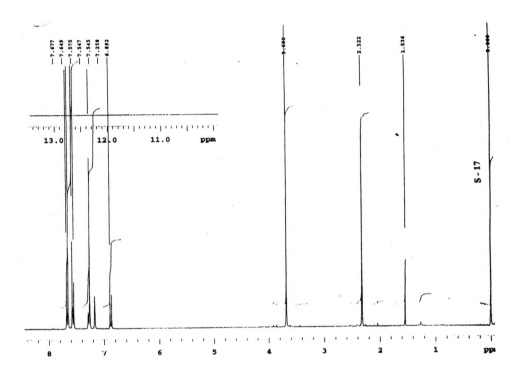

S-17

Table 14.1 The ^{13}C Shifts for Some Linear and Branched-Chain Alkanes (ppm from TMS)

COMPOUND	C-1	C-2	C-3	C-4	C-5
Methane	−2.3				
Ethane	5.7				
Propane	15.8	16.3	15.8		
Butane	13.4	25.2	25.2		
Pentane	13.9	22.8	34.7	22.8	13.9
Hexane	14.1	23.1	32.2	32.2	23.1
Heptane	14.1	23.2	32.6	29.7	32.6
Octane	14.2	23.2	32.6	29.9	29.9
Nonane	14.2	23.3	32.6	30.0	30.3
Decane	14.2	23.2	32.6	31.1	30.5
Isobutane	24.5	25.4			
Isopentane	22.2	31.1	32.0	11.7	
Isohexane	22.7	28.0	42.0	20.9	14.3
Neopentane	31.7	28.1			
2,2-Dimethylbutane	29.1	30.6	36.9	8.9	
3-Methylpentane	11.5	29.5	36.9	(18.8, 3-CH$_3$)	
2,3-Dimethylbutane	19.5	34.3			
2,2,3-Trimethylbutane	27.4	33.1	38.3	16.1	
2,3-Dimethylpentane	7.0	25.3	36.3	(14.6, 3-CH$_3$)	

From R. M. Silverstein, G. C. Bassler, and T. C. Morrill, *Spectrometric Identification of Organic Compounds,* 5th ed., New York: John Wiley and Sons, Inc., 1991. Reprinted with permission of John Wiley and Sons, Inc. p. 236.

Table 14.2   Incremental Substituent Effects (ppm) on Replacement of H by Y in Alkanes. Y is Terminal or Internal[a] (+ downfield, − upfield)

Y	α		β		γ
	TERMINAL	INTERNAL	TERMINAL	INTERNAL	
CH_3	+ 9	+ 9	+ 6	+ 6	−2
$CH=CH_2$	+20		+ 6		−0.5
C CH	+ 4.5		+ 5.5		−3.5
COOH	+21	+16	+3	+ 2	−2
COO^-	+25	+20	+5	+ 3	−2
COOR	+20	+17	+3	+ 2	−2
COCl	+ 33	28		+ 2	
$CONH_2$	+22		+ 2.5		−0.5
COR	+30	+24	+ 1	+ 1	−2
CHO	+31		0		−2
Phenyl	+23	+17	+ 9	+ 7	−2
OH	+48	+41	+10	+ 8	−5
OR	+58	+51	+ 8	+ 5	−4
OCOR	+51	+45	+ 6	+ 5	−3
NH_2	+29	+24	+11	+10	−5
NH_3	+26	+24	+ 8	+ 6	−5
NHR	+37	+31	+ 8	+ 6	−4
NR_2	+42		+ 6		−3
NR_3	+31		+ 5		−7
NO_2	+63	+57	+ 4	+ 4	
CN	+ 4	+ 1	+ 3	+ 3	−3
SH	+ 11	+11	+12	+11	−4
SR	+ 20		+ 7		−3
F	+ 68	+63	+ 9	+ 6	−4
Cl	+ 31	+32	+11	+10	−4
Br	+ 20	+25	+11	+10	−3
I	− 6	+ 4	+11	+12	−1

[a]Add these increments to the shift values of the appropriate carbon atom in Table 12.1.

Source: F. W. Wehrli, A. P. Marchand, and S. Wehrli, *Interpretation of Carbon-13 NMR Spectra.* 2nd ed., London: Heyden, 1983.

From R. M. Silverstein, G. C. Bassler, and T. C. Morrill, *Spectrometric Identification of Organic Compounds,* 5th ed., New York: John Wiley and Sons, Inc., 1991. Reprinted with permission of John Wiley and Sons, Inc. Table 5.3, p. 236.

Table 14.3   Chemical Shifts of Alkyne Protons

HC≡CR	1.73–1.88
HC≡C—COH	2.23
HC≡C—C≡CR	1.95
HC≡CH	1.80
HC≡CAR	2.71–3.37
HC≡C—C≡CR	2.60–3.10

From R. M. Silverstein, G. C. Bassler, and T. C. Morrill, *Spectrometric Identification of Organic Compounds,* 5th ed., New York: John Wiley and Sons, Inc., 1991. Reprinted with permission of John Wiley and Sons, Inc. Table D.3, p. 217.

Table 14.4   Calculation of ^{1}H NMR Chemical Shifts for Alkenes

See Figure 14.12 for more information.

	Z				Z		
SUBSTITUENT R	GEM	CIS	TRANS	SUBSTITUENT R	GEM	CIS	TRANS
—H	0	0	0	H⧵—C=O	1.03	0.97	1.21
—Alkyl	0.44	−0.26	−0.29				
—Alkyl-ring[a]	0.71	−0.33	−0.30				
—CH₂0, —CH₂I	0.67	−0.02	−0.07	N⧵—C=O	1.37	0.93	0.35
—CH₂S	0.53	−0.15	−0.15				
—CH₂Cl, —CH₂Br	0.72	0.12	0.07	Cl⧵—C=O	1.10	1.41	0.99
—CH₂N	0.66	−0.05	−0.23				
—C≡C	0.50	0.35	0.10	—OR, R:aliph	1.18	−1.06	−1.28
—C N	0.23	0.78	0.58	—OR, R:conj[b]	1.14	−0.65	−1.05
—C=C	0.98	−0.04	−0.21	—OCOR	2.09	−0.40	−0.67
—C=C conj[b]	1.26	0.08	−0.01	—Aromatic	1.35	0.37	−0.10
—C=O	1.10	1.13	0.81	—Cl	1.00	0.19	0.03
—C=O conj[b]	1.06	1.01	0.95	—Br	1.04	0.40	0.55
—COOH	1.00	1.35	0.74	—N⟨R,R R: aliph	0.69	−1.19	−1.31
—COOH conj[b]	0.69	0.97	0.39	—N⟨R,R R:conj[b]	2.30	−0.73	−0.81
—COOR	0.84	1.15	0.56				
—COOR conj[b]	0.68	1.02	0.33				
				—SR	1.00	−0.24	−0.04
				—SO₂	1.58	1.15	0.95

[a]Alkyl ring indicates that the double bond is part of the ring R |⟨C,C

[b]The Z factor for the conjugated substituent is used when either the substituent or the double bond is further conjugated with other groups.

From R. M. Silverstein, G. C. Bassler, and T. C. Morrill, *Spectrometric Identification of Organic Compounds,* 5th ed., New York: John Wiley and Sons, Inc., 1991; and C. Pascual, J. Meier, and W. Simon, *Helv. Chim. Acta,* 49, 164 (1966). Reprinted with permission of John Wiley and Sons, Inc. Appendix D, p. 215.

Table 14.5   Incremental Shifts of the Aromatic Carbon Atoms of Monosubstituted Benzenes (ppm from Benzene at 128.5 ppm, + downfield, − upfield). Carbon Atom of Substituents in parts per million from TMS[a]

SUBSTITUENT	C-1 (ATTACHMENT)	C-2	C-3	C-4	C OF SUBSTITUENT (ppm from TMS)
H	0.0	0.0	0.0	0.0	
CH_3	9.3	+0.7	−0.1	−2.9	21.3
CH_2CH_3	+15.6	−0.5	0.0	−2.6	29.2 (CH_2), 15.8 (CH_3)
$CH(CH_3)_2$	+20.1	−2.0	0.0	−2.5	34.4 (CH), 24.1 (CH_3)
$C(CH_3)_3$	+22.2	−3.4	−0.4	−3.1	34.5 (C), 31.4 (CH_3)
$CH=CH_2$	+9.1	−2.4	+0.2	−0.5	137.1 (CH), 113.3 (CH_2)
$C\equiv CH$	−5.8	+6.9	+0.1	+0.4	84.0 (C), 77.8 (CH)
C_6H_5	+12.1	−1.8	−0.1	−1.6	
CH_2OH	+13.3	−0.8	−0.6	−0.4	64.5
CH_2OCCH_3 $\overset{\|}{O}$	+7.7	~0.0	~0.0	~0.0	20.7 (CH_3), 66.1 (CH_2), 170.5 (C=O)
OH	+26.6	−12.7	+1.6	−7.3	
OCH_3	+31.4	−14.4	+1.0	−7.7	54.1
OC_6H_5	+29.0	−9.4	+1.6	−5.3	
$\overset{O}{\overset{\|}{O}}CCH_3$	+22.4	−7.1	−0.4	−3.2	23.9 (CH_3), 169.7 (C=O)
$\overset{O}{\overset{\|}{}}CH$	+8.2	+1.2	+0.6	+5.8	192.0
$\overset{O}{\overset{\|}{}}CCH_3$	+7.8	−0.4	−0.4	+2.8	24.6 (CH_3), 195.7 (C=O)
$\overset{O}{\overset{\|}{}}CC_6H_5$	+9.1	+1.5	−0.2	+3.8	196.4 (C=O)
$\overset{O}{\overset{\|}{}}CCF_3$	−5.6	+1.8	+0.7	+6.7	
$\overset{O}{\overset{\|}{}}COH$	+2.9	+1.3	+0.4	+4.3	168.0
$\overset{O}{\overset{\|}{}}COCH_3$	+2.0	+1.2	−0.1	+4.8	51.0 (CH_3), 166.8 (C=O)
$\overset{O}{\overset{\|}{}}CCl$	+4.6	+2.9	+0.6	+7.0	168.5
$C\equiv N$	−16.0	+3.6	+0.6	+4.3	119.5
NH_2	+19.2	−12.4	+1.3	−9.5	
$N(CH_3)_2$	+22.4	−15.7	+0.8	−11.8	40.3
$\overset{O}{\overset{\|}{}}NHCCH_3$	+11.1	−9.9	+0.2	−5.6	

Table 14.5 (Continued)

SUBSTITUENT	C-1 (ATTACHMENT)	C-2	C-3	C-4	C OF SUBSTITUENT (ppm from TMS)
NO_2	+19.6	−5.3	+0.9	+6.0	
N=C=O	+5.7	−3.6	+1.2	−2.8	129.5
F	+35.1	−14.3	+0.9	−4.5	
Cl	+6.4	+0.2	+1.0	−2.0	
Br	−5.4	+3.4	+2.2	−1.0	
I	−32.2	+9.9	+2.6	−7.3	
CF_3	+2.6	−3.1	+0.4	+3.4	
SH	+2.3	+0.6	+0.2	−3.3	
SCH_3	+10.2	−1.8	+0.4	−3.6	15.9
SO_2NH_2	+15.3	−2.9	+0.4	+3.3	
$Si(CH_3)_3$	+13.4	+4.4	−1.1	−1.1	

[a]See D. E. Ewing, *Org. Magn. Reson.,* 12, 499 (1979) for chemical shifts of 709 monosubstituted benzenes.

From R. M. Silverstein, G. C. Bassler, and T. C. Morrill, *Spectrometric Identification of Organic Compounds,* 5th ed., New York: John Wiley and Sons, Inc., 1991. Reprinted with permission of John Wiley and Sons, Inc. Table 5.9, p. 240.

Table 14.6 Chemical Shifts of Protons on Monosubstituted Benzene Rings

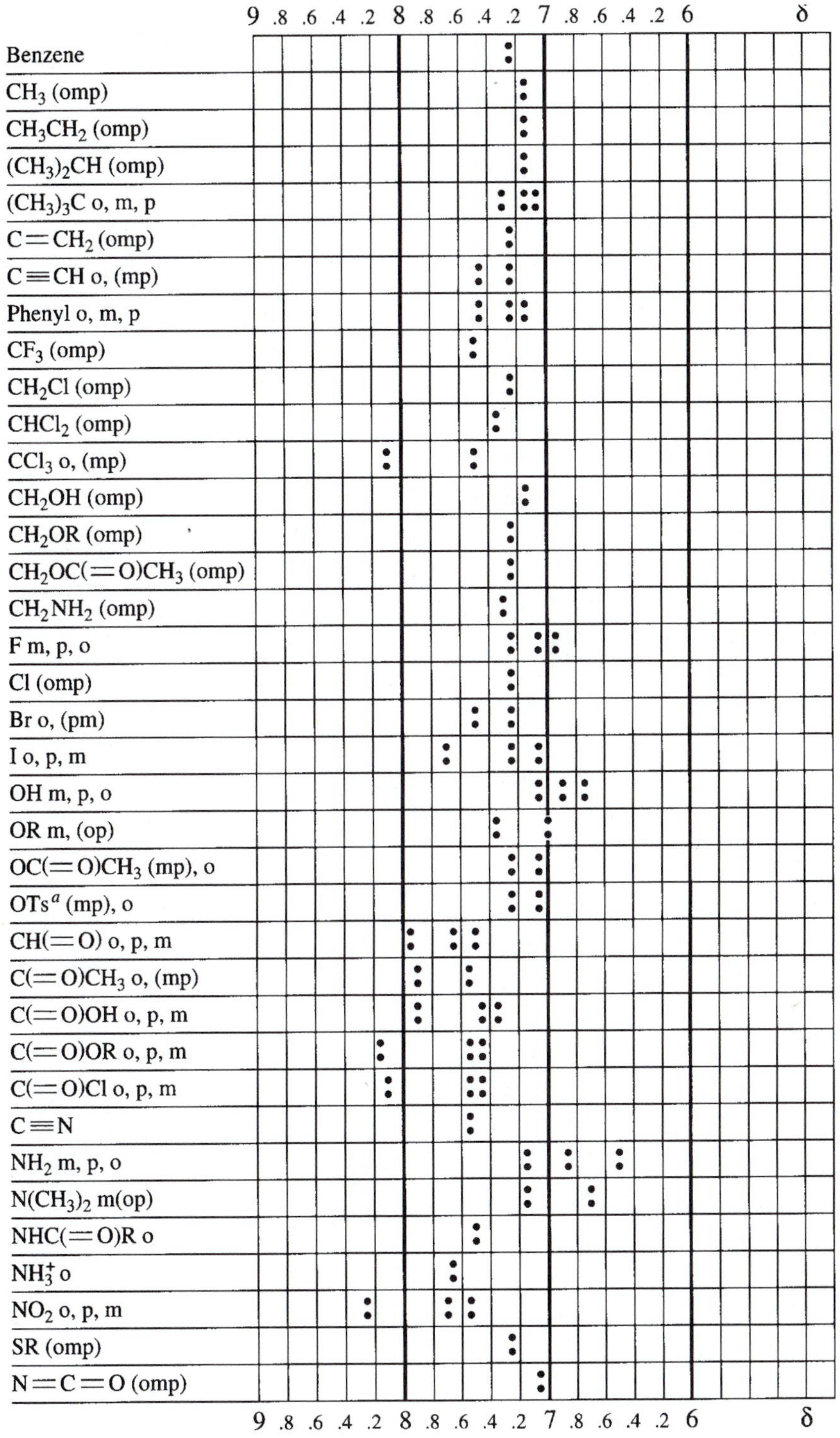
9 .8 .6 .4 .2 8 .8 .6 .4 .2 7 .8 .6 .4 .2 6 δ
Benzene
CH_3 (omp)
CH_3CH_2 (omp)
$(CH_3)_2CH$ (omp)
$(CH_3)_3C$ o, m, p
$C{=}CH_2$ (omp)
$C{\equiv}CH$ o, (mp)
Phenyl o, m, p
CF_3 (omp)
CH_2Cl (omp)
$CHCl_2$ (omp)
CCl_3 o, (mp)
CH_2OH (omp)
CH_2OR (omp)
$CH_2OC({=}O)CH_3$ (omp)
CH_2NH_2 (omp)
F m, p, o
Cl (omp)
Br o, (pm)
I o, p, m
OH m, p, o
OR m, (op)
$OC({=}O)CH_3$ (mp), o
OTs[a] (mp), o
$CH({=}O)$ o, p, m
$C({=}O)CH_3$ o, (mp)
$C({=}O)OH$ o, p, m
$C({=}O)OR$ o, p, m
$C({=}O)Cl$ o, p, m
$C{\equiv}N$
NH_2 m, p, o
$N(CH_3)_2$ m(op)
$NHC({=}O)R$ o
NH_3^+ o
NO_2 o, p, m
SR (omp)
$N{=}C{=}O$ (omp)
9 .8 .6 .4 .2 8 .8 .6 .4 .2 7 .8 .6 .4 .2 6 δ

[a] OTs = *p*-Toluenesulfonyloxy group.

From R. M. Silverstein, G. C. Bassler, and T. C. Morrill, *Spectrometric Identification of Organic Compounds,* 5th ed., New York: John Wiley and Sons, Inc., 1991. Reprinted with permission of John Wiley and Sons, Inc. Chart D.1, p. 218.

The Literature
of Organic Chemistry

The one facet of behavior that distinguishes chemistry from alchemy is that chemists share information (data), whereas alchemists were sublimely secretive; in fact, they went to great lengths to hide their results from others. For a variety of reasons, they were determined to keep experimental details to themselves for fear of persecution or worse. The coming of the research societies in Italy (1657), England (1660), and France (1666), with their attention to the scientific method and their attendant publications, made an immediate impact on science, and chemistry in particular. No longer was it in vogue to hide results; rather, free communication was the proper thing to do. The real beginnings of chemistry lay in the assembling of experimental facts and interpretations, along with the publication thereof. The continuous record of experimental fact and interpretation that has been evolving ever since is called the **chemical literature.**

One of the major elements in this body of knowledge is the description of the individual compounds that have been encountered and characterized by chemists in the past 150 years. Students of organic chemistry quickly become aware that a very large number of organic compounds can exist, and they learn to write structures and devise synthetic routes without particular concern as to whether the compound in question has actually been prepared or isolated and described somewhere in the literature. This point becomes of crucial importance, however, if an actual sample of the compound or information on some of its properties is needed. This chapter is an introduction to the chemical literature.

Computers have aided in the task of coordinating data on new as well as previously known compounds in a systematic way. The sheer bulk of this material is difficult to comprehend. All new compounds that appear in the current literature are now entered in a computer file by the Chemical Abstracts Service of the American Chemical Society. Each compound is given a unique registry number that permits rapid access to bibliographic data by computer search. This computer registry file includes data on almost twenty-one million compounds, and the number is increasing by about 1.7 million per year. This computer registry provides the base for both Chemical Abstracts and Beilstein, the two most widely used systems for retrieving chemical information.

SEARCHING THE LITERATURE FOR COMPOUNDS

The primary source of chemical information of any kind is the original report of the data in a journal or patent. In order to make this information more accessible, numerous encyclopedias, indexes, and reviews have been and are continuing to be published. Several of the secondary sources that are useful in locating syntheses and properties of organic compounds are described here. An excellent text that provides a broad view of the field of chemical literature is *Chemical Information. A Practical Guide to Utilization,* 2nd ed. by Yecheskel Wolman (John Wiley & Sons: New York, 1988).

HANDBOOKS

For quick reference to physical properties (melting and boiling points, solubility, density) of simple organic compounds, several compilations are available. In most of these, references are given to the source of the data. The following list includes a few of the most common and useful references for specific physical properties:

1. Physical Constants of Organic Compounds. In *Handbook of Chemistry and Physics,* 80th ed.; Lide, D. R., Jr., Ed. (CRC Press: Boca Raton, FL, 2000). Contains mp, bp, solubility, and other data for approximately 14,000 organic compounds.
2. Physical Constants of Organic Compounds. In *Lange's Handbook of Chemistry,* 15th ed.; Dean, J. A., Ed. (McGraw-Hill: New York, 1999). Similar to the preceding handbook but less extensive in coverage (approximately 6500 compounds are listed).
3. *Handbook of Tables for Identification of Organic Compounds,* 3rd ed. (Chemical Rubber Co.: Cleveland, 1967). Contains mp and bp data for more than 4000 compounds and melting points of derivative compounds for each. Arranged by functional groups.
4. *The Merck Index,* 12th ed.; Budvari, S. et al., Eds. (Merck & Co. Inc: Whitehouse Station, NJ, 1996). An encyclopedia with data on nearly 10,000 compounds. Contains information on occurrence, synthesis, and pharmacology of many compounds of importance in biochemistry and medicine. Also provides an extensive cross-list of synonyms and trade names.
5. *Dictionary of Organic Compounds,* 5th ed.; I. Heilbron (Oxford University Press: New York, 1982; vols. 1–7 and annual supplements). The Heilbron Dictionary is an alphabetical listing of more than 190,000 compounds with selected reactions, derivatives, and literature references.

BEILSTEIN

The German treatise, *Beilstein Handbuch der Organische Chemie* (Springer-Verlag: Berlin, 1918 to the present), is the ultimate secondary source on all organic compounds in the earlier literature (before 1950). The fourth edition, in 31 volumes (*Bande*), was published in the period 1918 to 1938, and this *Hauptwerk* (H) covers exhaustively the literature through the year 1909. This edition has been updated periodically by supplements (E) (Table 15.1).

With a rudimentary understanding of German nomenclature, it is possible to locate a compound in Beilstein using the cumulative formula index (*Gesamtregister Formelregister*) for the fourth supplement. This index gives references to entries for the same compound in the H through E-IV series. Alternatively, one can make use of the organization of the volumes to seek out the compound directly. To do the latter, some understanding of the classification system is necessary.

All of the compounds in Beilstein are classified in one or another of three main categories: Aliphatic (*Acyclische*), Carbocyclic (*Isocyclische,* including both alicyclic and aromatic rings), and Heterocyclic (*Heterocyclische*). Heterocyclic compounds are further subdivided according to the type and number of heteroatoms in the ring(s). Thus *Band* XX (Vol. 20), *Heterocyclische Verbindungen, Heteroclasse* 2 O *bis* 9 O, contains heterocyclic compounds with from two to nine oxygen atoms in the ring(s).

Within each of the main series, compounds are arranged according to the functional group present. The major functional group categories are *Kohlenwasserstoffe* (hydrocarbons), *Oxyverbindungen* (alcohols), *Oxoverbindungen* (aldehydes and ketones), *Carbonsauren* (carboxylic acids), *Sulfinsauren* (sulfinic acids), *Sulfonsauren* (sulfonic acids), *Amine, Hydroxylamine, Hydrazine, Azoverbindungen,* and *Phosphine.* The *Stammkerne* sections in the *Heterocyclische Band* contain heterocyclic compounds with none of the foregoing functional groups; that is, they are analogous to the *Kohlenwasserstoffe* sections of the *Acyclische* and

Table 15.1 The Organization of Beilstein

SERIES	ABBREVIATION	PERIOD OF LITERATURE COMPLETELY COVERED	COLOR OF LABEL ON SPINE
Basic Series	H	Up to 1909	Dark green
Supplementary Series I	E-I	1910–1919	Red
Supplementary Series II	E-II	1920–1929	White
Supplementary Series III	E-III	1930–1949	Blue
Supplementary Series III/IV	E-III/IV*	1930–1959	Blue/black
Supplementary Series IV	E-IV	1950–1959	Black
Supplementary Series V†	E-V	1960–1979	Red on blue binding
Cumulative Subject and Formula Indexes through E-IV			Light green

*Volumes 17–27 of Supplementary Series III and IV, covering the heterocyclic compounds, are combined in a joint issue.

†Supplementary Series V is in English.

Isocyclische categories. In addition to the preceding functional group categories, there are separate sections for compounds containing more than one major functional group. The titles of these are self-explanatory: *polycarbonsauren, oxocarbonsauren, oxyoxoverbindungen, oxoamine,* and so forth.

Compounds containing halogen atoms, nitrogen, nitroso, or azide groups are listed under the parent compound; for example, chloroacetone under acetone. Esters, primary amides, anhydrides, nitriles, and acid halides are listed under the parent carboxylic acids. Secondary and tertiary amides are given under the parent amine. Ethers, peroxides, mercaptans, sulfides, and sulfones are given following the corresponding alcohols.

Within any functional group category, the compounds are ordered according to the number of oxygen, sulfur, or nitrogen atoms contained. These subgroups are further ordered according to the degree of unsaturation (number of rings and double bonds) and, finally, according to the number of carbon atoms present. Isomeric compounds are generally arranged so that the simpler (less branched, fewer rings, and so forth) isomers precede the more complex. The preceding classification is outlined in the Table of Contents (*Inhalt*) at the beginning of each volume.

Finding methyl 5-chloro-2-hydroxybenzoate (methyl 5-chlorosalicylate) can be used as an example of how to use the Beilstein system of organization.

The compound is first recognized as a carbocyclic (*Isocyclische*) hydroxyacid (*Oxycarbonsauren*) derivative and should be found in *Band X,* which is so titled. The parent compound, salicylic acid, has a molecular formula of $C_7H_6O_3$, that is, $C_nH_{2n-8}O_3$ ($n = 7$). Looking in the Table of Contents of *Band X,* we first locate the subheading *oxycarbonsauren mit 3 sauerstoffchatomen* (hydroxyacids with three oxygen atoms). In this section, we then look for compounds with five units of unsaturation (i.e., $C_nH_{2n-8}O_3$) and, finally, the subheading for $n = 7$. At this point we are directed to *Seite* (page) 43. Turning to page 43, we find salicylic acid itself; in fact, pages 43 to 59 discuss salicylic acid. Continuing, we find salts of salicylic acid, ethers of salicylic acid, esters of salicylic acid, and so forth. On page 101 the listing of substituted salicylic acids begins, the first compounds being chlorosalicylic acids. After the 5-chloro isomer is a paragraph on page 103 describing its methyl ester. This is shown in Figure 15.1. Similarly, a supplemental entry for this compound can be found in E-II, *Band* X, *Seite* 62 (Fig. 15.2). There is no information on this compound in E-I, one entry in E-III, *Band* X, *Seite* 166 (Fig. 15.3), and an entry in Band X of E-IV, *Seite* 209.

5-Chlor-salicylsäure-methylester $C_8H_7O_3Cl=HO \cdot C_6H_3Cl \cdot CO_2 \cdot CH_3$. *B.* **Beim Einleiten von Chlorwasserstoff in eine methylalkoholische Lösung der 5-Chlor-salicylsäure** (VARNHOLT, *J. pr.* **[2] 36, 21**). **Bei der Einw. von Methyljodid auf das Silbersalz der 5-Chlor-salicylsäure** (LASSAR-COHN, SCHULTZE, *B.* **38, 3300**).–Nadeln (aus Alkohol). F: 48° (SMITH, *B.* **11, 1227; V.**), 50° (L.-C., SCH.). **Siedet unter teilweiser Zersetzung bei 249° (V.). Ziemlich leicht löslich in Alkohol (SM.; V.).**

Figure 15.1   Beilstein, *Hauptwerk*, Band X, p. 103.

5-Chlor-salicylsaure-methylester $C_8H_7O_3Cl=HO \cdot C_6H_3Cl \cdot CO_2 \cdot CH_3$ **(H 103).** *B.* **Bei der Belichtung einer Lösung von Chlorpikrin in Salicylsäuremethylester** (PIUTTI, BADOLATO, *R. A. L.* **[5] 33 I, 477**). −F: 48°.

Figure 15.2   Beilstein, E-II, Band X, p. 62.

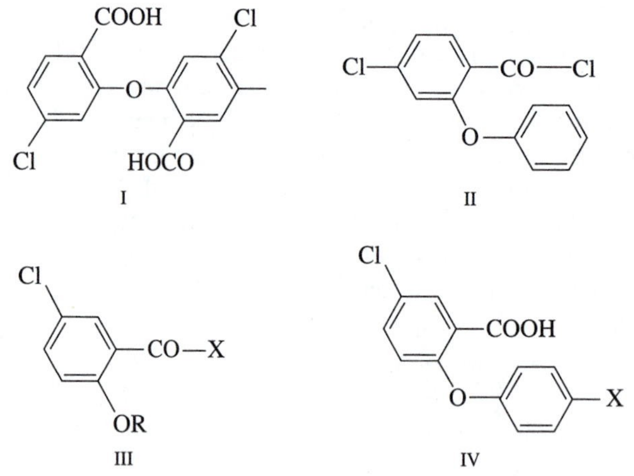

Figure 15.3   Beilstein, E-III, Band X, p. 166.

The preceding entries can be located equally well using the *Gesamtregister Formelregister* of E-IV. Nomenclature based on trivial or unsystematic names used in the H, E-I through E-III Series has been replaced by the IUPAC system in E-IV. For further information, consult O. Weissbach, *The Beilstein Guide* (Springer-Verlag: New York, 1976).

CHEMICAL ABSTRACTS

Publication of *Chemical Abstracts* was begun in 1907, and since 1945 the volume and collective indexes have become the "key to the chemical literature." All new compounds appearing in chemical journals (over 8000 periodicals are now abstracted) and patents are recorded in these indexes, as well as entries on most citations of previously known compounds. The most recent collective index (1987–1991) occupies 115 bound volumes. Despite this huge bulk, a compound mentioned in any abstract published in this five-year period can be found in a few minutes' time after a little practice, and traced back from index to abstract and thence to the original primary journal description. Ten-year collective indexes were published for the first five decades; the collective indexes are now on a five-year basis. Only these collective indexes need be consulted for the period before 1991, but individual volume indexes (now semiannual) must be consulted until the next collective index is published.

Compounds are indexed both by molecular formula and by the chemical name. If a specific compound is wanted, the formula index is generally consulted first, and after the compound in question is located by name out of a list of perhaps 10 or 20 isomeric compounds, an abstract number, the subject index name, and the registry number are obtained. This procedure may immediately turn up references to the desired compound, but it is usually advantageous to go from the formula index to the subject index, using the name obtained in the formula index. The subject index may lead to additional entries, and some indication of the information available in the reference will be given.

Since 1972 (Vol. 76), the subject indexes of *Chemical Abstracts* have been divided into a *Chemical Substance Index* and a *General Subject Index*. The former contains references to specific compounds and the latter to other subjects such as classes of compounds, reactions, concepts, applications, and so on. It should also be mentioned that in the early volumes of *Chemical Abstracts,* the formula indexes did not provide as complete a coverage as the subject index.

A great strength of the subject index is its correlative function. The system used for index names is designed to group closely related compounds together; thus, a series of esters of a complex acid will be found under the main entry for the acid; for example, *salicylic acid, 5-chloro-, methyl ester* is the subject index "inversion name," and other derivatives of the carboxyl group will be listed similarly under this heading. Some experience is required to find a compound in the subject index; a detailed discussion of the nomenclature system used up to 1972 is given in a preface to the index of Volume 56 (1962). Since 1972, *Chemical Abstracts* has used rigidly systematic nomenclature in its indexes, abandoning nearly all trivial names. Thus, toluene is now found as benzene, methyl-; anisole as benzene, methoxy-; and acetone as 2-propanone. An Index Guide, part of the index to Volume 76 and reissued periodically (latest 1989), provides a complete description of the new indexing nomenclature. For this reason, except for simple compounds, the formula index provides a surer way to find a specific compound, but if any of several related compounds would be equally useful, each must be searched for individually in the formula index. Selected *Chemical Abstracts* references to methyl 5-chlorosalicylate are shown in Figures 15.4, 15.5, and 15.6.

Another guide to Chemical Abstracts is an audio course, *Chemical Abstracts: An Introduction to Its Effective Use,* by J. T. Dickman, M. O'Hara, and O. B. Ramsay (American Chemical Society: Washington, DC, 1980).

SEARCHING FOR PREPARATIVE METHODS

When the preparation of a known compound is required, the first step is to locate the substance in the literature. All sources or preparations of a compound recorded in the time period covered by a volume of Beilstein will be cited therein. One way to check systematically for more recent

C$_8$H$_7$CIO$_3$ Acetic acid, (chlorophenoxy)-, **16**:3483[7]; **21**:1096[7]; **24**:1344[7]; **25**:930[2]; **30**:2963[5]; **36**:6199[7], 6512[5]; **37**:4426[3], 5637[4]; **38**:2327[6]; **39**:1199[4], 5286[8]; **40**:1474[1], P 2264[5], 3846[8], 6197[4, 7]; *Na salt,* **40**:7491[6].

Acetophenone, chlorodihydroxy-, **18**:675[3]; **23**:2161[3]; **24**:4012[9]; **29**:3338[1], 5839[5]; **30**:159[9], 3421[9]; **34**:401[6]; **37**:101[3], 2358[2]; **39**:1471[4].

Anisic acid, 3-chloro-, **38**:3629[9]; **40**:2131[1].

Benzoic acid, 3-chloro-4-hydroxy-, methyl ester, **20**:3712[8].

—, chloromethoxy-, **18**:386[6]; **20**:1065[3]; **21**:3189[8]; **23**:1128[9]; **24**:1859[3]; **34**:7867[4]; **35**:2125[6]; **38**:5495[7]; *and salts,* **33**:2124[6,7].

Carbonic acid, chloromethyl phenyl ester, **14**:739[9].

Cresotic acid, chloro-, **25**: P 4558[6]; **26**: P 1130[8], P 1946[4], P 2604[3]; **27**: P 852[5], P 2044[1]; **33**:6897[6].

4-Cyclohexene-1,2-dicarboxylic anhydride, 1-chloro-, **40**: P 3136[5].

2-Furoic acid, 2-chloroallyl ester, **32**: P 7925[5].

Homoprotocatechuyl chloride, **32**:851[7].

Mandelic acid, chloro-, **15**:2632[2]; **22**:2746[4]; **25**:3637[4]; **32**:4969[2]; **35**:5637[2]; **40**:1156[6], 3044[3].

p-Orsellinaldehyde, 3-chloro-, **28**:6130[5]; **29**:1078[7].

Piperonyl alcohol, 6-chloro-, **33**:1294[1].

Salicylic acid, chloro-, Me ester, **31**:8599[6]; **33**:2124[5]; **37**:2010[3]; **38**:3256[1]. ◀——

Sorbic acid, γ-chloroacetyl-δ-hydroxy-, δ-lactone, **35**:7405[1].

Toluic acid, chlorohydroxy-, **25**:3325[5]; **37**:3420[7].

Vanillin, chloro-, **19**:2494[8]; **20**:1980[6]; **21**:906[1]; **23**:4456[5]; **25**:94[6, 7]; **40**:3896[6].

Figure 15.4 *Chemical Abstracts,* Collective Formula Index (1920–1946).

Salicylic acid

droxyphenylazo)-3-hydroxy-4-biphenylylazo]-5-hydroxy-3-methyl-1-pyrazolyl}phenylazo}-5-hydroxy-3 -methyl-1-pyrazolyl}-, **40**: P 6265[3].

—, **chloro-,** effect on dermatophytes, **34**:5873[8].

—, **3-chloro-, 40**:5423[4].

—, **3**(and **5**)-**chloro-,** and derivs., **33**:2124[3].

—, **4**(and **5**)-**chloro-, 40**:71[3].

—, **5-chloro-, 31**:8117[7]; **33**:8181[7].

4-ethoxybutyl ester, **35**: P 3738[6].

methyl ester, **37**:2010[3]; **38**:3256[1].

and methyl ester, **31**:8599[6].

methyl ester, *N*-(*p*-methoxyphenyl)benzimidate, **32**:1666[7].

p-nitrophenyl ester, **33**:1296[3].

—, **4-*p*-chlorobenzoyl-, 35**: P 4218[5]; **36**: P 2732[1].

Figure 15.5 *Chemical Abstracts,* 4th Decennial Subject Index (1937–1946).

syntheses is a search of *Chemical Abstracts* subject indexes. This entails some labor, since a number of entries may be found, and not all of those pertaining to synthesis will necessarily be identified as such. Some entries may refer to uses, analytical methods, and the like and can be ignored; abstracts corresponding to the other entries must then be scanned. Alternatively, the online service Beilstein Online can be used. An input of the CAS registry number followed by the simple command "d fpre" (for "display preparations") will lead to an up-to-date (within the last few months) listing of every known preparative method for the compound in question.

N, *N*-**Dichlorocarbamates; chlorination reactions.** J. Bougault and P. Chabrier. *Compt. rend.* **213,** 400–2 (1941); cf. *C.* A. **37,** 87[7].—PhOH is transformed by a small excess of NCl_2CO_2Me (**I**) in AcOH into $2,4,6-Cl_3C_6H_2OH$, and *o*-$HOC_6H_4CO_2Me$ into its 5-Cl compd. **I** and $(ClCH_2CH_2)_2S$ in C_6H_6 give unstable *bis(1,2-dichloroethyl) sulfide*, $b_{15}°$, which readily yields HCl and $CHCl:CHSCHClCH_2Cl$. Carbazole and a small excess of **I** in AcOH yield *tetrachlorocarbozole*, m. 213°. $BzNH_2$ and **1** in aq. suspension afford BzNHCl, and $PhCH_2$-$CONH_2$ gives *N=chlorophenylacetamide*, m. 120°. In alk. soln. *2,4-dichloro-3,5-diketo-6-benzyl-*, m. 119°, and *-6-phenylethyl-*, m. 130°, *tetrahydro*-1,2,4-*triazine* are obtained from the Cl-free parents. *2-Chloro-3,5-diketodibenzyltetrahydro-1,2,4-triazine* m. 153°. **I** and diphenylhydantoin in alk. soln. afford *1,3-dichloro-5, 5-diphenylhydantoin*, m. 166° B. C. P. A.

Figure 15.6   *Chemical Abstracts.* **1943,** *37,* **2010**[3].

The method or methods described for the compound must then be *evaluated* from the standpoint of two primary criteria: (1) simplicity and practicality and (2) economy in terms of yield from readily available starting materials. Under the first heading come such items as the need for special apparatus, high pressures, or temperatures, and ease of isolating the product. These requirements may not be completely satisfied for any of several reasons. The literature reference may deal with a catalytic process suitable for industrial equipment but impractical for laboratory work. On the other hand, the previous source may not have been a deliberate synthesis; for example, the compound may have been obtained from a more complex molecule or as a minor product in the investigation of a reaction for some other purpose. A practical synthetic method requires that the compound be obtained in useful yield from more readily available or cheaper precursors. Ultimately, of course, the synthesis must go back to commercially available chemicals.

An important consideration, particularly when the literature references are not contemporary, is the possibility that the aforementioned criteria could be better met by a method not previously described for the compound. Improved synthetic procedures and reagents are constantly being developed. For example, the preparation of a primary alcohol may have been carried out before 1948 by reduction of a carboxylic acid ester with sodium in ethanol. The use of metal hydrides has largely displaced this method for laboratory work.

A number of monographs and compendia deal with preparative reactions, and reference to these sources may be necessary even if a previous preparation is to be repeated, since sufficient detail may not be given in the original paper. In applying a reaction to a compound for which it has not been used previously, general textbook knowledge is seldom sufficient, since reaction conditions and isolation procedures are needed. Some of the more important sources for laboratory procedures are the following:

1. *Houben-Weyl Methoden der Organische Chemie*, 4th ed.; Miller E., Ed.; G. Thieme Verlag: Stuttgart, 1952–.

This is an encyclopedic multivolume series, in German, dealing with methods of general laboratory practice and procedures and also with the preparation and reactions of classes of compounds. The fourth edition provides comprehensive coverage for many major functional group classes.

2. *Organic Reactions;* John Wiley & Sons: New York, 1942–.

In this series (currently 54 volumes), more than 170 general reactions of preparative utility are discussed in detail, with typical experimental procedures and extensive tables of examples with references. Cumulative subject indexes in the more recent volumes can be scanned for a desired reaction.

3. Theilheimer, W.; *Synthetic Methods of Organic Chemistry;* S. Karger: Basel; Interscience: New York, 1948–.

The annual volumes in this series contain brief descriptions of useful synthetic transformations taken from the literature of the year covered. The reactions are indexed according to a special system based on the type of bond formed. Five-year indices are included.

4. *Organic Syntheses;* John Wiley & Sons: New York, 1920–.

The annual volumes and six collective volumes of *Organic Syntheses* contain detailed procedures for the preparation of more than 1500 compounds. Apparatus, conditions, and workup procedures are specified, and many of the syntheses illustrate general methods that can be applied to related compounds.

5. In addition to these compendia, a number of books have been published devoted to synthetic organic chemistry. They vary widely in approach and breadth of coverage, but each of the following can provide references to specific procedures in the literature:

Buehler, C. A.; Pearson, D. E. *Survey of Organic Syntheses;* Wiley-Interscience: New York, 1970, 1977; vols. 1 and 2.

Carey, F. A.; Sundberg, R. J. *Advanced Organic Chemistry,* 3rd ed.; Parts A and B, Plenum: New York, 1990.

Fieser, L. F.; Fieser, M. *Reagents for Organic Synthesis;* John Wiley: New York, 1968–; vols. 1 to 18.

Harrison, I. T.; Harrison, S.; et al. *Compendium of Organic Synthetic Methods;* Wiley-Interscience: New York, 1971–1992; vols. 1–7.

House, H. O. *Modern Synthetic Reactions,* 2nd ed.; W. A. Benjamin: Menlo Park, CA, 1972.

Larock, R. C. *Comprehensive Organic Transformations,* 2nd ed.; John Wiley & Sons: New York, 1999.

Rodd's Chemistry of Carbon Compounds, 2nd ed.; Coffey, S., Ed.; Elsevier: New York, 1990.

Sandler, S. R.; Karo, W. R. *Organic Functional Group Preparations,* 2nd ed.; Academic Press: New York, 1983, 1986, 1989; vols. I, II, and III.

To illustrate the process of searching for a preparative method, methods for preparing methyl 5-chlorosalicylate can be examined. Three methods are given in Beilstein (see Figs. 15.1 to 15.3). The first entry cites the preparation by HCl-catalyzed esterification of 5-chlorosalicylic acid in methanol and gives the recrystallization solvent (needles from ethanol) and melting point (48°C).

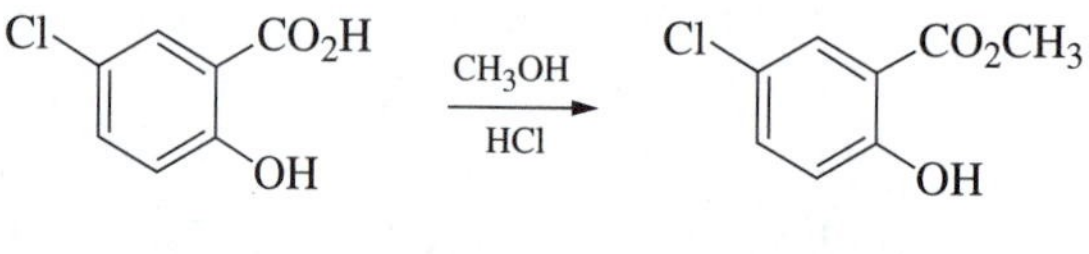

The original reference, however, to Series 2 of *Journal fur Praktische Chemie,* is not easily accessible. In the second Beilstein reference, a preparation is indicated by irradiation of a solution of trichloronitromethane (chloropicrin) in methyl salicylate:

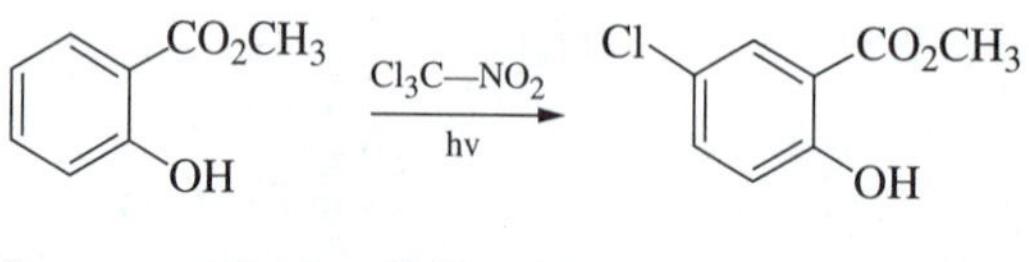

The third Beilstein reference and one of the *Chemical Abstracts* citations (see Fig. 15.6) indicates a preparation by chlorination of the hydroxy ester with *N,N*-dichlorocarbamate esters. The other abstracts cited in the subject index (see Fig. 15.5) lead to papers dealing with the bactericidal properties of chlorophenol derivatives.

At this point, we can assess the possibilities. Without consulting the original references (which appear in Italian and French journals), it is apparent that the two chlorination procedures require rather special reagents, neither of which is very commonly used. Further checking on these reagents in contemporary sources reveals that a chlorination step is required for their preparation. On the other hand, the main starting material is a very cheap compound, methyl salicylate (oil of wintergreen).

The alternative is esterification of 5-chlorosalicylic acid. This acid, but not the ester, is listed in chemical supply catalogs. The cost is about eight times that of methyl salicylate. The other reagents, however, are common laboratory chemicals, and the esterification is a standard operation (procedures using sulfuric acid are given in Chapter 28). For preparation of a small sample of the ester, this method would probably be the best choice.

For specific information on the procedure for carrying out the esterification and isolation of the product, many examples of acid-catalyzed esterifications can be found in the indexes of *Organic Syntheses Collective Volumes.* Another useful step at this point is to consult the section on acid-catalyzed esterification in one of the books on synthetic methods. References to esterification of closely similar hydroxybenzoic acids may be given, and from these a procedure in a conveniently accessible journal usually can be found.

KEEPING UP-TO-DATE

A number of journals are designed to allow the reader to scan listings that might contain very recent citations of keywords relating to the area of interest.

Published weekly are the seven editions of *Current Contents,* three of which relate to Chemistry: *Physical Chemical and Earth Sciences, Engineering Technology and Applied Science,* and *Life Sciences.* Each edition lists the contents of approximately 1000 journals as well as a subject index and an author index and is also available on diskette. Usually, no more than two or three weeks elapse between the time the original journal article appears and the time of its appearance in *Current Contents.* The articles are not abstracted.

Published biweekly is *Chemical Titles,* containing Keyword in Context Index, Author Index, and Bibliography sections, and this journal covers topics relating to chemistry and chemical engineering. Each issue represents approximately 700 primary journals.

ELECTRONIC METHODS

ONLINE SEARCHES

Along with the advent of computers and computer networking has come the ability to conduct literature searches of a large number of databases using online subscription services. Many of these are relatively inexpensive, and the cost is reasonable compared to the expense and expenditure of time necessary to conduct manual searches. Most modern libraries subscribe to one or more automated systems such as DIALOG, CAS ON-LINE, and JTN International, which include the ability to search *Chemical Abstracts, Biological Abstracts, World Patent Index,* and about 300 other databases. The cost of these searches depends on the file being searched, and searches that are well planned are well worth the cost. An excellent text that covers online databases, retrieval systems, and other useful information is *Communication, Storage and Retrieval of Chemical Information,* by J. Ash, P. Chubb, S. Ward, S. Welford, and P. Willett (John Wiley & Sons: New York, 1985).

Beilstein is now accessible online. Also available is CASREACT, an online reaction-searching tool from Chemical Abstracts Service that enables one to search for specific reactions, preparations of specific compounds or substructures, and information on reagents, solvents, and catalysts.

Many chemical supply houses have made their inventory of chemicals available for electronic searching, particularly relative to searches that include data on toxicity or disposal methods. These databases may be online or may be in the form of diskettes for microcomputers. These companies also upgrade the data periodically; thus the information remains current.

A service with great potential involves the use of CD-ROM, Compact Disk–Read Only Memory, in which a large amount of information is stored in a form readable by a laser device similar to a compact disk player. Although a compact disk reader is required, the possibility of having an enormous number of references available at a laboratory work station with minimum additional cost seems to have an advantage over the other methods. Among those now available are *Chemical Abstracts* and *Dissertation Abstracts*.

For detailed information concerning the electronic search capabilities of your library, consult the reference librarian or one of the references cited below:

Dialog Database Catalog 1987; Dialog Information Services, Inc.: 3460 Hillview Avenue, Palo Alto, CA 94304.

Keller, R. J. *The Sigma Library of FT-IR Spectra;* Nicolet Instruments: Madison, WI, 1985 (Computer Search Program).

Maizell, R. E. *How to Find Chemical Information,* 3rd ed.; John Wiley & Sons: New York, 1998.

Pouchert, C. J. *The Aldrich Library of FT-IR Spectra;* Nicolet Instruments: Madison, WI, 1993 (Computer Search Program).

Many of the references in this chapter are able to be accessed directly on the Internet.

LITERATURE PROBLEMS

1. **Initial Exercise:** As a first exercise in using the chemical literature, you will be asked to locate a reference to an organic substance of your choosing. You are to use the following procedure:

    a. Draw and name your molecule. Keep it simple, but it should have a molecular weight of at least 150 and contain at least six carbon atoms.
    b. Secure the approval of your teaching assistant or laboratory instructor for your choice, and for your naming.
    c. Locate a reference to your compound in the Chemical Abstracts indices. Make a copy of the actual abstract (it may not mention your compound in the abstract).
    d. Using the abstract, locate the journal and page number for the article. Make a copy of the first page of the article, and a copy of a page in which your compound is mentioned.
    e. Turn in all three pages (CA abstract and two journal pages) to your instructor.

2. **Advanced Exercise:** For a more advanced exercise in using the chemical literature, you will be asked to find a practical laboratory-scale synthesis for a compound. The compound can be prepared in one or more steps from simple starting materials and without special apparatus. Obtain an assignment of a compound from your instructor, consult the literature, and write up a complete experimental procedure for the synthesis on a scale to provide 50 to 100 mg of the final product. Your report should include all relevant literature references; the experimental descriptions should be sufficiently detailed that the preparation could be carried out without reference to the original literature.

At your instructor's direction, this preparation can then be carried out in the laboratory as a supplemental experiment. The starting materials for the synthesis should be reasonably priced commercial chemicals. Your instructor will determine at what point in the sequence your synthesis should begin; this will depend on the cost of various precursors and the amount of laboratory time available.

PRELABORATORY QUESTIONS

This exercise is designed to familiarize you with the relevant sections of your library that pertain to chemistry. You should become familiar with the reference sections and the periodicals locations before attempting the laboratory assignment.

1. Describe the locations in your library of each of the following publications; list the floor, aisle, and call number for each.
    a. Beilstein, *Handbuch der Chemie*
    b. *Chemical Abstracts,* bound copies
    c. *Chemical Abstracts,* loose copies
    d. *Heilbron's Dictionary of Organic Compounds*
    e. *Lange's Handbook of Chemistry*
    f. *The Journal of Organic Chemistry*
    g. *The Journal of the American Chemical Society*
2. Which of the publications in question 1 can undergraduate students borrow overnight?
3. List five journals abstracted by *Chemical Abstracts.*

Properties of Hydrocarbons

Hydrocarbons are those organic compounds composed only of carbon and hydrogen. There are *three main categories of hydrocarbons:* saturated, unsaturated, and aromatic. Saturated hydrocarbons have only carbon–carbon single bonds, whereas unsaturated hydrocarbons have carbon–carbon double or triple bonds. Aromatic hydrocarbons are cyclic compounds whose chemical properties are related to benzene.

Saturated hydrocarbons (alkanes and cycloalkanes) are relatively inert and do not react with common laboratory reagents. Unsaturated hydrocarbons (alkenes and alkynes), however, readily undergo addition reactions and oxidation reactions. Benzene and other aromatic compounds do not readily undergo addition reactions but are characterized by substitution reactions in which another atom or group of atoms replaces a ring hydrogen.

Although alkanes are relatively inert, they do undergo combustion in the presence of air if ignited. The fact that these reactions are highly exothermic and that huge quantities of alkanes are available as petroleum and natural gas has resulted in their extensive use as fuels. While the chemistry of alkanes is relatively straightforward, their economic impact can hardly be overestimated. Consequences resulting from this economic importance include oil spills, air pollution from automobiles, the greenhouse effect, and the threat of war in certain parts of the world.

EXPERIMENTS

Performing the following tests will illustrate general properties of hydrocarbons as well as differences in chemical reactivity due to the type of hydrocarbon (saturated, unsaturated, or aromatic) being considered. Perform all six tests on cyclohexane, cyclohexene, toluene, and two unknowns. One unknown will be an alkane or cycloalkane and a second will be an alkene or alkyne. Using the information that a lack of reactivity is characteristic of saturated hydrocarbons and addition reactions are common for unsaturated hydrocarbons, decide the identity of each unknown as to type of hydrocarbon.

SAFETY NOTE

Concentrated sulfuric acid, concentrated nitric acid, and bromine can cause burns. If any one of these is spilled on the skin, wash immediately with water. Exposure to the vapors of bromine or methylene chloride should be avoided; use these reagents in a hood.

Disposal. Dispose of all waste from these test tube experiments in the appropriate waste containers as directed by your instructor.

PROCEDURE

A. **Solubility**   Place 2 mL of water in a 13 × 100-mm test tube and add 2 or 3 drops of the hydrocarbon to be tested. Shake the mixture to determine whether the hydrocarbon is soluble (a colorless second layer may be hard to see). Record your results and save the mixture for Test B.

B. **Relative Density**   Reexamine the mixtures prepared above and decide in each case whether the hydrocarbon is more dense (sinks) or less dense than water (floats).

C. **Flammability**   Test the flammability of each hydrocarbon by placing 2 or 3 drops on an evaporating dish in the hood and igniting it with a match or microburner. Note the nature of the flame: sooty flames are characteristic of unsaturated compounds.

D. **Addition of Bromine**   Dissolve 3 or 4 drops of the hydrocarbon in 1 mL of methylene chloride (dichloromethane) in a 13 × 100-mm test tube. Add dropwise a 2% solution of bromine dissolved in methylene chloride with shaking. The loss of bromine color is an indication of an unsaturated compound (do not confuse diminution of color due to dilution of the Br_2/CH_2Cl_2 solution with an actual loss of color).

E. **Reaction with Potassium Permanganate**   Dissolve 3 or 4 drops of the hydrocarbon in 1 mL of reagent-grade acetone and then add dropwise a 1% solution of potassium permanganate with shaking. A loss of the purple color of the permanganate solution indicates that a reaction has taken place and that the hydrocarbon is unsaturated.

F. **Reaction with Sulfuric Acid**   Place 1 mL of concentrated sulfuric acid in a 13 × 100-mm test tube and then add 3 or 4 drops of the hydrocarbon one drop at a time. (Caution!) Reaction is indicated not only by the dissolution of the sample but also by changes in color, production of heat, or the formation of insoluble material. If no reaction occurs, warm the mixture (be careful that the hydrocarbon does not evaporate). If the unknown is a simple aromatic hydrocarbon it will react with the warm sulfuric acid to form a sulfonic acid which will dissolve.

At this point, information from Tests D, E, and F should allow you to determine which unknown is an alkane or cycloalkane and which is an alkene or alkyne. Record your conclusions about each unknown.

It should be pointed out that while the reagents in Tests D, E, and F may be used to distinguish saturated and unsaturated hydrocarbons, they will also react with other organic compounds. Potassium permanganate reacts with easily oxidized compounds (alcohols, aldehydes, amines, etc.), while concentrated sulfuric acid reacts with a variety of Lewis bases (such as alcohols and other oxygen-containing compounds and amines). Care should be taken not to use these tests with compounds other than hydrocarbons and then misinterpret the results.

QUESTIONS

1. Considering your results in the solubility tests, what do you conclude about the solubility of hydrocarbons in water? Predict the solubility of gasoline and motor oil in water.

2. Are hydrocarbons less dense or more dense than water? Does this have any practical importance when oil spills occur?

3. Write equations using 1-butene to illustrate the reactions occurring in Tests D, E, and F of the experimental procedure (use your text).

4. Write equations using cyclohexene to illustrate the reactions in Tests D, E, and F.

5. Write a balanced equation for the complete combustion of octane. What other product(s) are produced when incomplete combustion of octanes occurs in an automobile engine?
6. How could one distinguish octane from 1-octene by a simple chemical test; 1-octanol from 1-octene; and toluene from 1-octene?

PRELABORATORY QUESTIONS

1. Draw skeletal (line-bond) structures for cyclohexane, cyclohexene, and toluene.
2. What color is potassium permanganate solution? What is the formula for potassium permanganate?
3. Would you expect a significant difference between the reactivity of hexane and cyclohexane? Explain your answer.
4. The general formula for an alkane is C_nH_{2n+2}. What is the corresponding general formula of a cycloalkene?
5. List four hazardous chemicals that are used in this experiment.

Nucleophilic Substitution of Alkyl Halides

Alkyl halides, alkyl sulfates, and alkyl sulfonates serve as reactants or "substrates" in many nucleophilic substitution reactions. These substitution (or displacement) reactions are useful preparative methods, for example, the Williamson synthesis of ethers. For our purpose the term "nucleophilic" will refer to a reaction in which a nucleophile (HO^-, X^-, S^{2-}, HS^-, etc.) attacks a carbon atom and displaces a "leaving group," another nucleophile. Although these reactions proceed in a continuum of slightly different pathways, they can be viewed as occurring in a combination of two limiting mechanisms:

1. **Substitution: Nucleophilic, Unimolecular (S_N1)**

This mechanism involves more than one step, the first of which is the ionization of the alkyl halide substrate, yielding a carbocation and an anion. The planar carbocation then reacts with a nucleophile, forming the product (in the case of chiral compounds, racemic product generally results). Since the nucleophile is often a solvent molecule, the process is sometimes called **solvolysis.** A typical S_N1 reaction is the solvolysis reaction of *tertiary*-butyl chloride with water to form *tert*-butyl alcohol.

2. **Substitution: Nucleophilic, Bimolecular (S_N2)**

This reaction involves one so-called concerted step in which the nucleophile attacks the carbon atom of the substrate on the side opposite the leaving group. An inversion of configuration of chiral compounds results.

Studies of the mechanisms of these reactions were undertaken in the early 1930s in order to explain the vast differences in the rates of seemingly very similar reactions. Since that time, these findings have been refined to encompass the effects of solvent, ionic strength, added electrolytes, the nature of the alkyl group, and other factors. Some of these factors will be illustrated in this experiment. In addition, the complex methodology of determining the rate of a reaction will be undertaken.

EXPERIMENTS

A.   STRUCTURAL EFFECTS ON S_N1 AND S_N2 REACTIVITY

In this experiment the relative reactivities of a series of halides are observed under two sets of conditions, one of which favors the S_N1 mechanism and the other the S_N2. The S_N2 reaction is observed by the displacement of chloride or bromide by an iodide ion in acetone solution. The iodide ion is a good nucleophile for the S_N2 reaction, whereas acetone is a relatively poor ionizing solvent, and S_N1 dissociation is minimized. Sodium iodide is very soluble in acetone, but sodium chloride and sodium bromide have very low solubilities; so the course of the reactions can be followed by the formation of crystalline NaCl or NaBr (Eq. 17.1).

$$R\!-\!X \; + \; NaI \; \xrightarrow{\text{acetone}} \; R\!-\!I \; + \; NaX \downarrow \qquad\qquad (17.1)$$

The S_N1 reaction can be observed by treating the alkyl halide with a solution of silver nitrate in aqueous ethanol. Nitrate ion is a very poor nucleophile so there is little opportunity for S_N2 displacement. Dissociation of the alkyl halide by the S_N1 process is followed by the precipitation of the insoluble silver halide (Eq. 17.2); the carbocation is then captured by alcohol or water

$$R\!-\!X \longrightarrow R^+ + X^- \xrightarrow{\text{AgNO}_3} \; AgX \downarrow \; +NO_3^- \qquad\qquad (17.2)$$

PROCEDURE

SAFETY NOTE

Alkyl halides are toxic and flammable; avoid breathing them or spilling them on the skin. Insure that there is proper ventilation.

Label two series of five clean, thoroughly dry test tubes with numerals 1 to 5. In each series of tubes, place 0.2 mL of the following halides: (1) *n*-butyl chloride, (2) *n*-butyl bromide, (3) *sec*-butyl chloride, (4) *tert*-butyl chloride, and (5) crotyl chloride [$CH_3CH{=}CHCH_2Cl$]. Keep the tubes stoppered with corks or parafilm and leave them covered at all times, before and after adding reagents. Obtain 15 mL of 15% NaI-acetone solution and 15 mL of 1% ethanolic AgNO$_3$ solution from the side shelf.

Arrange one series of tubes in order, from 1 to 5. Add 2 mL of the NaI solution to tube number 1 and note the time. (Add the solution from a pipet as rapidly as possible—not dropwise.)

After 2 to 3 minutes, add 2 mL of NaI solution to the second tube and again note the time. Continue the addition at 2- to 3-minute intervals with the remaining tubes. After each addition, watch for any rapid reaction and then inspect the other tubes for signs of a precipitate. Note the time as closely as possible when precipitation begins to occur, recording the data in tabular form in your notebook. Cover the tubes and allow them to stand, observing them periodically while the next series is run.

Arrange the second series of tubes, and in the same way add 2-mL portions of the AgNO₃ solution to each tube at 2-minute intervals. Again, watch closely for any that change rapidly and then observe the others periodically. If possible, note the time both for the first appreciable turbidity and also for a definite precipitate. If any tubes in the NaI series are still clear at this point, loosen the covers slightly and place the tubes in a water bath at 50°C; note any changes that occur. Record the data in tabular form with column headings: tube number, time turbidity appears, and time of definite precipitate.

When you have completed these test tube reactions, pour the contents of all of the test tubes into the organic waste container.

B.   EFFECTS OF SOLVENT ON S_N1 REACTIVITY

In addition to structural features of the halide, reaction conditions can have a large effect on the rate of nucleophile substitution. For example, the solvent plays a major role in both S_N1 and S_N2 reactions, and in this experiment the effect of solvent on the rate of the S_N1 solvolysis of *tert*-butyl chloride will be studied. For this purpose, the most suitable method for comparing solvolysis rates is based on the fact that a strong acid is liberated in the reaction (Eq. 17.3). To determine the extent to which the reaction has proceeded, enough base is added to the reaction mixture to neutralize a small fraction of the acid produced. The solution becomes acidic after that fraction of the *tert*-butyl chloride has reacted, and the change in pH is detected with phenolphthalein indicator. The time required for neutralization is inversely proportional to the rate constant of the reaction, k_1, as shown in Equation 17.4.

$$t\text{-Bu}\!-\!\text{Cl} + \text{ROH} \rightarrow t\text{-Bu}\!-\!\text{OR} + \text{HCl} \qquad (17.3)$$

The rate of formation of HCl by S_N1 solvolysis of an alkyl chloride is equal to the rate of disappearance of alkyl chloride and is proportional to the concentration of the alkyl halide (Eq. 17.4).

$$\frac{d[\text{HCl}]}{dt} = k_1[\text{RCl}] = \frac{-d[\text{RCl}]}{dt} \qquad (17.4)$$

Integration of this equation from 0 to 50% reaction (100 to 50% of the RCl) provides Equation 17.5 where $t_{1/2}$ is the time required for 50% of the alkyl halide to react. If 0.50 equivalents of hydroxide ion were present at the start of the reaction, $t_{1/2}$ will be the time at which the solution becomes acidic.

$$\int_{1.0}^{0.5} \frac{d[\text{RCl}]}{[\text{RCl}]} = -\int_{0}^{t_{1/2}} k_1 dt$$

$$2.3 \log 0.5 = -k_1 t_{1/2} \qquad (17.5)$$

$$k_1 = -2.3 \log 0.5/t_{1/2} = 0.69/t_{1/2}$$

In the procedure to be used, the amounts of alkyl halide and hydroxide are not accurately measured, since only relative rates are desired. If the alkyl halide used were weighed and the NaOH solution were standardized and accurately dispensed, actual rate constants could be obtained.

Suitable solvent systems for this study are mixtures of water with methanol, ethanol, and acetone. The volume percent compositions indicated below will give conveniently measurable reaction rates, with the exception noted for 70% acetone. The following solutions will be available.

COMPOSITION (SOLVENT: WATER)	
55:45	(all solvents)
60:40	(all solvents)
65:35	(all solvents)
70:30	(not acetone)

PROCEDURE

Select three to five solvent mixtures, using either different ratios of one solvent or the same ratio with the three solvents, and label each of three to five clean, dry 13 × 100-mm test tubes accordingly. To each test tube add 2.0 mL of the solvent system and 3 drops of 0.5 M NaOH solution containing phenolphthalein indicator. Cover the tubes with corks or parafilm, and place them in a bath containing a thermometer and water at 30 ± 1°C. A Styrofoam cup placed in an empty beaker for stability makes a convenient insulated container for the bath. After 4 to 5 minutes in the bath to bring the solvent to bath temperature, add 3 drops of *tert*-butyl chloride to each test tube. Note the time of addition, shake or swirl the test tubes to mix the solutions, and replace them in the bath, swirling intermittently to insure good mixing. Add a few mL of hot water as needed to the bath to maintain the temperature at 30 ± 1°C. Record the time required for the pink (basic) color to disappear in each solvent mixture. Tabulate your results and compare them with those of others in the class to obtain a more complete picture of the solvent effects.

C.   EFFECT OF TEMPERATURE ON REACTION RATES

The rate of any reaction increases with increasing temperature because of the greater kinetic energy of the reacting molecules. The effect of temperature on the rate constant of a reaction is generally expressed in terms of the Arrhenius Equation (Eq. 17.6) where E_a is the activation energy, T is the temperature in K, A is a term related to the probability of a reaction occurring if the molecules have sufficient energy, and R is the gas constant (1.99 cal/mole deg). The activation energy can be obtained if the rate constant is known at two or more temperatures. Using the logarithmic form of Equation 17.6 (Eq. 17.7), it is seen that a plot of log k versus $1/T$ gives a straight line with slope equaling $-E_a/2.3R$.

$$k = A \exp\left(-\frac{E_a}{RT}\right) \qquad (17.6)$$

$$\log k = \log A - \frac{E_a}{2.3R}\left(\frac{1}{T}\right) \qquad (17.7)$$

The effect of temperature on the rate of solvolysis of *tert*-butyl chloride can be studied using the procedure and solvent systems of Part B at different temperatures. To obtain conveniently measurable rates, a solvent should be used that gives an end point at about 10 minutes in Part B. Although numerical values of the rate constants are not obtained by the comparative procedure in Part B, the neutralization times measured are inversely proportional to the rate constants (Eq. 17.5). Thus, a plot of log t values versus $1/T$ gives a straight line with slope $+ E_a/2.3R$ from which the value of E_a can be obtained.

PROCEDURE

From your results in Part B, select the solvent system for which the neutralization time at 30°C is closest to 10 minutes. Obtain two 2-mL mixtures of this solvent composition and add 3 drops of phenolphthalein indicator solution as in Part B. (In this part of the experiment, duplicate samples should be run.)

Adjust the temperature of the water bath to 20 ± 1°C, insert both tubes, and allow 4 to 5 minutes for temperature equilibration. Note the time, add 3 drops of *tert*-butyl chloride to each tube, mix, and measure the time required for the color to disappear. Repeat the procedure with two other samples after adjusting the bath temperature to 40 ± 1°C.

Using the times required for the samples at 20°C and 40°C and the time at 30°C from Part B, make a plot of log t versus $1/T$ (°K), and from the slope, calculate the activation energy for the solvolysis in the solvent used.

D.   REACTION OF ETHYL BROMIDE WITH HYDROXIDE ION, AN S_N2 REACTION

This is a variation of the classic experiments of Grant and Hinshelwood involving the reaction of ethyl bromide and hydroxide ion:

$$EtBr + HO^- \rightarrow EtOH + Br^- + \text{other products}$$

If this reaction goes by an S_N2 mechanism, the following equations should apply:

$$\text{rate} = k[EtBr][HO^-]$$

$$\text{if } [EtBr] = [HO^-], \text{ then rate} = k[HO^-]^2$$

$$\frac{1}{[HO^-]} = kt + \frac{1}{[HO^-]_0}$$

A plot of the hydroxide ion concentration versus time then should give a straight line with the slope equal to k (rate constant).

PROCEDURE

Students should work in pairs. For each pair of students, prepare the following solutions:

   a.   200 mL of 0.10 M HCl
   b.   50 mL of 0.40 M ethyl bromide in ethanol
   c.   50 mL of 0.40 M potassium hydroxide (CO_2 free) in ethanol.

Place the stock solutions of ethyl bromide and potassium hydroxide in a water bath at 40°C and allow them to come to temperature equilibrium.

In a 250-mL Erlenmeyer flask, pipet exactly 50 mL of the ethyl bromide solution and exactly 50 mL of the KOH solution. Mix the contents of the flask thoroughly and place it securely in the 40°C water bath (a 1-liter beaker may be used for each pair of students; add ice or hot water as needed). Immediately take a 10-mL aliquot and deliver it into a 250-mL Erlenmeyer flask containing 50 mL of ice water. This will quench the reaction so that a titration can be performed. Take as the starting time when the pipet is half empty. Titrate rapidly with 0.10 M HCl to the phenolphthalein end point. Record the number of mL of acid necessary to allow a faint pink color to remain in the solution and record this as the volume of NaOH at zero time. Repeat, taking aliquots and titrating them at 20-minute intervals up to 2 hours. Record the data in tabular form with these headings: clock, time (minutes), time (seconds), initial buret reading, final buret reading, volume HCl used, $[HO^-]$, and $1/[HO^-]$.

Plot the values of $1/[HO^-]$ versus the time in seconds. The slope of the line is the rate constant (units are liter/mole•sec.). Compare the value obtained with those of others in the class.

QUESTIONS

1. Explain the relative reactivities observed in Part A for butyl chloride, *sec*-butyl chloride, and *tert*-butyl chloride with NaI in acetone in terms of the structure of the transition state.
2. Similarly, explain the relative reactivities of the three butyl chlorides in ethanolic $AgNO_3$.
3. Based on the data obtained by your class in Part B, arrange the following solvents in order of increasing S_N1 solvolytic power: water, acetone, methanol, and ethanol. Explain in terms of the mechanism how the properties of the solvents account for the observed trend.
4. With the procedure used in Part B, would the color change occur sooner, later, or at the same time if: (a) twice as much *tert*-butyl chloride were used; (b) twice as much NaOH-phenolphthalein solution were used?
5. Sketch the reaction coordinate diagram for the solvolysis of *tert*-butyl chloride, labeling the maxima and minima with appropriate structures and indicating the activation energy measured in Part C.

PRELABORATORY QUESTIONS

1. Give structures of the products of the following nucleophilic substitution reactions:
    a. ethyl bromide and KOH (aq.)
    b. butyl chloride and NaI (acetone)
    c. *tert*-butyl iodide and ethanol
2. Illustrate the S_N2 mechanism for the reaction of methyl bromide with iodide ion.
3. How many mL of a 0.10 *M* solution of HCl are necessary to neutralize 10.00 mL of 0.16 *M* KOH?
4. Which should have more room for S_N2 attack by a nucleophile: methyl chloride or *tert*-butyl chloride? Explain.

REFERENCES

Grant, G. H.; Hinshelwood, C. N. *J. Chem. Soc.,* 1933, 258.

March, J. *Advanced Organic Chemistry,* 4th ed.; Wiley-Interscience: New York, 1992; Chapter 10.

Moore, J. W.; Pearson, R. G. *Kinetics and Mechanism;* John Wiley and Sons: New York, 1981.

Properties of Alcohols

Alcohols are an interesting and important group of organic compounds. They are used in beverages, in medicines, as solvents, and as synthetic intermediates. The characteristic functional group of alcohols is the hydroxyl group (OH) attached to a saturated carbon.

Alcohols are classified into three categories: primary (1°), secondary (2°), and tertiary (3°); the classification is based on the type of carbon to which the hydroxyl group is attached. If the carbon is attached to only one other carbon, it is a primary carbon and the alcohol is a primary alcohol. If the carbon is attached to two other carbons, it is a secondary carbon and the alcohol is a secondary alcohol, and so on.

Alcohols have much higher boiling points than ethers or alkanes of similar molecular weight. This is due to the association of alcohol molecules through hydrogen bonding. For example, the boiling point of butyl alcohol is 118°C, whereas the boiling point of the isomeric diethyl ether is 36°C.

A number of alcohols are common materials of commerce. Ethyl alcohol (grain alcohol) has been consumed in wines and other alcoholic beverages for thousands of years. Most methyl alcohol was once produced by the destructive distillation of wood, hence the name "wood alcohol." It is the major component of a number of commercial preparations used to avoid gasoline line freeze-up in automobiles. Ethylene glycol has a high boiling point and low freezing point and is miscible with water; therefore, it finds wide use as an antifreeze in automobile radiators. Glycerol is frequently used in cosmetics and medicinal preparations.

EXPERIMENTS

Some important reactions or properties of alcohols are demonstrated in the following experiments. The alcohols to be used in these experiments are ethanol (ethyl alcohol), 2-propanol (isopropyl alcohol), 1-butanol (*n*-butyl alcohol), 2-butanol (*sec*-butyl alcohol), 2-methyl-2-propanol (*tert*-butyl alcohol), and cyclohexanol.

A. REACTION WITH SODIUM (DEMONSTRATION)

SAFETY NOTE

Sodium is very reactive and must be treated with caution. Handle it with forceps and use only small pieces. Sodium reacts violently with water. Carry out all reactions in a hood. Sodium is normally stored under mineral oil.

PROCEDURE

1. Drop a small piece (2- to 3-mm cube) of sodium metal into a beaker of water (use a hood). After the reaction is complete, add 1 or 2 drops of phenolphthalein indicator to the remaining solution. Write an equation for the reaction of sodium with water.
2. Place 2 mL of ethanol in a large, dry test tube and add a small piece of sodium. Test the resulting solution with phenolphthalein. Write an equation for the reaction of sodium with ethanol. Is this reaction more or less vigorous than with water?
3. Place 2 mL of 1-butanol in a large, dry test tube and add a small piece of sodium. Note the rate of this reaction compared with the previous ones. Test the resulting solution with phenolphthalein. Write an equation for the reaction of sodium with 1-butanol.

Note that many organic liquids give a reaction with sodium metal even if the compound does not have a reactive hydrogen such as alcohols do. This is normally due to small amounts of water in the sample.

B.   SOLUBILITY IN WATER

To 2 mL of water in a 13×100-mm test tube, add ethanol, drop by drop, and shake the tube to mix the two materials. Note how many drops must be added until the alcohol is no longer soluble in water, but do *not* use more than 10 drops.

Repeat this experiment using in turn 2-propanol, 1-butanol, and cyclohexanol. Record your results as very soluble (6 to 10 drops), soluble (2 to 5 drops), or insoluble (1 drop).

C.   LUCAS TEST

The OH group of an alcohol is a poor leaving group and is not readily displaced by nucleophiles. However, in strongly acidic solutions, protonation of the OH group can occur to give R—$^+OH_2$ and water may be displaced. The **Lucas reagent,** prepared by dissolving 16 g of anhydrous zinc chloride in 10 mL of concentrated hydrochloric acid with cooling, will cause conversion of alcohols to insoluble alkyl halides if the alcohol is sufficiently soluble in the reagent. Thus, a second phase of droplets of alkyl halide will form in the test tube. With this reagent, tertiary alcohols react immediately and secondary alcohols in 3 to 10 minutes, while primary alcohols may take an hour or longer.

SAFETY NOTE

In the following experiments concentrated hydrochloric acid and sulfuric acid, glacial acetic acid, and acetic anhydride are used. These are corrosive materials and should be used with care, even though small amounts are required. Chromic acid and its salts are classified as suspected carcinogens. If any of the above materials are spilled on the skin, rinse thoroughly with water, then wash with soap and water.

Disposal

Dispose of all waste from these test tube experiments in the appropriate waste containers as directed by your instructor.

PROCEDURE

Place 2 mL of the Lucas reagent in a 13×100-mm test tube, add 4 drops of alcohol, shake the tube to mix the materials, and note the time required for the mixture to become cloudy or to separate into two layers. Try the test on 1-butanol, 2-butanol, 2-methyl-2-propanol, and your unknown. (*Note:* The alcohol must dissolve initially.) Then, try using just concentrated HCl on 2-methyl-2-propanol.

D.   OXIDATION USING CHROMIC ACID

Primary and secondary alcohols are oxidized rapidly to acids and ketones, respectively, by **chromic acid reagent.** This reagent, also called the **Jones reagent,** is prepared by dissolving 13.4 g of CrO_3 in 10 mL of water, adding 12 mL of concentrated H_2SO_4, and diluting the solution to 50 mL with water to give 2.68 M CrO_3. Tertiary alcohols are not oxidized by this reagent; however, they dehydrate in the acid environment and then the dehydration products oxidize after 2 or 3 minutes.

PROCEDURE

Place 1 mL of acetone in a 13 × 100-mm test tube (acetone that is free of traces of alcohol must be used); add 1 drop of the alcohol and then 2 drops of the chromic acid reagent. Note whether the mixture stays clear and orange [CrO_3 color] or turns turbid blue-green [formation of Cr(III)]. Make the decision about the test within 5 seconds after the reagent is added. Try this test on 1-butanol, 2-butanol, 2-methyl-2-propanol, and your unknown.

Based on the result of the Lucas test and the chromic acid test, what can you conclude about the nature of your unknown alcohol (1°, 2°, or 3°)?

E.   ESTER FORMATION

Alcohols react with carboxylic acids to yield esters in a condensation reaction in which water is eliminated. This reaction is catalyzed by mineral acids. Many low molecular weight, aliphatic esters have fragrant odors and are the materials primarily responsible for the odor of flowers and ripe fruits.

PROCEDURE

1. **Acid and Alcohol.** Mix 5 drops of ethanol and 5 drops of glacial acetic acid in a 13 × 100-mm test tube; add 1 drop of concentrated sulfuric acid, warm the mixture gently (3 to 4 minutes in a hot water bath), and then add 2 mL of cold water. Note the odor (if you have trouble detecting the odor or can only smell traces of acetic acid, pour the mixture into a small beaker and make it slightly basic with dilute NaOH). Write an equation for the reaction of acetic acid and ethanol.
2. **Acid Anhydride and Alcohol.** Place 5 drops of 1-butanol and 5 drops of acetic anhydride in a 13 × 100-mm test tube, stir well, and warm briefly. Add 2 mL of cold water, make the mixture slightly basic with dilute NaOH, and note the odor. Write an equation for the reaction of acetic anhydride and 1-butanol.

QUESTIONS

1. How does the size of the alkyl group in an alcohol affect the vigor of its reaction with sodium metal?
2. Why does the length of the carbon chain in an alcohol affect its solubility in water?
3. Write an equation for the reaction of 2-propanol with the Lucas reagent (ignore $ZnCl_2$).
4. Do primary or tertiary alcohols react more rapidly with the Lucas reagent? Why?

5. Which of the following alcohols would not be oxidized by chromic acid reagent within 5 seconds: 1-pentanol, cyclohexanol, 2-methyl-1-propanol, 2-methyl-2-butanol?
6. Using the tests described in this experiment, how could one distinguish between the following pairs of compounds? Tell what reagent or test you would use and what you would see.
    a. 1-butanol and 2-butanol
    b. isobutyl alcohol and *tert*-butyl alcohol
    c. 1-propanol and 1-heptanol
    d. cyclohexanol and cyclohexane

PRELABORATORY QUESTIONS

1. Ethanol, C_2H_6O, has a much higher boiling point than dimethyl ether, C_2H_6O, despite having the same molecular formula. Explain this phenomenon.
2. What are the components of the Lucas reagent?
3. What is the chemical name of "wood alcohol"?
4. Write an equation for the initial reaction of *tert*-butyl alcohol with the Jones reagent.
5. What class of organic compounds is responsible for the odor of flowers and ripe fruit?

The Grignard Reaction

Grignard reagents (organomagnesium halides, RMgX) play a commanding role in organic synthesis. These compounds can be adapted to the preparation of a large variety of functional groups, so the reaction of organomagnesium derivatives is one of the major uses of alkyl halides in organic synthesis.

The reaction of a halide and magnesium occurs on the metallic surface and is formally an oxidation of the metal. The reaction is usually carried out in an ether solution, often diethyl ether or tetrahydrofuran. The ether solvent functions as a Lewis base by solvating the Grignard reagent and permitting it to diffuse away from the metal.

$$RX \ + \ Mg \ \xrightarrow{\ Et_2O\ } \ \overset{\displaystyle OEt_2}{\underset{\displaystyle OEt_2}{\vdots}}RMgX\vdots$$

The formation of the organometallic reagent requires an active metal surface, and some difficulty may be encountered in getting the reaction started. Grinding the magnesium is usually effective in starting the reaction by providing a clean surface. Other tricks include exposure to an ultrasonicator, the addition of a small crystal of iodine, the addition of a very reactive halide such as ethyl iodide, or the addition of a small amount of 1,2-dibromoethane (suspected carcinogen), which reacts rapidly with the magnesium to form ethylene.

$$BrCH_2CH_2Br + Mg \longrightarrow CH_2{=}CH_2 \ + \ MgBr_2$$

A major pitfall in a Grignard reaction is the presence of water, alcohol, or any other "active hydrogen" compound that can act rapidly as an acid with the organomagnesium compound.

$$R'OH \ + \ RMgX \longrightarrow RH \ + \ R'OMgX$$

This reaction consumes and wastes the Grignard reagent by the formation of a hydrocarbon and may seriously complicate the preparation since the product R'OMgX can coat the surface of the metal and prevent further reaction with the alkyl halide. For this reason, the apparatus, reagents, and solvents must all be scrupulously dry.

Grignard reagents are most frequently prepared from alkyl bromides, which are usually more costly than the chlorides but are also somewhat more reactive with magnesium. The order of reactivity of halides with magnesium is $RI > RBr > RCl$. Aryl Grignard reagents are normally prepared using aryl bromides or aryl iodides since aryl chlorides react very slowly.

EXPERIMENTS

A. TRIPHENYLMETHANOL AND BENZOIC ACID

SAFETY NOTE

The major potential hazard in this experiment is fire. Ether is highly flammable, and there should be **no flames** in the laboratory at any time during this experiment.

(Macroscale)

PROCEDURE

Clamp a 100-mL single-neck flask fitted with a Claisen adapter to a ring stand, leaving room to slide an ice bath under the flask. (Do not support the flask on a ring.) Place a reflux condenser in the side neck of the Claisen adapter (see Fig. 19.1) and a dropping funnel in the vertical neck. Because all of the apparatus must be thoroughly dry, you should place it in a drying oven before assembly. All ground glass joints should be lightly greased. Fit a drying tube containing calcium chloride to the top of the condenser.

Obtain 10.0 g of bromobenzene and 1.50 g of magnesium turnings. Grind the magnesium in a mortar to provide a fresh surface, place it in the flask, and add 20 mL of *anhydrous* ether. With a pipet, add approximately 2 mL of the bromobenzene below the surface of the ether, to obtain a high concentration of halide close to the metal. Replace the dropping funnel, and warm the flask gently with warm water or the palm of your hand.

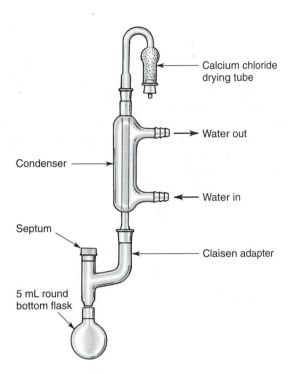

Figure 19.1   Setup for the microscale Grignard reaction.

If no visible sign of reaction (bubbling, turbidity) is seen after 5 minutes, add a small crystal of iodine and carefully crush it against the magnesium with a stirring rod. Continue the warming, and if the reaction again does not begin after 5 minutes, a "starter" solution should be added. To prepare a **starter solution,** place 0.2 to 0.3 mL of the bromobenzene, 1 mL of ether, and two to three chips of magnesium in a small test tube. With a glass rod, gently grind the metal against the glass to expose fresh surface. If this mixture does not begin to react, try using an ultrasonicator or one of the methods mentioned previously. When a vigorous reaction occurs (solution begins to bubble and becomes warm), quickly pour the contents of the test tube into the reaction flask.

When the reaction is actively in progress (heat evolved, turbulence), begin the flow of condenser water. Mix the remaining bromobenzene with 23 mL of anhydrous ether and place this solution in the dropping funnel. Add the bromobenzene solution dropwise, with occasional swirling of the flask, at a rate to maintain a steady reflux. After the halide has been added, warm the mixture on the steam bath as needed to maintain reflux for another 20 minutes. If the solution is black at this point, it is due to finely dispersed impurities from the magnesium; a few bits of metal may also remain. Proceed from this point with Experiment B or C.

(Microscale)

PROCEDURE

On a ring stand, set up the apparatus shown in Figure 19.1 consisting of a 5-mL round-bottom flask fitted with a Claisen adapter. Cap the shorter arm of the adapter with a septum, and place a condenser with a drying tube containing calcium chloride in the other. The apparatus must be absolutely dry to assure a successful reaction, and you should dry the pieces in an oven before assembly.

Weigh out about 50 mg of magnesium turnings that had been previously ground in a mortar to expose a fresh surface. Because of the size of the individual turnings, it may be difficult to obtain exactly 50 mg; however, this is not important. Get an accurate weight, calculate the millimoles, and use an equal millimolar amount of bromobenzene where it is called for in the procedure below. Place the magnesium in the 5-mL flask, and assemble the apparatus as shown in Figure 19.1. All ground glass joints (except the one capped by the septum) must be properly greased to avoid their fusion in the presence of the strongly basic Grignard reagent. With a syringe, add 0.5 mL of *anhydrous* ether through the septum. In a separate container mix 0.5 mL of *anhydrous* ether and the equimolar amount of bromobenzene needed to react completely with the magnesium. Add approximately 10 drops of this mixture through the septum to the magnesium and gently shake the apparatus to mix the contents. If the reaction does not start in several minutes, remove the septum and gently rub the magnesium against the bottom of the flask with a dry stirring rod. If the reaction does not begin after several additional minutes, one of the methods mentioned in the Macroscale procedure should be tried. When the reaction is vigorously under way, turn on the condenser water and add the remainder of the bromobenzene–ether mixture through the septum with the syringe at such a rate that the reaction remains at a moderate reflux. If the reaction becomes too vigorous, it may be moderated by immersing the flask in a beaker of cold water. (Should reaction cease completely, it may be necessary to warm the flask to start it again.) When the reaction begins to slow, place it in a beaker of warm water (40°C) and continue the gentle reflux until the magnesium is essentially consumed. Replace any anhydrous ether that may have been lost due to evaporation, and proceed with Experiment B or C.

B.   PREPARATION OF TRIPHENYLMETHANOL

In synthetic chemistry, the Grignard reaction is one of the most useful and versatile, particularly when it involves carbonyl systems. For example, reaction with formaldehyde gives primary alcohols, with other aldehydes it yields secondary alcohols, and with ketones it produces tertiary alcohols. The reaction with esters also yields tertiary alcohols, but with two groups the same, both arising from the Grignard reagent (Table 19.1).

(Macroscale)

PROCEDURE

While the Grignard solution is being refluxed, weigh out methyl benzoate equivalent in moles to one-half the amount of bromobenzene used. (Remember that the reaction of 1 mole of ester requires 2 moles of Grignard reagent.) Mix the ester with three volumes of anhydrous ether and place it in the dropping funnel. When the formation of the Grignard reagent is complete, cool the flask and its contents briefly in an ice bath. Add the ester solution dropwise with swirling of the apparatus to insure mixing. The addition should be at a rate to prevent excessive refluxing, and if necessary, an ice bath can be used to control the reaction. When the addition is complete, heat the mixture at reflux for 30 minutes or let stand overnight.

Pour the reaction mixture into a 125-mL Erlenmeyer flask containing 25 mL of 10% sulfuric acid and approximately 15 g of ice. Rinse the flask with a small amount of ordinary ether, followed by a small amount of 10% sulfuric acid, and add these rinses to the flask containing the reaction mixture. Swirl the mixture well to promote hydrolysis and then transfer the contents with a pipet to a separatory funnel, leaving the magnesium and remaining ice behind. Separate the layers, wash the ether layer with 10% sulfuric acid, followed by saturated sodium chloride solution, and let it dry over magnesium sulfate. Filter the mixture to remove the dry-

Table 19.1   The Reaction of a Grignard Reagent with Various Carbonyl Systems

PREPARATION OF A	GRIGNARD REAGENT UTILIZED	CARBONYL COMPOUND REQUIRED		PRODUCT OBTAINED

Primary alcohol $RMgX \;+\; HCHO \longrightarrow RCH_2OMgX \xrightarrow{H_3O^+} RCH_2OH$

Secondary alcohol $RMgX \;+\; R'CHO \longrightarrow R\!-\!\underset{\underset{R'}{|}}{CH}\!-\!OMgX \xrightarrow{H_3O^+} R\underset{\underset{R'}{|}}{CH}\!-\!OH$

Tertiary alcohol $RMgX \;+\; R'\!-\!\overset{O}{\overset{\|}{C}}\!-\!R'' \longrightarrow R\!-\!\underset{\underset{R'}{|}}{\overset{\overset{R''}{|}}{C}}\!-\!OMgX \xrightarrow{H_3O^+} R\!-\!\underset{\underset{R'}{|}}{\overset{\overset{R''}{|}}{C}}\!-\!OH$

$RMgX \;+\; R'\!-\!\overset{O}{\overset{\|}{C}}\!-\!O\!-\!R'' \longrightarrow R\!-\!\underset{\underset{R'}{|}}{\overset{\overset{R}{|}}{C}}\!-\!OMgX \xrightarrow{H_3O^+} R\!-\!\underset{\underset{R'}{|}}{\overset{\overset{R}{|}}{C}}\!-\!OH$

ing agent, add 15 mL of high-boiling petroleum ether, and evaporate the solution slowly on a steam bath until crystallization of the triphenylmethanol begins. Cool in an ice bath to complete the crystallization and collect, dry, and weigh the product. Dispose of solvent in the solvent waste container.

(Microscale)

PROCEDURE

In a test tube mix 1.0 mL of anhydrous ether and an amount of methyl benzoate equivalent in moles to one-half the amount of bromobenzene used (recall that two moles of Grignard reagent react with each mole of ester). When the formation of the Grignard reagent is complete, let the mixture cool and add the ester solution dropwise through the septum with a syringe. Because of the low vapor pressure of ether, the ester solution may tend to dribble or even squirt out of the syringe—hold the plunger tightly and work carefully so as not to lose any of the material. The addition rate should be such that the reaction always remains under control. If the mixture begins to reflux excessively, the reaction can be moderated by cooling in a beaker of water or even briefly in ice. When the addition is complete, heat the mixture at gentle reflux on a steam bath for 15 minutes or let it stand overnight.

Transfer the mixture to a centrifuge tube that contains 1.0 mL of 10% sulfuric acid and is approximately half filled with ice. Rinse the reaction flask with approximately 0.5 mL of ordinary ether and add this to the tube containing the reaction mixture. Stopper the tube and shake gently to hydrolyze the alkoxide salt. Unstopper the tube and let it stand. When the ice has melted, remove the lower (aqueous) layer with a pipet and bulb and discard. Wash the ether layer with water, followed by saturated sodium chloride solution, and dry it over magnesium sulfate. Add 0.5 mL of high-boiling petroleum ether and a boiling stone, and evaporate some of the solvent on a steam bath until the triphenylmethanol crystallizes. Complete the crystallization by cooling the tube in ice. Filter the mixture using a Hirsch funnel, dry the product, and weigh. Dispose of solvent in the solvent waste container. Determine the melting point and IR of the product and submit it in an appropriately labeled container.

C.   PREPARATION OF BENZOIC ACID

Carbonation of a Grignard reagent with CO_2 is a general method for preparing carboxylic acids. The initial product is the halomagnesium salt of the acid, which undergoes further addition of Grignard reagent at a very much slower rate than does CO_2. Carbon dioxide can be bubbled into the Grignard solution from a cylinder of liquid CO_2 under pressure, or more conveniently for small reactions, the Grignard solution can simply be poured onto dry ice.

benzoic acid
mp 121°C

(Macroscale)

PROCEDURE

While the Grignard solution is maintained at reflux, obtain approximately 30 g of crushed dry ice in a towel. Cool the Grignard solution and remove the condenser and dropping funnel. Blot the dry ice to remove condensed ice crystals, place it in a dry 250-mL Erlenmeyer flask, and pour the Grignard solution with care into the flask. Stopper the flask *loosely* with a cork and allow it to stand until the excess dry ice has sublimed. (If necessary, the reaction mixture can be stored at this point until the next laboratory period.)

Add 20 mL of ether to the residue and then add 20% hydrochloric acid in small portions, with shaking, until all solid has dissolved. Separate the layers, and wash the ether solution with water. To isolate the acid, extract the ether solution thoroughly with two 15-mL portions of 10% NaOH solution and then with 10 mL of water. Combine the aqueous extracts, back-extract once with ether, and then acidify with hydrochloric acid. Extract the mixture with ether, dry the extract with Na_2SO_4, and evaporate the ether in a tared round-bottom flask in a hood. Remove the last traces of ether with aspirator vacuum, and weigh the crude acid. Cool the crude acid in ice to crystallize it if the product is not solid. The crude benzoic acid contains a small amount of biphenyl (C_6H_5—C_6H_5, mp 69–72°C), which can be removed by recrystallizing the product from aqueous acidic solution. Determine the weight and melting point of the product. Dispose of ether in the solvent waste container.

(Microscale)

PROCEDURE

Wipe the frost from several small pieces of dry ice, crush them in a cloth towel, and add enough to a centrifuge tube to fill it about one-fourth full. With a syringe, squirt the Grignard reagent directly onto the dry ice. Rinse the reaction flask with approximately 1 mL of anhydrous ether, add this to the dry ice, and allow the excess dry ice to sublime. (At this point, the reaction mixture can be stored until the next period, if required.)

Add 1 mL of ordinary ether to the residue, and then add dropwise, with shaking, 10% hydrochloric acid until all of the solid has dissolved. With a syringe or pipet, remove and discard the aqueous layer, and wash the ether solution with water. The benzoic acid is purified by extracting the ether solution with two 1-mL portions of 10% aqueous NaOH and then with 1 mL of water. Combine the aqueous solutions, back-extract with 1 mL of ordinary ether, and then acidify with 10% hydrochloric acid. Extract the acidified mixture with three 1-mL portions of ordinary ether, and dry the combined ether layers over Na_2SO_4. With a syringe or pipet, carefully remove the ether solution from the drying agent, and place it in a tared 10-mL round-bottom flask. Evaporate the ether on a steam bath, preferably in a hood, and remove the last traces of solvent under aspirator vacuum. Weigh the crude acid, and place the flask in ice to effect crystallization if the product is not solid. The crude acid can be recrystallized from a minimum amount of slightly acidified water if required. Determine the weight and melting point of the product. Dispose of ether in the solvent waste container.

D. PREPARATION OF MALACHITE GREEN

SAFETY NOTE

The major potential hazard in this experiment is fire. Ether is highly flammable, and there should be **no flames** in the laboratory at any time during this experiment.

This experiment demonstrates the use of the Grignard method in preparing Malachite Green, which is used for directly dyeing silk and wool and for dyeing cotton after mordanting (see Chapter 45). *p*-Bromo-N,N-dimethylaniline is reacted with magnesium in tetrahydrofuran to form a Grignard reagent [I]. (The synthesis is possible with this amine because there are no protons on the nitrogen.) The Grignard reagent is then reacted with methyl benzoate to form the magnesium complex [II] which then is neutralized to form the dye [III].

(Microscale)

PROCEDURE

Dry the following equipment in a 110°C oven for 30 minutes: 50-mL round-bottom flask, condenser, 100-mL graduated cylinder, 50-mL beaker, Pasteur pipet, and a spatula.

The Grignard is water sensitive, so once the glassware has been dried, it is to be loaded and assembled directly to minimize exposure to the moisture in the air.

Allow the beaker to cool. Add 0.5 gram of *p*-bromo-N,N-dimethylaniline to the beaker and add 15 mL of tetrahydrofuran. Rinse the solution into the cooled round-bottom flask. Rinse the beaker twice with the 5-mL portions of tetrahydrofuran and add the rinsate to the round-bottom flask. Then add 0.1 gram of fresh, dry magnesium turnings to the reaction flask. Assemble the flask and condenser, with cooling water tubes attached, add a stirring bar, and clamp the assembly in place over a heating well that is held over a magnetic stirrer. Through the top of the condenser add a crystal of iodine. Begin stirring the mixture. If a spontaneous reaction does not occur, slowly bring the solution to a mild reflux. Continue heating for 30 minutes. The dark color will slowly change to a milky-gray. Allow the system to cool to room temperature.

Using the Pasteur pipet, add 0.09 grams (4 drops) of dry methyl benzoate to the mixture through the top of the condenser. Rinse the methyl benzoate down the condenser with about 1 mL of ether. A green solution will result. If residual magnesium is present, add 5% HCl dropwise until the magnesium is consumed.

The purity of this dye is not suitable for dyeing clothing or other routine applications. Avoid contact with the dye. Dye a small piece of cloth or paper to record the color you obtained. Your report will include observations of color changes the reaction goes through, as well as a cloth or paper sample. You will be graded on the quality of your sample rather than percentage yield.

Malachite Green is potent! That is, a little goes a long way! Wear old clothes to lab, avoid contact with the dye, and wash promptly in case of contact.

Note. Crystal violet can be made by substituting 0.06 gram of diethyl carbonate for methyl benzoate.

Disposal. Excess dye must be disposed of in the chemical waste container.

QUESTIONS

1. Give equations for a detailed procedure showing the conversion of 1-bromoheptane to 3-decanol using a Grignard reaction.
2. Consider the reaction of bromobenzene with magnesium to give phenylmagnesium bromide and its subsequent reaction with benzophenone to give triphenylmethanol. What weight of triphenylmethanol could be prepared if 3.14 g of bromobenzene and 3.80 g of benzophenone are used?
3. In a reaction of phenylmagnesium bromide with 2-butanone, the Grignard reagent was prepared from 25 g of bromobenzene and was then treated with a solution of 10 g of the 2-butanone in 50 mL of ether that contained 1% by weight of water. Calculate the grams of benzene that would be produced and the maximum theoretical yield of tertiary alcohol, corrected for loss due to benzene formation.
4. Triphenylmethanol can also be prepared by the reaction of phenylmagnesium bromide with diethyl carbonate. Write a stepwise mechanism for this overall reaction.

PRELABORATORY QUESTIONS

1. Which react more readily to form Grignard reagents, alkyl chlorides or alkyl bromides?
2. What is a common solvent for Grignard reactions?
3. Sometimes difficulty is encountered in getting a Grignard reaction started. What "tricks" can be used to get the reaction to start?
4. Show the reaction of butylmagnesium bromide with water.
5. Show the reaction of phenylmagnesium bromide with acetone.

REFERENCES

Clough, S., et al. *J. Chem. Ed.* 1986, *63*, 176.

March, J. *Advanced Organic Chemistry,* 4th ed.; Wiley: New York, 1992, pp. 622–624, 920–928.

Taber, D. F., et al. *J. Chem. Ed.* 1996, *73*, 259.

Dehydration of Alcohols

lcohols may be dehydrated under acid-catalyzed conditions to form alkenes. In the presence of Bronsted acids, such as sulfuric or phosphoric acid, dehydrations involve the initial protonation of the hydroxyl group (Eq. 20.1), converting it to an excellent leaving group. Loss of water then normally takes place by a unimolecular process, resulting in a highly reactive carbocation that can stabilize itself by the elimination of a proton from an adjacent carbon to give the alkene.

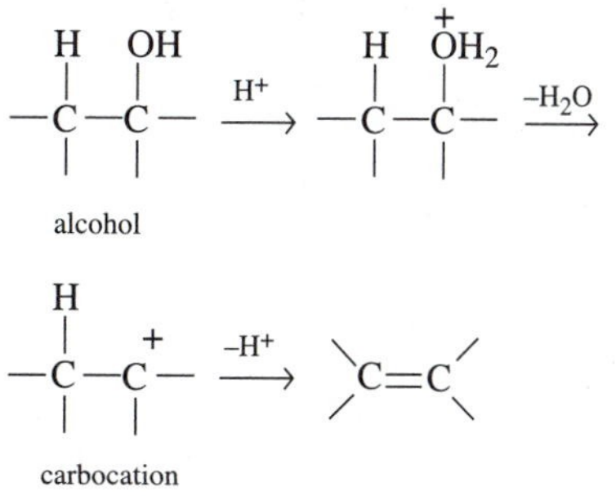

$$
\underset{\text{alcohol}}{-\overset{\displaystyle H}{\underset{\displaystyle |}{C}}-\overset{\displaystyle OH}{\underset{\displaystyle |}{C}}-} \xrightarrow{H^+} -\overset{\displaystyle H}{\underset{\displaystyle |}{C}}-\overset{\displaystyle \overset{+}{O}H_2}{\underset{\displaystyle |}{C}}- \xrightarrow{-H_2O}
$$

(20.1)

$$
\underset{\text{carbocation}}{-\overset{\displaystyle H}{\underset{\displaystyle |}{C}}-\overset{+}{\underset{\displaystyle |}{C}}-} \xrightarrow{-H^+} \overset{\diagup}{\underset{\diagdown}{}}C=C\overset{\diagdown}{\underset{\diagup}{}}
$$

Carbocations can also rearrange, by alkyl and/or hydride shifts, to form more stable cationic species that then yield other alkenes. A typical example of such a multiple-product reaction is shown in Equation 20.2.

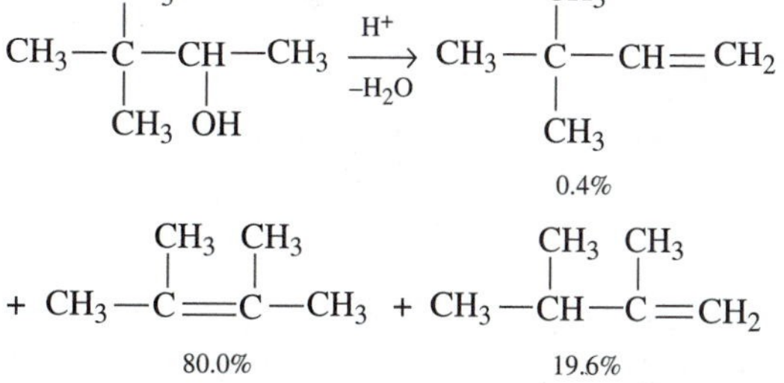

$$
\underset{}{CH_3-\overset{\displaystyle CH_3}{\underset{\displaystyle CH_3}{C}}-\overset{}{\underset{\displaystyle OH}{CH}}-CH_3} \xrightarrow[-H_2O]{H^+} \underset{0.4\%}{CH_3-\overset{\displaystyle CH_3}{\underset{\displaystyle CH_3}{C}}-CH=CH_2}
$$

(20.2)

$$
+ \underset{80.0\%}{CH_3-\overset{\displaystyle CH_3}{C}=\overset{\displaystyle CH_3}{C}-CH_3} + \underset{19.6\%}{CH_3-\overset{\displaystyle CH_3}{CH}-\overset{\displaystyle CH_3}{C}=CH_2}
$$

The ratios and types of alkenes formed depend on the catalyst used and the reaction conditions. Many types of catalysts other than Bronsted acids have been used in alcohol dehydrations.

These include iodine, alumina, molecular sieves, and ion exchange resins. Dehydrating agents such as $POCl_3$ and acetic anhydride have also been used, in which case the elimination usually proceeds via the corresponding ester intermediate.

With secondary and tertiary alcohols, direct Bronsted acid–catalyzed dehydration is often a simple and convenient procedure. If the elimination product is sensitive to further reaction with acid, as may be the case if the alcohol contains other functional groups, procedures that avoid strong acids must be used.

Dehydration of alcohols using Bronsted acids is illustrated in the following experiments. The procedures described use phosphoric acid; sulfuric acid can cause charring and can give off sulfur dioxide fumes.

EXPERIMENTS

SAFETY NOTE

The dehydration products formed in these reactions are highly flammable. If a burner is used in the distillation, take care that vapors of the product are not exposed to the flame. Handle phosphoric acid with care. Should phosphoric acid be spilled on your skin, wash it off with copious amounts of water. If droplets of phosphoric acid are splashed in your eyes, use the eyewash fountain to properly wash them.

Disposal. Dispose of all waste products from these experiments in the appropriate waste containers as directed by your instructor.

A.   PREPARATION OF CYCLOHEXENE

(Macroscale)

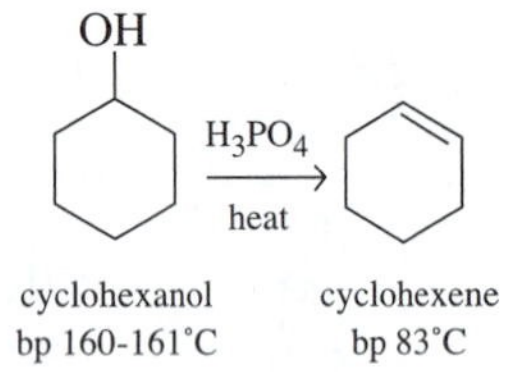

cyclohexanol
bp 160-161°C

cyclohexene
bp 83°C

PROCEDURE

Weigh 10.0 grams of cyclohexanol into a 25-mL round-bottom flask. Carefully add 5 mL of 85% phosphoric acid, then a boiling stone, and mount the flask for fractional distillation (see Fig. 5.7; either use no column packing or a low hold-up packing). Slowly heat the mixture until it comes to a gentle boil. After 10 minutes of gentle boiling, increase the heat sufficiently to cause distillation (the temperature of the distilling vapor should not exceed 100°C) and collect the distillate in a cooled receiver.

To the distillate, add 2 mL of 10% sodium carbonate solution to neutralize any traces of acid that may have been carried over. Transfer the liquids to a separatory funnel, add 10 mL of cold water, swirl the mixture gently, and drain off the lower aqueous layer. Pour the cyclohexene into a small, dry Erlenmeyer flask and dry it over anhydrous sodium sulfate for 10 to 15 minutes.

Decant the dried cyclohexene into a small dry distilling flask, add a boiling stone, attach the flask to a simple distillation assembly, and distill carefully. Collect the material distilling at 80 to 85°C. Determine the weight of the product and calculate the percentage yield.

Test the reactions of the product with bromine and with potassium permanganate (see Chapter 16, Procedure Tests D and E). Also examine the purity of product by use of vapor phase chromatography, IR, or NMR as designated by your instructor. Submit the product in an appropriately labeled container to your instructor.

B.   DEHYDRATION OF 2-METHYLCYCLOHEXANOL

(Macroscale)

2-methylcyclohexanol
bp 165-168°C

1-methylcyclohexene
bp 110°C

3-methylcyclohexene
bp 104°C

The dehydration of 2-methylcyclohexanol can occur in two directions, giving rise to 1-methylcyclohexene or 3-methylcyclohexene. In this experiment, gas chromatography (GC) will be used to determine the product composition. A disadvantage of using sulfuric acid in this experiment is that it is a sufficiently strong acid to reprotonate the double bond of the initial product and cause isomerization to other alkenes; therefore, phosphoric acid is used. In addition to the reactants it may be advantageous to use a "chaser" such as bromobenzene or decalin. These relatively high-boiling substances are inert under the reaction conditions and allow a more complete distillation of the product.

PROCEDURE

To a 25-mL round-bottom flask, add 6 mL of 2-methylcyclohexanol, 5 mL of 85% phosphoric acid, and a boiling stone. Attach the flask to a fractional distillation assembly (see Fig. 5.7; either use no column packing or a low hold-up packing). Slowly heat the contents of the flask to boiling and distill out the product. The distillation temperature should be kept below 96°C by regulating the rate of heating. Continue the distillation until 4 to 6 mL of liquid have been collected (a test tube or distillation receiver [centrifuge tube] is a convenient collection vessel).

Separate the product from the water layer by using a Pasteur pipet. Transfer the product to a clean, dry test tube and dry it over Na_2SO_4 for 10 minutes.

Inject 0.5 μL of the dried liquid onto a nonpolar gas chromatograph column. Before doing so, be sure to fill and empty the syringe several times, discarding the sample each time, to clean out the previous contents. Draw approximately 1 μL of air into the syringe after the liquid when you are ready to analyze the sample. Insert the needle as far as possible into the injection port before injecting the sample to insure placing the sample directly onto the column.

Mark the point of injection on the recorder chart paper and measure accurately the distance (or time) from this point to the olefin peak(s). By comparing the results with the retention times of 1-methylcyclohexene and 3-methylcyclohexene, determine the identity of the dehydration product(s). If both are present, determine the approximate composition by triangulation of peak areas (see Fig. 6.4). Explain your results. If additional peaks are present, attempt to identify these as well.

If GC analysis is unavailable, measure the refractive index of the product and assume a linear relationship between the molar concentration and the refractive index. Refractive index values for 1-methylcyclohexene and 3-methylcyclohexene are 1.4503 and 1.4414, respectively.

C.   DEHYDRATION OF 2-METHYLCYCLOHEXANOL

(Microscale)

See the previous discussion in Experiment B. In the following procedure, phosphoric acid is used as a catalyst; however, the quantity used also allows it to function as a "chaser." Because of the relatively high ratio of phosphoric acid, a separate water phase may not be seen in the distillate.

PROCEDURE

To a 5-mL round-bottom flask, add 1 mL of 2-methylcyclohexanol, 2 mL of 85% phosphoric acid, and a boiling stone. Mount the flask for simple distillation (see Fig. 5.6), slowly heat the contents of the flask to boiling, and distill out the product keeping the distillation temperature below 96°C. Continue the distillation until 0.5 to 0.7 mL of product is collected (a small test tube or centrifuge tube is a convenient receiver). Dry the product for 10 minutes over anhydrous Na_2SO_4, then transfer the dried liquid to a clean test tube so as not to get Na_2SO_4 in the syringe. Examine the dried liquid by gas chromatography as described in Experiment B.

QUESTIONS

1. What alkene(s) will be produced when each of the following alcohols is dehydrated:
   a.   cyclohexanol
   b.   2-methylcyclopentanol
   c.   3-methylcyclopentanol
   d.   2,3-dimethyl-2-butanol
2. Illustrate the mechanism for the acid-catalyzed dehydration of cyclohexanol.
3. Illustrate the mechanism for the acid-catalyzed dehydration of 2-methylcyclohexanol. How many products are possible?
4. Give equations to show how sulfuric acid can convert 4-methylcyclohexene into a mixture of four isomeric C_7H_{12} compounds (including starting material). Predict the relative amounts of each isomer.
5. Explain how the methylcyclohexene(s) can be distilled from the reaction mixture if the distillation temperature is kept well below 100°C.
6. The 2-methylcyclohexanol used in Experiments B and C is actually a mixture of two isomers. What are they? Is it likely that this makes a difference in the product composition? Explain your answer.
7. Write equations to illustrate the reaction of cyclohexene with $KMnO_4$.
8. Given the following data, calculate the theoretical yield and percentage yield of cyclohexene:
   Cyclohexanol used: 20.0 mL
   Cyclohexene obtained: 12.0 g

PRELABORATORY QUESTIONS

1. What alkene(s) will be produced when each of the following alcohols is dehydrated?
   a.   2-propanol
   b.   2-butanol

2. Why is phosphoric acid preferred over sulfuric acid for dehydration of alcohols?
3. What measures should be taken if droplets of phosphoric acid are splashed in the eyes?
4. Give an example of a skeletal rearrangement during a dehydration experiment.
5. What is the purpose of adding anhydrous sodium sulfate to the dehydration product in Experiment B?

REFERENCES

Banthrope, D. V. *Elimination Reactions;* Elsevier: New York, 1963.

Taber, R. L.; Champion, W. C. *J. Chem. Ed.,* 1967, *44*, 620.

Taber, R. L.; Grantham, G. D.; Champion, W. C. *J. Chem. Ed.,* 1969, *46*, 849.

Oxidation and Reduction

OXIDATION

Oxidation is a common and widely used reaction in organic chemistry. For example, primary alcohols may be oxidized to aldehydes or carboxylic acids, secondary alcohols to ketones, aldehydes to carboxylic acids, alkenes to glycols, aromatic side chains to carboxylic acids, and so forth. Alkenes, alkynes, and 1,2-glycols may be cleaved by oxidizing agents. These and numerous other reactions make oxidation of organic compounds an important category.

Two of the most important and general oxidizing agents are chromic acid and potassium permanganate. Both are strong oxidizing agents and can be used for a variety of purposes. Other oxidizing agents include hypochlorous acid, nitric acid, Tollens reagent, Benedict's reagent, and various reagents containing chromium trioxide such as the Jones reagent (CrO_3 in aqueous H_2SO_4), the Collins reagent (CrO_3 in pyridine), and the Corey reagent (pyridinium chlorochromate).

Alcohol oxidations are usually carried out with chromic acid, H_2CrO_4, generated either from the reaction of $Na_2Cr_2O_7$ or CrO_3 (chromic anhydride) with sulfuric or acetic acid. Permanganate in alkaline solution is more often the preferred reagent for side-chain oxidations or cleavage of a carbon chain at a double bond or carbonyl group (the latter via the enol). In both cases the oxidations proceed by the formation of intermediate esters with the reagent followed by breakdown to the product and a reduced inorganic species.

In the chromic acid oxidation of an alcohol, the intermediate chromate ester gives initially an unstable Cr(IV) species that reacts with Cr(VI) to give two equivalents of a Cr(V) species, $HCrO_3$. The $HCrO_3$ oxidizes additional alcohol and is finally reduced to Cr(III). The overall electron change for chromium is thus VI → III, or a gain of three electrons. The color change (orange to turbid green) in the oxidation provides a sensitive test to detect a primary or secondary alcohol; aliphatic aldehydes also give a positive test. Under controlled conditions, other functional groups, including tertiary hydroxyl groups, amines, aromatic aldehydes, and double bonds, do not react within the time-frame. The reaction is the basis for the determination of ethanol in exhaled breath in devices used by law-enforcement officials.

$$
\begin{array}{c}
\underset{R}{\overset{R}{}} \!\! \overset{OH}{\underset{H}{C}} + H_2CrO_4 \xrightarrow[-H_2O]{} \underset{R}{\overset{R}{}} \!\! \overset{O-CrO_3H}{\underset{H}{C}} \longrightarrow \underset{R}{\overset{R}{}}C{=}O + H_2CrO_3
\end{array}
$$

(unstable)
Cr(IV)

$$
\underset{R}{\overset{R}{}}C{=}O + Cr(III) \longleftarrow \overset{HCrO_3}{\underset{}{R_2CHOH}}
$$

Permanganate oxidation of an alkene or enol involves a cyclic permanganate ester. Under mild conditions, a glycol or ketol can be obtained; more vigorous oxidation gives carboxylic acids. The stable reduction product of permanganate in alkaline solution is MnO_2, which forms an insoluble brown precipitate. The electron change for this process is thus a gain of three; however, the actual stoichiometry may be more complex, since molecular oxygen is often formed in a side reaction during permanganate oxidations:

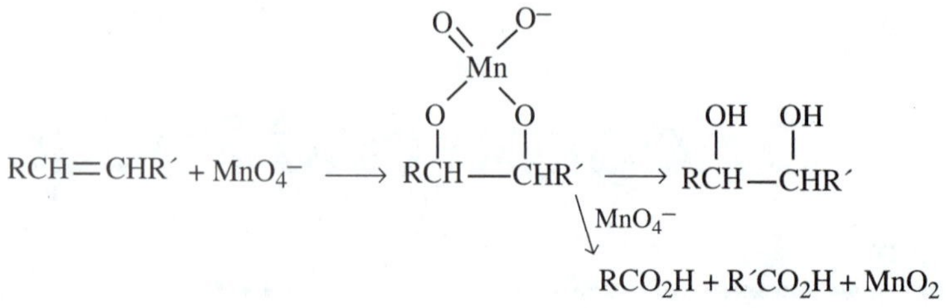

An important consideration in carrying out an oxidation reaction is the amount of oxidant required. For this, of course, a balanced equation for the oxidation is needed. To balance such a reaction, keep in mind that for any C—H or C—C bond that is broken (and replaced by a C—O bond), the molecule loses two electrons. For example, in the chromic acid oxidation of 2-propanol to acetone, one C—H bond is broken, with loss of two electrons. Since chromic acid gains three electrons, one then requires a ratio of three moles of alcohol (3×2) to two of chromic acid (2×3). The equation is completed by adding acid to balance charges and water to balance oxygen atoms:

$$3(CH_3)_2CHOH + 2H_2CrO_4 6H^+ \rightarrow 3(CH_3)_2C=O + 2Cr^{3+} + 8H_2O$$

In the vigorous oxidation of ethylbenzene with $KMnO_4$, assuming complete oxidation of the CH_3 group to CO_2, there are five C—H bonds and one C—C bond broken for an overall electron change of 12 electrons, requiring four moles of permanganate.

$$C_6H_5CH_2CH_3 + 4MnO_4^- \rightarrow C_6H_5CO_2^- + 4MnO_2 + 2H_2O + OH^- + CO_3^{2-}$$

In this chapter, procedures for a medium-scale oxidation and several smaller-scale oxidations are described. It should be noted that many oxidation reactions are exothermic and caution must be exercised in keeping the reaction under control. This is particularly important in larger-scale reactions.

In the oxidation of cyclohexanol (Experiment A), hypochlorous acid is used. Chromic acid has been widely used for this and similar reactions; however, chromic acid and its salts are suspected carcinogens so their use in other than small-scale reactions is not advisable. The procedure employs household bleach, a relatively safe material, as a source of the hypochlorite ion.

In the second reaction (Experiment B), cyclohexanone is oxidized further by permanganate with ring opening to give a dicarboxylic acid. The reaction is carried out on a microscale. Since the reaction is exothermic, careful addition of one of the reactants would be necessary to keep the reaction under control if it were being carried out on a large scale.

Pyridinium chlorochromate (PCC), used to convert primary alcohols into the corresponding aldehydes without contamination by the corresponding carboxylic acid, is prepared from CrO_3, pyridine, and HCl, and is commercially available. Experiment C is an example of this.

In the last oxidation experiment (Experiment D), the haloform reaction of acetophenone is illustrated. This interesting reaction results in the oxidation of the side chain to a carboxylic acid.

REDUCTION

The complex hydrides NaBH₄ and LiAlH₄ are among the most useful reagents for the conversion of carbonyl compounds to alcohols. Sodium borohydride is the less reactive of the two; for example, it does not reduce esters and acids. Also, sodium borohydride is more convenient to use, since reactions can be carried out in aqueous or alcoholic solutions. Lithium aluminum hydride, on the other hand, requires strictly water- and alcohol-free conditions. The experiments E and F included here use NaBH₄ as the reducing agent.

EXPERIMENTS

SAFETY NOTE

Alkaline solutions, strongly acid solutions, and chromic acid solutions (suspected carcinogen) are used in the following experiments. Take care to avoid contact with these reagents; if some is spilled on the hands, rinse thoroughly with water and wash with soap and water. Do not inhale the vapors of methylene chloride. Compounds containing chromium in the +6 oxidation state are carcinogens; handle them carefully.

Disposal. Dispose of all chromium and manganese waste products in a properly labeled container that will be provided by your instructor. All methylene chloride residues should be poured into the proper organic liquid container. In each case the proper notation of quantity and identity should be made on the container.

A.   PREPARATION OF CYCLOHEXANONE

(Macroscale)

cyclohexanol
bp 161°C

cyclohexanone
bp 155°C

PROCEDURE

In a 250-mL Erlenmeyer flask, place 10.0 g of cyclohexanol (10.4 mL, 0.1 mole) and then 25 mL of glacial acetic acid. Place the flask in an ice bath, and using a separatory funnel, add 75 mL of 5% NaOCl solution (Clorox) dropwise, with occasional swirling, while keeping the temperature at 30 to 35°C. Swirl the mixture well, return it to the ice bath, and add another 75 mL of 5% NaOCl, dropwise. At this point the mixture should give a positive reaction with potassium iodide–starch paper. If it does not, add more NaOCl solution (1 to 5 mL) until a positive reaction is obtained. Let the mixture stand 15 minutes at room temperature. Add saturated sodium bisulfite solution until the mixture gives a negative potassium iodide–starch test.

Transfer the mixture to a 250-mL distilling flask and distill until the distillate appears free of droplets (approximately 75 mL). It is convenient to collect the distillate in a 250-mL Erlenmeyer flask.

To free the distillate of any acetic acid, add anhydrous sodium carbonate carefully with stirring until foaming ceases. Then add 8 g of sodium chloride, and stir approximately 10 minutes to further salt out the cyclohexanone. Decant the mixture into a separatory funnel, and separate the layers; place the cyclohexanone in a 25-mL Erlenmeyer flask. Return the aqueous layer to the separatory funnel and extract it with 5 mL of ether. Separate the two layers, and add the ether extract to the cyclohexanone. Dry the product over anhydrous magnesium sulfate, decant through a cotton plug into a 25-mL distilling flask, and after distilling off the ether (CAUTION: fire hazard), collect the fraction boiling at 150 to 155°C in a tared receiver. Record the weight of product obtained, and calculate the percent yield. Examine the product by gas chromatography and infrared spectroscopy.

B.   PREPARATION OF ADIPIC ACID

(Microscale)

d 0.95
bp 155°C

adipic acid
bp 153-154°C

PROCEDURE

Place 245 mg (0.0025 mole) of cyclohexanone in a 50-mL Erlenmeyer flask, add a mixture of 790 mg of $KMnO_4$ (0.0050 mole) in 15 mL of water, and then add 0.1 mL (2 to 3 drops) of 10% NaOH. Let the mixture stand for 10 minutes with occasional swirling, and then place it in a boiling water bath for 20 minutes. Test for residual potassium permanganate by placing a drop of the reaction mixture from the tip of a stirring rod on a piece of filter paper. Unreacted potassium permanganate will appear as a purple ring around the brown manganese dioxide. If unreacted permanganate remains, decompose it by adding a small portion of solid sodium bisulfite.

Vacuum filter the mixture through a small Hirsch funnel to remove most of the manganese dioxide, and wash the manganese dioxide with approximately 2 mL of hot water. Place the filtrate in a 25-mL beaker, add a boiling stone, and boil it gently on a hot plate until the volume is reduced to approximately 5 mL. If the solution is colored at this stage, add decolorizing charcoal (after letting the solution cool), boil briefly, and remove the charcoal by filtration through a filter pipet. Add concentrated HCl to the clarified solution until it is acid to litmus, and then add an additional 10 drops of HCl. Allow the solution to cool to room temperature, and collect the product by vacuum filtration. Dry the product, and determine its weight and melting point.

C.   OXIDATION OF *PARA*-NITROBENZYL ALCOHOL

(Microscale)

Oxidant Preparation

Prepare the oxidant by combining equal weights of pyridinium chlorochromate, anhydrons, sodium acetate, and activated 4 A molecular sieve and grinding to a fine powder with a mortar and pestle.

p-nitrobenzyl alcohol   p-nitrobenzaldehyde

Oxidation. In a 50-mL Erlenmeyer flask with magnetic stirring, suspend 1.70 g (2.64 mmol) of the oxidant in 10 mL of methylene chloride, then add 200 mg *para*-nitrobenzyl alcohol. Stir the mixture for 30 minutes at room temperature. Check to see that the reaction is complete by TLC monitoring (30% ethyl acetate in petroleum ether, UV visualization). Add 2 grams of Florisil to the reaction mixture and stir vigorously for 5 minutes. Filter the reaction mixture and wash with 10 mL of ether. This removes the oxidant waste, which can be disposed of in the appropriate waste bottle. Transfer the filtrate to a round-bottom flask, add one gram of coarse (60–200 mesh) silica gel, and evaporate the suspension to dryness on a rotary evaporator. Retain the dry solid for column chromatography.

Pack a small amount of glass wool into the bottom of a 2-cm (i.d.) by 20-cm chromatography column, then secure it with a clamp. To the column add one gram of sand to cover the glass wool followed by 5 grams of silica gel (230–400 mesh). Tap the column to settle the packing and add the one gram of the coarse silica gel with the product on it to the top of the column. Cover the top with an additional gram of sand. Elute the column with 15% methylene chloride in petroleum ether. Collect 10 mL fractions in test tubes and check for the presence of the *para*-nitrobenzaldehyde by TLC as before. Combine the fractions containing the pure aldehyde in a tared round-bottom flask and concentrate them on a rotary evaporator to give the product. A typical yield is about 95% of a white solid (mp 185–187°C).

D.   HALOFORM REACTION

(Microscale)

benzoic acid
mp 122°C

PROCEDURE

Place 100 mg of acetophenone in a 15 × 125-mm test tube, add 4.0 mL of 5% NaOCl solution (Clorox or other household bleach), and then add 0.30 mL of 10% NaOH. Heat the mixture in an 80 to 90°C water bath for 15 to 25 minutes with frequent shaking; note when all the acetophenone appears to have dissolved. Add 4 drops of acetone to destroy excess hypochlorite, and cool to room temperature. Add concentrated HCl dropwise until the mixture is acid to litmus, and then add 2 drops in excess. Chill the acidified solution in an ice bath for 10 to 15 minutes to fully precipitate the benzoic acid. Vacuum filter the product, allow it to dry, and then determine the weight and melting point. Calculate the percent yield. If necessary, the benzoic acid may be recrystallized from water.

E.   REDUCTION OF BENZOPHENONE

(Microscale)

benzophenone   diphenylmethanol

Note: Dispense $NaBH_4$ from a small (15 to 25 g) bottle. Store the bottle in a dessicator or dessi-cator jar when not in use.

     In a 25-mL Erlenmeyer flask, place 0.64 g (0.0035 mole) of benzophenone and 4 mL of methanol. Swirl to dissolve. Slight warming might be necessary to dissolve the benzophenone; if so, allow the solution to come back to room temperature. Add 0.064 g (0.0017 mole) of sodium borohydride ($NaBH_4$) in one portion. Swirl to dissolve. Let the reaction mixture stand at room temperature with occasional swirling for 20 minutes (the solution should bubble or fizz slightly when swirled).

     After the reaction period, add 2 mL of cold water and heat intermittently with swirling on a steam bath or hot water bath for 5 minutes. Do not overheat and boil off solvent (mixture should boil gently when heated). Next, cool the mixture to room temperature and chill in an ice bath for 4 to 5 minutes; if appreciable solid does not separate, add 1 or 2 flakes of ice to the mixture. Filter the solid with vacuum in a small Hirsch funnel, suck dry, and allow to air dry. Determine the weight and the melting point. The melting point of diphenylmethanol is 68–69°C. If the product melts appreciably below this range, recrystallize it from 50% methanol-water.

     Determine the infrared spectrum of your product via a KBr pellet. Is the carbonyl absorption absent? Compare your spectrum with a spectrum of diphenylmethanol. Calculate the percent-age yield and submit the product to your instructor in an appropriately labeled container.

F.   REDUCTION OF *PARA*-NITROBENZALDEHYDE

(Microscale)

     Dissolve 30 mg (0.2 mmol) of *para*-nitrobenzaldehyde in 0.5 mL of ethanol and add 30 mg of sodium borohydride to the reaction mixture. Stir the reaction mixture and monitor it by TLC as in Experiment C. The reaction goes to completion very quickly. Compare the R_f values of the pure aldehyde and alcohol to the R_f values of those in the reaction mixture. Note the time when the aldehyde TLC spot no longer is visible.

QUESTIONS

1. Considering that the Cr in CrO_3 and $Cr_2O_7^{2-}$ is in the 6+ oxidation state and that each step in the sequence $CH_4 \rightarrow CH_3OH \rightarrow CH_2=O \rightarrow HCOOH \rightarrow CO_2$ corresponds to a two-electron change, complete and balance the following reactions:
   a. $CH_3CH_2CHOHCH_3 + Na_2Cr_2O_7 + H_2SO_4 \rightarrow CH_3CH_2COCH_3 + Cr^{3+}$
   b. $C_6H_5CH_2OH + CrO_3 \rightarrow C_6H_5COOH + Cr^{3+}$
   c. $CH_3CH_2{-}\underset{\underset{CH_3}{|}}{C}{=}CH_2 + H_2Cr_2O_7 \rightarrow CH_3CH_2COCH_3 + CO_2 + Cr^{3+}$

2. Write a balanced equation for the oxidation of cyclohexanone to adipic acid, and calculate the theoretical amount of permanganate needed for this procedure.

3. In an oxidation of cyclohexanol in which the temperature was allowed to rise above 50°C, the yield of cyclohexanone was low, and an acid by-product, with a melting point of 153 to 154°C, was isolated. Suggest the structure of this product.

4. In basic solution, sodium bisulfite reduces permanganate to manganese dioxide, whereas in acidic solution the product is manganous ion (Mn^{+2}). Write balanced equations for these reactions.

5. What organic product would you expect from the permanganate oxidation of the following ketones:
   a. acetophenone, **b.** 2-pentanone, **c.** 1-phenyl-2-propanone?

6. What organic product would you expect from $NaBH_4$ reduction of the following ketones:
   a. acetophenone, **b.** 2-pentanone, **c.** 1-phenyl-2-propanone?

PRELABORATORY QUESTIONS

1. When oxidations are carried out with $H_2Cr_2O_7$, usually excess oxidant is used. It is important to quench this excess reagent before the reaction is worked up. What might you use to quench the excess oxidant?

2. Why is $Cr_2O_7^=$ always used in acidic, not alkaline, solution?

3. Write a balanced equation for the oxidation of cyclohexanol by hypochlorous acid, HClO. (Note: The reduction product of hypochlorite ion is chloride ion.)

4. What is the structure of the organic product of the reaction of benzophenone with $NaBH_4$?

5. What safety precautions should be taken with $NaBH_4$?

REFERENCES

Chinn, L. J. *Selection of Oxidants in Synthesis;* Marcel Dekker: New York, 1971.

Corey, E. J.; Suggs, J. W. *Tetrahedron Letters,* 1975, 2647.

Jones, A. G. *J. Chem. Ed.,* 1988, *65*, 611.

Kafarski, P.; Ottenbreit, B.; Weiczorek, P.; Pawjowicz, P. *J. Chem. Ed.,* 1988, *65*, 549.

Perkins, R. *J. Chem. Ed.,* 1984, *61*, 551.

Perkins, R. A.; Chau, F. *J. Chem. Ed.,* 1982, *59*, 981.

Trahanovesky, W. S., Ed. *Oxidation in Organic Chemistry,* Parts B, C, D; Academic Press: New York, 1973, 1978, 1982.

Wiberg, K. B., Ed. *Oxidation in Organic Chemistry,* Part A; Academic Press: New York, 1965.

Zuczek, N. M.; Furth, P. S. *J. Chem. Ed.,* 1981, *58*, 824.

22

Aldehydes and Ketones

Aldehydes and ketones are two of several types of compounds that contain the carbonyl group. Reactions that occur because of the presence of the carbonyl group include nucleophilic addition reactions and base-catalyzed condensations. Aldehydes are also easily oxidized, which provides a convenient means to distinguish them from ketones.

The carbonyl group in aldehydes and ketones is highly polarized; the carbonyl carbon bears a substantial partial positive charge and is susceptible to nucleophilic attack. Further, since it is sp^2 hybridized it is relatively open to attack. Because the carbonyl carbon contains no good leaving group, nucleophilic addition occurs rather than nucleophilic substitution. Hydrogens attached to a carbon adjacent to the carbonyl group (α-hydrogens) are unusually acidic because of the polarization of the carbonyl group and the resonance stabilization of the resulting anion. The acidity of these α-hydrogens leads to a variety of base-catalyzed condensation reactions.

Because of the carbonyl group, aldehydes and ketones are polar compounds; however, the pure compounds do not undergo hydrogen bonding as alcohols do. The boiling points of aldehydes and ketones are therefore lower than alcohols of comparable molecular weight but higher than alkanes or ethers. The carbonyl oxygen will form hydrogen bonds with water so that low molecular weight aldehydes and ketones are appreciably soluble in water.

Aldehydes and ketones are compounds of considerable commercial and biological interest. Formaldehyde, acetaldehyde, and acetone are important industrial chemicals and the annual production of each is in the range of 1- to 3-million tons. Naturally occurring compounds include vanillin, camphor, cinnamaldehyde, and carvone (spearmint oil).

Each of the tests in the Experiments section will be performed on known compounds to familiarize you with the test procedure and the interpretation of results. You will also perform these tests on an unknown, to determine whether the compound is an aldehyde or ketone and, if it is a ketone, whether it is a methyl ketone. You may be instructed to identify the specific compound, in which case the data in Table 22.1 will be helpful.

EXPERIMENTS

SAFETY NOTE

Although only small amounts of reagents are used in this experiment, some are strongly acidic or basic and appropriate care should be taken in handling them.

Table 22.1   Derivatives of Aldehydes and Ketones

COMPOUND	2,4-DNP (°C)	SEMICARBAZONE (°C)
Acetone	126	187
Propanal	154	154
Butanal	123	106
2-Butanone	117	146
2-Pentanone	144	112
3-Pentanone	156	139
Cyclopentanone	146	203
Cyclohexanone	162	167
Heptanal	108	109
Benzaldehyde	237	222
Acetophenone	250	198
p-Tolualdehyde	234	215

Disposal. Care should be taken, as always, not to dispose of reaction mixtures or waste therefrom down the drain. Properly labeled waste containers will be provided by your instructor. Acidic or basic mixtures should be neutralized before disposal. Special containers should be provided for chromium waste (may be carcinogenic) and for the silver waste from the Tollens test.

A.   REACTIONS WITH NITROGEN COMPOUNDS

Aldehydes and ketones react with a number of nitrogen-containing compounds through nucleophilic addition and subsequent loss of water to give products that have a carbon–nitrogen double bond. These reactions are useful in distinguishing aldehydes and ketones from other functional groups and in the identification of specific aldehydes and ketones. Two reagents that frequently yield crystalline derivatives are 2,4-dinitrophenylhydrazine and semicarbazide.

1.   **2,4-Dinitrophenylhydrazones**   Prepare a solution of 4 drops of sample, or a comparable amount of solid, in 1 mL of ethanol. Add 1 mL of 2,4-dinitrophenylhydrazine reagent (see p. 90), mix thoroughly, and allow to stand. Many aldehydes and ketones give a solid derivative immediately. If no precipitate appears within 5 minutes, heat the solution in a hot water bath for 5 minutes and then allow it to cool to room temperature. The solid 2,4-dinitrophenylhydrazone (2,4-DNP) may be isolated by vacuum filtration, rinsed with a little cold water, and recrystallized from ethanol. Perform this test on benzaldehyde, methyl ethyl ketone, and your unknown.

2. Semicarbazones Add 4 drops of sample, or a comparable amount of solid, to 1 mL of semicarbazide reagent (see Chapter 46) and mix thoroughly. Allow the mixture to stand until the product has crystallized well; it may be necessary to chill it in an ice bath. Vacuum filter the solid semicarbazone. Prepare this derivative for benzaldehyde, methyl ethyl ketone, and your unknown.

$$
\underset{\substack{|\\R}}{\overset{\substack{H(R')\\|}}{C}}=O + H_2N-NH-\overset{\overset{\textstyle O}{\|}}{C}-NH_2 \longrightarrow \underset{\substack{|\\R}}{\overset{\substack{H(R')\\|}}{C}}=N-NH-\overset{\overset{\textstyle O}{\|}}{C}-NH_2 + H_2O
$$

B. OXIDATION REACTIONS

Aldehydes are very easily oxidized to yield carboxylic acids or their salts if the oxidation is done in basic solution. Since ketones are not readily oxidized, this difference provides a convenient means to distinguish aldehydes from ketones. Strong oxidizing agents such as chromic acid are frequently employed; however, much less vigorous oxidizing agents such as Tollens reagent have been used.

1. Chromic Acid Reagent Dissolve 1 drop of sample in 1 mL of reagent-grade acetone, and to this solution, add 1 drop of chromic acid reagent (Chapter 18, Experiment D). Rapid formation of a green to blue-green precipitate constitutes a positive test for an oxidizable compound, that is, in this case an aldehyde. Aliphatic aldehydes react usually within 15 seconds, whereas aromatic aldehydes require approximately 1 minute. Ketones react only after several minutes. Note that other relatively easily oxidized compounds, such as primary and secondary alcohols, also react readily. Perform this test on propionaldehyde, benzaldehyde, methyl ethyl ketone, and your unknown.

$$
3R-\overset{\overset{\textstyle O}{\|}}{C}-H + 2CrO_3 + 6H^+ \longrightarrow 3R-\overset{\overset{\textstyle O}{\|}}{C}-OH + 2Cr^{3+} + 3H_2O
$$

2. Tollens' Silver Mirror Test

SAFETY NOTES

1. This reagent must be prepared immediately before use because it can form silver fulminate if stored. All reagent residues and the silver mirror should be destroyed by adding nitric acid.
2. Silver nitrate can stain the skin black. Use gloves!
3. The residual silver ions can be precipitated as silver chloride by hydrochloric acid, filtered, and discarded as directed by your instructor.

Reagent Preparation: In a clean dry test tube add 1.0 mL of 0.05 M silver nitrate. Add dropwise approximately 5 drops of 3.0 M sodium hydroxide until a gray precipitate of silver oxide (Ag_2O) appears. Mix the solution thoroughly and add another drop of the NaOH solution to see if further precipitate forms. Do not add excess NaOH. When the silver oxide precipitation is complete, add dropwise and with vigorous stirring, 2% aqueous ammonia until the silver oxide precipitate just disappears. DO NOT ADD EXCESS aqueous ammonia or the reagent will not work properly. Dilute the mixture (1:1) with deionized or distilled water.

Add 2 drops of sample to 1 ml of the reagent in a thoroughly clean test tube, and let stand for 10 minutes. (For water-insoluble aldehydes, dissolve 2 drops of sample in 0.5 mL of 95% ethanol, then add 1 mL of the reagent.) If the test is negative, place the test tube in a boiling water bath for 1 minute. The formation of a silver mirror on the wall of the tube through reduction of silver ion constitutes a positive test for aldehydes. Discard all unused reagent and test solutions *immediately* after use, as these solutions can become explosive on standing. Excess Tollens' reagent can be destroyed by the addition of dilute nitric acid. Perform this test on propionaldehyde, benzaldehyde, methyl ethyl ketone, and your unknown.

$$\underset{\substack{|| \\ R-C-H}}{\overset{O}{}} + 2Ag(NH_3)_2\,OH \longrightarrow \underset{\substack{|| \\ R-C-O^-NH_4^+}}{\overset{O}{}} + 2Ag + H_2O + 3NH_3$$

C. IODOFORM TEST

Methyl ketones can be distinguished from other ketones by their reaction with iodine in basic solution to yield iodoform. Alcohols, such as 2-propanol and 2-butanol, that can be oxidized to methyl ketones also give a positive reaction.

In a small test tube, dissolve 3 drops of liquid unknown, or about 100 mg of solid, in 2 mL of water (if insoluble in water, add 1 to 2 mL of 1-propanol and warm) and then add 1 mL of 10% NaOH. Add dropwise, with shaking, a solution of iodine in aqueous potassium iodide until a dark iodine color persists. Note the formation of any yellow precipitate (CHI_3). If no precipitate appears within 3 to 5 minutes, warm the solution slightly, and if the color fades, add more I_2/KI until the color remains for 1 to 2 minutes. Then add a few drops of NaOH to remove the excess iodine, and dilute the mixture with cold water. Iodoform, if present, will precipitate as a yellow solid, mp 119 to 121°. Perform this test on methyl ethyl ketone and your unknown.

$$\underset{\substack{|| \\ R-C-CH_3}}{\overset{O}{}} + 3I_2 + 4NaOH \longrightarrow \underset{\substack{|| \\ R-C-O^-Na^+}}{\overset{O}{}} + CHI_3 + 3NaI + 3H_2O$$

D. INFRARED SPECTRA

PROCEDURE

1. Record the infrared spectrum of benzaldehyde (neat) between salt plates. Annotate the bands due to (a) the carbonyl stretch, (b) the aromatic C—H stretch, and (c) the two bands due to the aldehydic C—H stretch. Also examine the spectrum to determine whether benzoic acid is present. Aldehydes are rapidly air oxidized.
2. Record the infrared spectrum of your unknown (neat) between salt plates. Annotate the absorption band due to the presence of a carbonyl group. Is the unknown an aldehyde or is it a ketone? Is the unknown aromatic or is it aliphatic? Annotate your spectrum to answer these questions.

E. CROSSED ALDOL CONDENSATION

(Microscale)

In a centrifuge tube, combine 640 mg of benzaldehyde (0.006 mole), 170 mg of acetone (0.003 mole), and 3 mL of 95% ethanol. Shake to dissolve the components, and then add 1 mL of 10% sodium hydroxide solution. Shake the mixture until a precipitate begins to form, and then let it stand with occasional shaking for 20 minutes.

$$2 \; \text{C}_6\text{H}_5{-}\text{CHO} + \text{CH}_3{-}\overset{\overset{\text{O}}{\|}}{\text{C}}{-}\text{CH}_3 \xrightarrow{\text{HO}^-}$$

$$\text{C}_6\text{H}_5{-}\text{CH}{=}\text{CH}{-}\overset{\overset{\text{O}}{\|}}{\text{C}}{-}\text{CH}{=}\text{CH}{-}\text{C}_6\text{H}_5 + 2\text{H}_2\text{O}$$

dibenzalacetone
mp 110–111°C

Cool the mixture in an ice bath for 5 to 10 minutes, and then remove the liquid by using a Pasteur pipet with some cotton in the tip pressed against the bottom of the test tube or centrifuge the mixture and remove the supernatant liquid using a Pasteur pipet. Add 2 mL of ice water to the centrifuge tube, stir to wash the crystals thoroughly, and remove the water using a Pasteur pipet. Repeat with another 2 mL of ice-water wash. After removing as much of the water as possible, recrystallize the solid in the centrifuge tube using 95% ethanol.

Collect the product by vacuum filtration, and allow it to dry. Determine the weight, melting point, and percent yield. Submit the product to your instructor in a properly labeled container.

QUESTIONS

1. Using the tables at the end of this text, list the following compounds in order of increasing boiling point: butanal, 2-butanone, butyl alcohol, and diethyl ether. Explain the molecular basis of this order.
2. Write a balanced equation for the oxidation of tolualdehyde with CrO_3.
3. Write a stepwise mechanism to illustrate the "crossed aldol condensation" of benzaldehyde and acetaldehyde.
4. Distinguish between the following pairs by simple chemical tests. Tell what you would do and see.
   a. pentanal and 2-pentanone
   b. 2-pentanone and 3-pentanone
   c. pentanal and 1-pentanol
5. How would you distinguish between the following pairs by use of infrared spectroscopy only? (See Chapter 11.) Tell what absorption would be present or absent in each case.
   a. 2-pentanone and 2-pentanol
   b. acetophenone and p-tolualdehyde
   c. 3-pentanone and benzophenone
6. How would you distinguish between the pairs in Question 5 (above) using only proton NMR spectroscopy? (See Chapter 13.) Sketch a possible spectrum for each.

PRELABORATORY QUESTIONS

1. Write equations for the reaction of 2,4-dinitrophenylhydrazine with
   a. butanal
   b. cyclohexanone
2. Write an equation for the reaction of semicarbazide with acetophenone to form a semicarbazone. What is the melting point of this derivative?
3. Write an equation for the reaction of the Tollens Reagent with p-tolualdehyde.
4. Write the structure of the organic product resulting from the reaction of basic KI/I_2 reagent with 2-pentanone.
5. What should be done with the waste from the Tollens test? How do you destroy excess reagent?

REFERENCE

Hathaway, B. A. *J. Chem. Ed.,* 1987, *64*, 367.

Determination of pK_a: Effects of Substituents on Acidity

23

The acidity or acid strength of carboxylic acids, not to be confused with acid concentration, is a property often used to examine the relationship between structure and chemical properties. In a series of acids, RCO_2H, with different R groups, measurement of the acidity provides a direct means of determining the influence of the group R. In this experiment, the effect of substituents on acidity will be determined for an unknown aliphatic acid and then for a series of benzoic acids.

The *acid strength* of a carboxylic is expressed quantitatively as the **equilibrium constant, K_a,** for the dissociation of the acid in water. (Another way of looking at this is to describe, in Bronsted–Lowry terms, the ability of the acid to donate a proton to the base, water.)

$$RCO_2H + H_2O \rightleftharpoons RCO_2^- + H_3O^+ \tag{23.1}$$

$$K_{eq} = \frac{[RCO_2^-][H_3O^+]}{[RCO_2H][H_2O]}$$

$$K_{eq}[H_2O] = \frac{[RCO_2^-][H_3O^+]}{RCO_2H} = K_a$$

or

$$K_a = \frac{[H_3O^+][RCO_2^-]}{[RCO_2H]} \tag{23.2}$$

A logarithmic form of Equation 23.2 is usually used, just as hydrogen ion concentration (actually hydrogen ion *activity*) is expressed in pH units. Recall that **pH** is defined as the negative log of the hydrogen ion activity. For purposes of this experiment it will be assumed that the activity and concentration of the hydrogen ion are the same. Equation 23.2 expressed in negative logarithmic terms becomes the following:

$$-\log K_a = pK_a = -\log\left(\frac{[H_3O^+][RCO_2^-]}{[RCO_2H]}\right) \tag{23.3}$$

Simplifying and rearranging:

$$pK_a = pH - \log\left(\frac{[RCO_2^-]}{[RCO_2H]}\right) \tag{23.4}$$

or

$$pH = pK_a + \log\left(\frac{[RCO_2^-]}{[RCO_2H]}\right) \tag{23.5}$$

Using Equation 23.5, it can be shown that the value of the pK_a of an acid is the pH at which the anion and the acid are at the same concentration (that is, the acid is half-neutralized). In other words, if

$$[RCO_2^-] = [RCO_2H]$$

then

$$\log\left(\frac{[RCO_2^-]}{[RCO_2H]}\right) = 0$$

and

$$pK_a = pH$$

The method for determining pK_a by the titration of an acid with a base is illustrated in Figure 23.1. The pH is measured with a pH meter. Note that the equivalent weight of an unknown acid can be determined at the same time if the weight of acid and the concentration of base are known with reasonable accuracy.

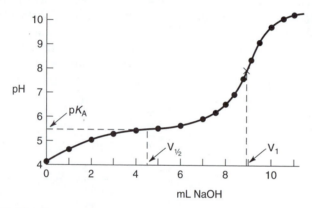

Figure 23.1   pK_a titration curve.

DETERMINATION OF THE pK_a OF AN UNKNOWN SUBSTITUTED BENZOIC ACID

(Microscale)

An unknown substituted benzoic acid will be provided.

PROCEDURE

From the sample of the substituted benzoic acid provided, weigh out approximately 20 mg and dissolve it in 25 mL of ethanol in a 100-mL beaker. Dilute the solution with 25 mL of water, and titrate with 0.01 M sodium hydroxide solution using a pH meter. Carry out duplicate analyses, and plot the results as shown in Figure 23.1. The instructor will demonstrate the use and standardization of a pH meter. Use it with care.

Rinse a 50-mL buret with small portions of standardized NaOH solution (0.01 M), and then fill the buret with the standard solution. Record the exact molarity of the NaOH and the initial buret reading.

Position the pH meter electrodes so that they dip sufficiently into the acid solution. Read and record the pH. Add approximately 1 mL of NaOH, swirl the solution gently, then accurately read and record the buret volume and the pH of the mixture being titrated. Repeat the addition of NaOH, swirling, reading, and recording using 1- to 1.5-mL additions of NaOH until the pH begins to change rapidly. At this point, add the NaOH in 3- to 5-drop increments until pH 9 is reached. Then add additional 1-mL increments to continue the titration approximately 5 mL past the end point where the curve starts to flatten out. Carefully graph the pH versus the volume of base. From the graph determine the equivalence point (V_1), which is the point where the curve is the steepest, and the half-equivalence point ($V_{1/2}$) of the titration. Using the half-equivalence point, determine the pK_a of the unknown acid, and from the volume of base at the equivalence point, calculate the equivalent weight (neutralization equivalent) of the acid.

After completing duplicate titrations that agree to within ±2%, empty the buret and rinse it with distilled water, and record your data on the Data Report Sheet.

Report the pK_a of your acid to your instructor, who will tabulate the results obtained on all the acids given out. Order this list of acids according to the acidity. (*Note:* Higher acidity = larger K_a = smaller pK_a.) Explain the order and any differences between the effect of *meta-* and *para*-substitution.

QUESTIONS

1. It is not necessary that the pH at half-neutralization be used to calculate the pK_a. Use Equation 23.5 and your titration curve to calculate the pK_a of your acid from the pH of the solution when the acid is 75% neutralized. How does the result compare with the value calculated in the experiment? What does any difference indicate?
2. A solution of acetic acid, pK_a 4.7, has been partially neutralized with base to a pH of 6.0. Calculate the percentage of the acid that is present.
3. Circle the stronger acid in each pair:
    a. *para*-chlorobenzoic acid or benzoic acid
    b. 2-chlorobutanoic acid or 3-chlorobenzoic acid
    c. butanoic acid or 3-bromobenzoic acid
4. The pH of normal blood is approximately 7.4. What is the hydrogen ion concentration?

PRELABORATORY QUESTIONS

1. Using the list of acids in the tables at the end of Chapter 46 and CRC *Handbook of Chemistry and Physics,* list four weak acids and the pK_a of each.
2. What is the major organic species present in an 0.1 *M* solution of acetic acid: unionized acetic acid or acetate ion?
3. Calculate the pK_a of an acid with K_a of 6.3×10^{-5}.
4. What is the K_a of an acid with pK_a of 3.6?
5. Which is the *stronger acid:* acid A, pK_a = 4.55, or acid B, pK_a = 2.33? Explain your answer.

REFERENCE

King, J. F. In *Techniques of Organic Chemistry;* Bentley, K. W., Ed.; Interscience: New York, 1963; Vol. XI, part 1; pp. 318–370.

Carboxylic Acid Mini-Unknown

This experiment is intended to serve as an introduction to qualitative organic analysis and to illustrate how a variety of techniques might be useful in determining the identity of a compound. In this experiment, the unknown is drawn from the set of mono and disubstituted benzene derivatives, where the substituent(s) can be methyl, methoxy, bromo, or chloro. With four possible substituents, and the possibility of 1,2-, 1,3-, or 1,4-substitution when two substituents are present, the list of potential unknowns is quite extensive.

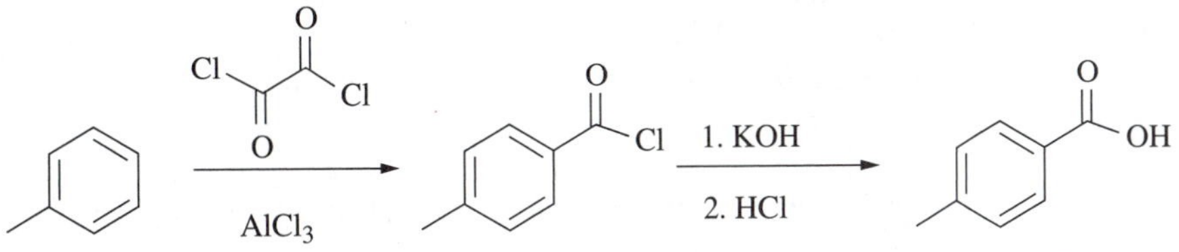

Figure 24.1   Conversion of toluene to p-toluic acid.

To deduce the structure of the unknown, the arene, illustrated by toluene **1**, is reacted with oxalyl chloride and AlCl$_3$, to give the acid choride **2**. Hydrolysis of **2** gives the crystalline acid **3**. By titration with 0.1 M aqueous NaOH (standardized by titration against benzoic acid), you will be able to establish the molecular weight of your product acid, and so determine which substituents were attached to the benzene ring in your initial unknown. For monosubstituted unknowns, the melting point of the acid will confirm the substituent deduced from the equivalent weight. For disubstituted unknowns, the melting point of the product acid will also allow the assignment of the pattern of the substituents on the benzene ring.

There are two limitations to this procedure. The first is that the initial unknown cannot be dimethoxy (an electron transfer reaction intervenes) or dihalogen (unreactive). The other limitation is that the acid **3** must be *water-free* before the titration. This requirement is easily accomplished by letting the acid oven-dry between laboratory periods. Although one might expect the initial electrophilic aromatic substitution to give mixtures of regioisomers, it in fact appears to be highly selective, converting monosubstituted aromatics to the 1,4-products.

EXPERIMENTS

Acid Preparation: To 1.0 g of the unknown aromatic compound in a 50-mL Erlenmeyer flask add 10 mL of CH_2Cl_2 containing 10 weight % of oxalyl chloride (in hood! **CAUTION:** Corrosive!). Stir the mixture while it cools in an ice-water bath. Add 1.0 g of anhydrous $AlCl_3$ (**CAUTION:** Corrosive! Do not breathe dust!) in portions with swirling. Allow the reaction to subside after each addition. If the $AlCl_3$ does not visibly react (dissolution, gas evolution), add the rest of it and then warm the mixture to room temperature until a reaction is observed. Most of the $AlCl_3$ should dissolve, to provide a clear orange-yellow solution.

When the reaction has stopped bubbling, pour the mixture over 50 mL of crushed ice in a beaker, and rinse the reaction flask with a little additional CH_2Cl_2. Swirl the mixture until the orange color fades to a pale yellow. Separate the layers, extract the aqueous layer with 10 mL of CH_2Cl_2, and concentrate the combined CH_2Cl_2 extracts *in vacuo* or by distillation.

To the concentrated residue, add 20 mL of 1 *M* aqueous KOH, and stir for 15 min at RT. Extract the KOH solution with 2 × 10 mL of $CHCl_3$, and set the extract aside to recover the unreacted unknown, if necessary. Acidify the aqueous phase with concentrated aqueous HCl to pH = 1, and filter the precipitated acid. Wash the solid acid with cold distilled water, and let it thoroughly air dry (drying it in an 80° C oven is even better) overnight before titration. A portion of the acid should be recrystallized from ethanol/water to constant melting point.

If the acid does not precipitate directly on acidification, extract the aqueous layer with 3 × 10 mL of $CHCl_3$, dry (Na_2SO_4), and concentrate on the rotary evaporator.

Acid Titration: Precisely standardize 0.1 *M* aqueous NaOH by titration of benzoic acid in ethanol, using a phenolphthalein (or bromthymol blue) endpoint. Dissolve your dry, carefully weighed unknown acid in ethanol, and titrate it using the standardized aqueous NaOH. From the titration, calculate the exact molecular weight of the unknown acid.

Spectroscopic Data: IR and/or NMR data may also be used to confirm the structural assignments both of the product benzoic acid and of the starting unknown substituted benzene.

QUESTIONS

1. What is the molecular weight of hexanedioic acid? What is its equivalent weight?
2. If a 0.200-g sample of a carboxylic acid requires 14.50 mL of 0.100 M NaOH for titration, what is its milliequivalent weight? What is its equivalent weight?
3. Without actual spectra of both compounds, would ^1H NMR be useful in distinguishing 4-bromobenzoic acid from 4-chlorobenzoic acid? What about ^{13}C NMR? Would determination of the equivalent weight differentiate between them? Explain.
4. Would determination of the equivalent weight be useful to distinguish between *p*-toluic acid and phenylacetic acid? Would proton NMR be useful? Explain.

PRELABORATORY QUESTIONS

1. Write an equation for the reaction of benzoic acid with sodium bicarbonate solution.
2. What is the equivalent weight of hexanoic acid?
3. If 15.00 mL of 0.100 N NaOH are needed to titrate an unknown acid sample, how many milliequivalents of NaOH are used? How many milliequivalents of acid were present?

4. Tell what absorptions you would look for in the infrared spectrum of an unknown to determine whether or not it was a carboxylic acid.

5. How can a water-insoluble unknown be determined to be a carboxylic acid or a phenol on the basis of solubility tests?

REFERENCES

Osman, M. A. *Helvetica Chim. Acta,* 1982, *65*, 2448.

Taber, D. F.; Nelson, J. D.; Northrop, J. P. *J. Chem. Educ.,* 1999, *76*, 828.

25

Esterification and Saponification: Effects of Substituents on Reactivity

In addition to having great utility in the understanding of reaction mechanisms, the study of rate constants also provides an insight into the effects (steric, inductive, or resonance) that substituent groups have on molecular interactions. By taking a particular reaction and by making changes in the nature of the substituents, a correlation of structure and reactivity may be made. Much of our present-day understanding of these phenomena is attributed to the pioneering work of such eminent chemists as Brönsted, Bartlett, Hinshelwood, Conant, and Norris. A landmark book by L. P. Hammett opened the door to further refinements by Taft, Leffler, and others. The correlation of thermodynamic and kinetic effects has led to a better understanding of molecular dynamics, especially in the field of enzyme-catalyzed reactions. Knowledge of the details of how molecules interact and the effect of a minute change in the structure has led to a wholly new way to approach the fields of medicinal chemistry and pharmacology.

Hammett proposed the correlation of the effect of substituents on the acidity of benzoic acids (see Chapter 23) with the effect of those substituents on the rate constants of the hydrolysis of benzoic acid esters. He found a reasonably good correlation of the *para* and *meta* acids and esters. The ortho derivatives, however, did not show the same relationship (the *ortho* effect).

In this experiment, different esters will be hydrolyzed by different students, and the results of all will be compared to see if results similar to those of Hammett can be obtained. If a good selection of methyl benzoates is not available, they may be synthesized by the method described in Experiment A.

$$\text{(benzene ring)}-\overset{\overset{\displaystyle O}{\|}}{C}-\text{OMe} + HO^- \overset{k}{\longrightarrow} \text{(benzene ring)}-\overset{\overset{\displaystyle O}{\|}}{C}\overset{-}{\underset{O}{}} + MeOH \qquad (25.1)$$

EXPERIMENTS

A. PREPARATION OF METHYL BENZOATES

(Microscale)

SAFETY NOTE

Concentrated sulfuric acid causes burns on contact with skin; rinse immediately with large quantities of cold water and then wash with soap and water. Get assistance from a physician if necessary. Ethers are flammable; use with caution.

PROCEDURE

Place 0.5 g of the acid assigned to you in a 25-mL Erlenmeyer flask, add 5.0 mL of methanol, and add 10 drops of BF_3 etherate. Warm the mixture for 10 minutes. The progress of the reaction can be checked by TLC on silica gel.

The solvent or solvent mixture required for this will depend on the acid being used and will have to be determined individually. Before taking the first sample, find a solvent that barely, but perceptibly, moves the acid from the origin. The ester, being less polar, will have a larger R_f (see Chapter 8) and will move well up the plate.

When the reaction is complete as measured by TLC analysis, cool the solution to room temperature and transfer it to a 25.0-mL separatory funnel. Add 5.0 mL of diethyl ether and 7.5 mL of water and shake well. Allow the layers to separate, and discard the aqueous layer. Rinse the organic layers successively with 2.5 mL of water, twice with 2.5 mL of 10% $NaHCO_3$ solution, and once with 2.5 mL of saturated NaCl solution. Dry the organic layer over Na_2SO_4 for 5 to 10 minutes, remove the drying agent by filtration, rinse it on the funnel with ether, and evaporate the ether from the filtrate. The aqueous extracts can be neutralized and flushed down the drain with ample water.

Most of the assigned esters [H, p-Br, p-Cl, m-Cl, p-Me, m-Me, p-MeO, m-MeO, p-NO_2, m-NO_2] can be obtained as solids and purified by recrystallization from hexane or a mixture of toluene and hexane (exceptions, of course, are the liquids methyl benzoate, methyl m-toluate, methyl m-chlorobenzoate, and methyl m-anisate).

Some of the solid esters have relatively low melting points, so their recrystallization must be done at ice bath temperatures. For use in the following experiment, the esters must be completely dried of the recrystallization solvent. Those that are oils should be distilled from a small distilling flask. Because of their high boiling points (190 to 260°C), a condenser will not be needed; the receiver should be cooled in ice. Report the observed and literature melting (or boiling) range of the ester and the percent yield.

B.   RATE MEASUREMENTS

PROCEDURE

Weigh out 0.0100 mole of the known methyl ester supplied by your instructor into a tared 125-mL Erlenmeyer flask; the weight should be measured accurately and should differ from 0.0100 mole by no more than 1%. Accurately weigh 60.0 mL of reagent grade acetone (d = 0.791) into the flask, and swirl the mixture to dissolve the ester completely. Suspend the flask with a buret clamp in an ice-water bath. During the reaction, ice should be added to the bath as necessary to keep the temperature at 0°C.

In a second Erlenmeyer flask accurately weigh out 40.0 mL of 0.25 M NaOH solution (d = 1.011) and cool to 0°C. To start the reaction, add this base to the ester solution and mix well by swirling. Note the time, using a watch or clock with a second hand, when the base is added. With a 10-mL pipet, remove 10-mL aliquots of the solution at regular intervals. (The length of the intervals will depend on which ester you are using. See your instructor for directions on this point.) Add each sample to a separate 100-mL flask containing 10 mL (measured by buret) of 0.100 M HCl, and note the time when the solution is added. This quenches the reaction.

Add 3 drops of phenolphthalein indicator solution to each flask, and titrate each with 0.010 M NaOH solution to a persistent pink end point. For each sample, record the time when it was taken and the volume of titrant required. The acid being titrated is the benzoic acid liberated in the hydrolysis.* Using the volume and molarity of titrant, calculate the number of moles of

---

*This is so because after x percent of the ester has reacted, there are in an aliquot $0.001(1 - x/100)$ mole each of ester and OH^-, and $0.001(x/100)$ mole of benzoate anion. Addition of 10 mL of 0.10 M H^+ gives $0.001(1 - x/100)$ mole of H_2O (from OH^-) plus $0.001(x/100)$ mole of the benzoic acid.

unreacted ester remaining. The concentration of ester can be calculated simply by dividing the number of moles by the sample volume.

C. CALCULATION OF THE RATE CONSTANT FOR HYDROLYSIS

PROCEDURE

Since the slow (rate-determining) step in the reaction in Equation 25.1 is the attack of HO^- on the ester (Eq. 25.2), the rate of disappearance of ester is proportional to the product of the concentrations of these two species (Eq. 25.3).

$$R-\overset{\overset{O}{\|}}{\underset{\underset{OMe}{}}{C}} + HO^- \xrightarrow{\text{slow}} R-\overset{\overset{O^-}{}}{\underset{\underset{OH}{|}}{\underset{OMe}{C}}} \xrightarrow{\text{fast}} R-\overset{\overset{O}{\|}}{\underset{\underset{O^-}{}}{C}} + MeOH \tag{25.2}$$

$$\text{rate} = -\frac{d[\text{ester}]}{dt} = k[HO^-][\text{ester}] \tag{25.3}$$

In this experiment $[HO^-] = [\text{ester}]$, so it is possible to simplify Equation 25.3 to Equation 25.4

$$\frac{-d[\text{ester}]}{dt} = k[\text{ester}]^2 \tag{25.4}$$

After integration of this differential form of the rate equation, Equation 25.5 is obtained: $[\text{ester}]_t$ is the concentration of ester after t seconds of reaction.

$$\frac{1}{[\text{ester}]_t} = kt + \frac{1}{[\text{ester}]_0} \tag{25.5}$$

The value of k can be obtained by substituting values of $[\text{ester}]_t$, t, and $[\text{ester}]_0$ (which equals 0.1 M) into Equation 25.5. However, it is simpler and more accurate to treat Equation 25.5 as that of a straight-line plot of $[\text{ester}]_t^{-1}$ versus t, whose slope is k. Therefore, to obtain k for your ester, plot the values of $[\text{ester}]_t^{-1}$ (Column 7 in Table 25.1) versus time (Column 2) on graph paper with appropriate scales, and then with a straightedge, draw a line that passes as close as possible to all of the points. (Because of experimental errors, you should not expect that the line will go through each of the points.) Using two well-separated points on the line (not necessarily data points), (X_1, Y_1) and (X_2, Y_2), calculate the slope of the line $k = (Y_2 - Y_1)/(X_2 - X_1)$. Since the ordinate (Y) is in units of M^{-1} (liters/mole) and the abscissa (X) is in units of sec, the rate constant has units of $M^{-1} \sec^{-1}$.

After everyone in the class has calculated a value of k for a different ester, the results can be collected and compared. For this purpose, it is useful to prepare a table of esters and rate constants in order of increasing values of k. How does this list compare with the corresponding list of acids and pK_as from the previous experiment? Why? If a more quantitative comparison of the two sets of quantities seems justified, you can make a graph of the values of $\log k$ for the esters versus the pK_as $(-\log K_a$s) of the corresponding acids.

Table 25.1 Data Format

Sample Number	Time Taken (X)	Vol of 0.01 M NaOH	mmoles of Acid	mmoles of Ester	Concentration of Ester (M)	(Concentration of Ester)$^{-1}$ Y

QUESTIONS

1. How many moles of benzoic acid are present in a sample that requires 30 mL of 0.200 M NaOH for titration?
2. How many grams of benzoic acid are present in a sample that requires 20 mL of 0.500 M NaOH for titration?
3. How many millimoles of ester are present in a 10-mL aliquot of the reaction mixture in this experiment if 10 mL of 0.010 M NaOH is needed for titration?
4. How would you expect the acid strength (as reflected in K_a values) of benzoic acid to compare with p-nitrobenzoic acid? Explain.

PRELABORATORY QUESTIONS

1. Write an equation for the basic hydrolysis (saponification) of methyl benzoate.
2. List three electron-donating groups and three electron-withdrawing groups.
3. Draw the structure of the following compounds:
    a. methyl p-nitrobenzoate
    b. methyl p-toluate
4. In the saponification of 0.0100 mole of methyl benzoate with 0.0100 mole of NaOH, how many moles of HO^- remain after 20% reaction?
5. Write the rate equation for a second-order reaction.

REFERENCES

Hammett, L. P. *Chem. Rev.,* 1935, *17,* 125.

Moore, J. W., Pearson, R. G. *Kinetics and Mechanism,* 3rd ed.; John Wiley & Sons: New York, 1981; pp. 12–26.

26

Friedel-Crafts Alkylation and Acylation

The attachment of alkyl groups to an aromatic ring is an extremely important reaction, and is carried out on an industrial scale to produce compounds such as cumene (isopropylbenzene) in multimillion-pound amounts each year. Alkylation involves the attack of an aromatic ring by an electrophilic carbon species, the choices of reagent and conditions depending on the alkyl group and the reactivity of the ring.

The Friedel-Crafts reaction using an alkyl halide and aluminum halide is a general method for such aromatic alkylation. Other important variations of this reaction involve the generation of the electrophile from an alcohol or alkene. However, there are some practical limitations. Since these reactions involve the development of a positive charge on the alkyl carbon atom, carbocation rearrangements may occur. Furthermore, the ring becomes more susceptible to substitution (more nucleophilic) as alkyl groups are introduced, and there is thus a tendency for polysubstitution. Finally, "scrambling" of ring-attached alkyl groups may occur, when a thermodynamically more stable product can arise from one that is formed initially in a faster reaction.

The first experiment in this chapter involves the preparation of an antioxidant. Such a compound inhibits the process of autoxidation, which is the reaction of organic compounds with oxygen in the air. Autoxidation leads to the incorporation of oxygen, oxidative degradation, and cross-linking and is the cause of deterioration of foods, plastics, rubber, and many other materials.

To slow down autoxidation, inhibitors have been developed that trap peroxy radicals in the propagation step of the radical chain. Since the process has a long chain length, one molecule of inhibitor can interrupt a cycle that would otherwise lead to many molecules of hydroperoxide. Among the most effective antioxidants are phenols with a *para* substituent and one or two bulky alkyl groups at the *ortho* positions. These substituents stabilize the initially formed aryl radical until it can react with a second radical to form a stable end product.

Two phenols that are nontoxic and are permitted for use as food additives are 2-*t*-butyl-4-methoxyphenol and 2,6-di-*t*-butyl-4-methylphenol. When used as antioxidants these compounds are called **BHA** and **BHT,** respectively, abbreviations for the very unsystematic names "butylated hydroxyanisole" and "butylated hydroxytoluene."

The first experiment involves the preparation of the antioxidant BHT (2,6-di-*t*-butyl-4-methylphenol) by alkylation of the highly activated ring in *p*-cresol with *tert*-butyl alcohol and sulfuric acid. The conditions and molar ratio of reactants are critical in this reaction, since by-products can be formed that interfere with the isolation of the di-*t*-butyl-*p*-cresol. One obvious possibility is the monosubstituted compound, 2-*t*-butyl-*p*-cresol, which becomes the major product if the concentration of the sulfuric acid is reduced from 96% to 75%. Disubstitution is favored by high acid strength and use of excess *t*-butyl alcohol. However, the excess alcohol can lead to further complications caused by dehydration and the formation of diisobutylene.

237

The second experiment describes the preparation of *p*-di-*t*-butylbenzene by the alkylation of *t*-butylbenzene with *t*-butyl chloride in the presence of aluminum chloride. This procedure parallels the method originally discovered by chemists Charles Friedel and James Crafts in 1877 at the Sorbonne University in Paris. Alkyl halides are insufficiently electrophilic by themselves to react with aromatic rings, but in the presence of Lewis acids, they produce highly reactive carbocations, which readily undergo electrophilic substitution reactions with activated rings. As in other such substitutions, the rate is enhanced by electron-donating substituents and retarded by those that are electron withdrawing. In fact, the reaction fails with ring substituents more highly deactivating than the halogens, and this constitutes yet another limitation. Nevertheless, despite these drawbacks, the Friedel-Crafts reaction continues to be a highly useful synthetic method.

The third experiment describes the chlorocarbonylation of *p*-xylene with oxalyl chloride in the presence of aluminum chloride, to give an intermediate acid chloride (see Chapter 24). The aluminum chloride still present in the reaction is then used to promote the acylation of *p*-methyl anisole by the intermediate acid chloride, to give the diaryl ketone. Friedel-Crafts acylation is another reaction that fails with ring substituents more highly deactivating than the halogens. Despite this limitation, Friedel-Crafts acylation is widely used as a synthetic method.

EXPERIMENTS

A.   PREPARATION OF BHT

(Macroscale)

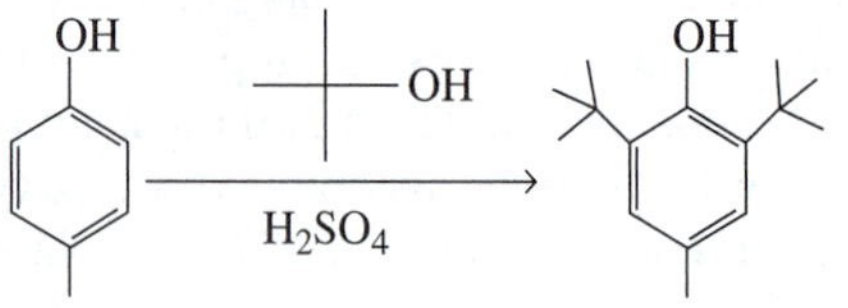

p-cresol

BHT
mp = 70°C

Figure 26.1   Preparation of Butylated Hydroxytoluene (BHT).

SAFETY NOTE

p-Cresol and concentrated sulfuric acid, especially the latter, will cause burns in contact with the skin. If either is spilled on the hand, rinse thoroughly with water and then wash with soap and water. Ether is highly flammable; there should be no flames in the room while it is in use.

PROCEDURE

In a 50-mL round-bottom flask equipped with a magnetic stir bar, place 2.16 g of *p*-cresol, 1.0 mL of glacial acetic acid, and 5.6 mL of *t*-butyl alcohol. (The alcohol has an mp of 26°C and should be warmed to between 30 and 35°C before measuring; it also helps to warm the graduated cylinder slightly to avoid solidification during transfer.) When the *p*-cresol has dissolved, cool the solution to 0°C in an ice bath while stirring magnetically. To this cold mixture,

SLOWLY add 5.0 mL of concentrated sulfuric acid with a disposable pipette over 15 minutes, and the color should remain pale yellow to pink. At this point an oily layer will separate.

Continue stirring the reaction mixture in the ice-water bath, recording any changes in appearance. At the same time, prepare a 125-mL separatory funnel containing 30 mL of distilled water. Remove the flask from the ice bath and carefully add a few pieces of ice to the reaction mixture. This will give a vigorous, exothermic reaction! Using a disposable pipet, slowly add the reaction mixture to the water in the separatory funnel. Rinse the residual material from the reaction flask into the separatory funnel first with 10 mL of water, then with 30 mL of ether. Stopper the separatory funnel and shake it, pausing every few seconds to vent built-up pressure. Remove the lower aqueous layer, and wash the ether layer with 10 mL of 2% aqueous sodium hydroxide and, finally, twice with 10-mL portions of water. Discard the aqueous layers as directed by your instructor. Transfer the ether layer to a dry 50-mL Erlenmeyer flask. Dry the ether solution with anhydrous sodium sulfate and filter through a fluted filter paper. Collect the filtrate in a dry 50-mL round-bottom flask, and evaporate the ether to approximately 10-mL volume.

Cool the solution in an ice-salt bath. To the viscous mass thus obtained, add 2 mL of methanol and stir the resulting solution with a glass rod. When a white solid precipitates, allow the mixture to stand in the ice-salt bath for 10 minutes. Collect the crystals via filtration, and allow them to dry. Weigh the dried product and determine its melting point. If the melting point is low, recrystallize again from methanol and determine the final yield and melting point.

Antioxidant Properties

A quick and direct way to evaluate antioxidant effectiveness is based on the drying properties of linseed oil. This oil, extracted from flax seed, contains a high proportion of linolenic acid, with the nonconjugated triene unit —CH=CH—CH$_2$—CH=CH—. For most applications, such as in varnishes and printing inks where rapid drying is desired, the oil is heated with a small amount of lead oxide. This treatment causes isomerization of the double bonds and preliminary polymerization. The resulting "boiled" oil dries in approximately one sixth of the time required for the "raw" oil. Drying occurs by autoxidation, with the formation of C—O and C—C bonds between low molecular weight polymer units. In the presence of an effective autoxidation inhibitor, the drying process is retarded.

PROCEDURE

In each of four small test tubes labeled 1 through 4, place approximately 0.5 mL of boiled linseed oil (simply divide 2 mL of oil evenly among the four tubes). Prepare 1% solutions of your di-*t*-butyl-*p*-cresol and of *p*-cresol in acetone (e.g., 50 mg in 5 mL of acetone). Dilute the linseed oil samples as follows:

Tube 1: Add 0.5 mL of acetone (blank).
Tube 2: Add 0.25 ml of di-*t*-butyl-*p*-cresol solution and 0.25 mL of acetone (0.5% BHT).
Tube 3: Add 0.5 mL of di-*t*-butyl-*p*-cresol solution (1% BHT).
Tube 4: Add 0.5 mL of *p*-cresol solution (1% *p*-cresol).

Place an open-end capillary tube in each solution and stir well until the solutions are mixed, and then place 2 to 3 drops from each tube on microscope slides (four samples can be placed on two slides). Place the slides on a paper towel, number them for identification, and place them on a hot plate at low heat or in an oven (70 to 80°C). Examine the slides after 20 minutes, compare the fluidity or stickiness of the four samples, and record the results.

If it is more convenient, the slides may be allowed to stand at room temperature and compared after several hours or on the following day.

B.   PREPARATION OF *p*-DI-*t*-BUTYLBENZENE

(Microscale)

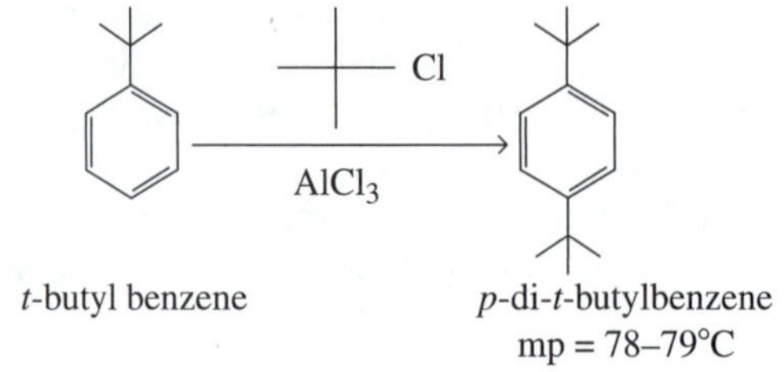

t-butyl benzene   *p*-di-*t*-butylbenzene
mp = 78–79°C

Figure 26.2   Preparation of p-di-*t*-butylbenzene.

SAFETY NOTE

Aluminum chloride is extremely hygroscopic and produces HCl in the presence of moisture. Avoid breathing the dust and immediately wash off any material that may get on the skin. Diethyl ether is highly flammable; there should be no flames in the room while it is in use.

PROCEDURE

In a dry 5-mL vial, place 1.0 mL of *t*-butyl chloride, 0.5 mL of *t*-butylbenzene, and a small stirring vane. Place the reaction vessel in an ice bath (a 100-mL beaker or a small crystallizing dish may be used for the bath), and place it on a stirring plate. HCl gas is produced in this reaction, and it should be trapped by passing the gas from the reaction vessel through a tube into a beaker (approximately 100 mL) containing water-moistened cotton. The HCl is readily absorbed by the wet cotton, which can then be disposed of by removing with forceps and rinsing under tap water. Alternatively, the reaction may be run in an efficient hood.

Weigh 0.05 g of anhydrous aluminum chloride and store in a capped vial until you are ready to use it. It may be difficult to weigh this material accurately because it reacts readily with the moisture in the air. To minimize exposure to air, keep the reagent bottle closed as much of the time as possible, have the weighing paper tared beforehand, and weigh and transfer the material as quickly as possible.

Add one half of the aluminum chloride to the chilled *t*-butylbenzene/*t*-butyl chloride mixture. Recap immediately. A vigorous reaction should ensue. Stir for 5 to 6 minutes, add the remainder of the aluminum chloride, and continue to stir. When the reaction begins to subside—a white product should be formed, although this may be difficult to detect—remove the flask from the ice bath and allow the mixture to warm to room temperature.

Add 1 mL of ice-cold water (from the bath), followed by 1 mL of diethyl ether. Swirl, and with a pipet transfer the contents to a centrifuge tube. Remove the ether layer with a clean pipet, and wash the remaining aqueous layer with two 1-mL portions of ether, adding these washes to the original ether layer. Dry the combined ether layers by filtering through a small column of

MgSO$_4$ prepared in a pipet with a cotton plug at the bottom to contain the drying agent. Collect the filtrate in a small Erlenmeyer flask, evaporate the solvent over a steam bath, and recrystallize the residue from 1 mL of methanol. Discard filtrate in the waste solvent container. Collect the product on a Hirsch funnel, let dry, weigh, and determine its melting point. If required, the product can again be recrystallized from methanol. Submit the product in an appropriately labeled container.

C. FRIEDEL-CRAFTS SYNTHESIS OF AN UNSYMMETRICAL DIARYL KETONE

(Microscale)

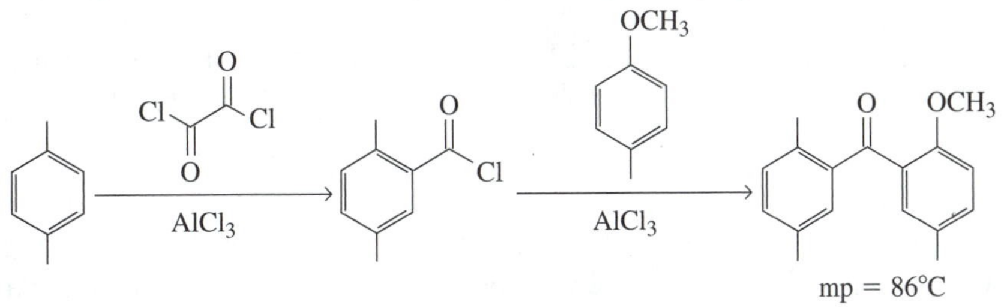

Figure 26.3 Preparation of an Unsymmetrical Diaryl Ketone.

Chlorocarbonylation of an arene followed by the addition of another arene, illustrated by *p*-xylene and anisole, leads smoothly to the unsymmetrical diaryl ketone. The product ketone can be purified by recrystallization or by column chromatography.

PROCEDURE

Combine 212 mg of *p*-xylene and 7 mL of CH$_2$Cl$_2$ in a 25-mL Erlenmeyer flask. Stir in an ice/water bath, then add 2.1 g of a 10 weight percent solution of oxalyl chloride (CAUTION: Corrosive!) in CH$_2$Cl$_2$. Add 0.53 g of solid AlCl$_3$ (CAUTION: Corrosive!) over 5 minutes. Remove from the ice bath, and let stir 20 minutes.

Add 0.22 g of anisole in 2 mL of CH$_2$Cl$_2$ dropwise over 5 minutes. Let stir for 40 minutes. Quench cautiously by adding the reaction mixture to cold water in a separatory funnel. Separate the CH$_2$Cl$_2$ layer, and extract the aqueous layer with two more 10-mL portions of CH$_2$Cl$_2$. Dry the CH$_2$Cl$_2$ over Na$_2$SO$_4$. Set aside 0.5 mL of the CH$_2$Cl$_2$ solution for TLC, and evaporate the rest of the CH$_2$Cl$_2$ solution onto 0.5 g of flash silica gel (see instructions for column chromatography in Chapter 7).

TLC: using 5% EtOAc/hexane, you should see two UV-absorbing spots that move, with more material at the baseline. The ketone that you want is the more slowly moving spot.

Column: Prepare a chromatography column using 5 g of flash silica gel. Add your sample to the top, followed by a layer of sand. Elute the column with 20 mL of 1% EtOAc/hexanes, 20 mL of 2% EtOAc/hexanes, and 20 mL of 5% EtOAc/hexanes. Collect 5-mL fractions in test tubes. (This could mean as many as 60 5-mL fractions. You may not have 60 test tubes. Collect ten fractions at a time, and check the TLC of each tube. If it does not have your desired product, rinse the contents of the tube into a 250-mL Erlenmeyer flask and reuse the tube. When you are sure that you have found your product, pour the contents of the 250-mL Erlenmeyer flask into the waste receptacle designated for it.) and check them by TLC. You should be able to spot at least four fractions across the bottom of a TLC plate. Combine the fractions that contain your pure ketone, and concentrate them on the rotary evaporator. The residue after evaporation should be solid. Filter it with cold hexane, collecting the crystalline product on a Hirsch funnel. Record the melting point, weight, and physical appearance of your product, and calculate the yield.

QUESTIONS

1. Explain why the Friedel-Crafts reaction of benzene plus *n*-propyl chloride in the presence of $AlCl_3$ gives mostly isopropylbenzene.
2. What monoalkylated product(s) would one expect to obtain if benzene were alkylated with *n*-butyl chloride?
3. Give a stepwise mechanism for the acylation of benzene with acetyl chloride in the presence of aluminum chloride.
4. Other than comparison of melting points, how could one demonstrate that the isolated product from Experiment A is BHT and not the monoalkylated product, 2-t-butyl-4-methylphenol?

PRELABORATORY QUESTIONS

1. Draw an arrow-pushing mechanism for the Friedel-Crafts reaction of phenol with t-butyl chloride in the presence of AlCl3.
2. The Friedel-Crafts reaction involves the electrophilic attack of a carbocation on an aromatic ring. Draw an arrow-pushing mechanism for the generation of a carbocation from 2-propanol with acid.
3. An aromatic ring more highly deactivated than a monohalobenzene cannot be alkylated by the Friedel-Crafts procedure. Which of the following will not undergo Friedel-Crafts alkylation: nitrobenzene, toluene, anisole, acetophenone?
4. Why is it important to add ice slowly to the reaction mixture in Experiment A?

REFERENCES

Taber, D. F.; Sethuraman, M. R. *J. Org. Chem.,* 2000, *65*, 254.

Aromatic Nitration

itration is one of the most important and thoroughly studied examples of electrophilic aromatic substitution. Benzene and benzene derivatives of widely different reactivities can be nitrated under suitable conditions; mononitro compounds can usually be obtained in good yields because the first nitro group deactivates the ring toward further substitution. Nitro compounds are useful as explosives (TNT), but their greater interest for organic chemists is in their usefulness as synthetic intermediates. The nitro group may be reduced to a nitroso, hydroxylamine, or amine function, and via the diazotization reaction amines can be converted to many other functional groups (—OH, —F, —I, —CN, —N=N—Ar, etc).

Nitration may be carried out by using fuming nitric acid, mixtures of concentrated nitric acid and sulfuric acid, and nitric acid in glacial acetic acid. Selection of the nitrating agent and the reaction conditions will depend upon the reactivity of the compound to be nitrated; however, nitration is most frequently carried out with a mixture of concentrated nitric acid and concentrated sulfuric acid. In the HNO_3/H_2SO_4 mixture (called mixed acid or nitrating acid), the electrophile is the nitronium ion, NO_2^+, which is generated by protonation of the nitric acid followed by the loss of water.

$$HONO_2 + H_2SO_4 \rightleftarrows H_2\overset{+}{O}NO_2 + HSO_4^-$$

$$H_2\overset{+}{O}NO_2 \rightarrow NO_2^+ + H_2O$$

$$H_2O + H_2SO_4 \rightarrow H_3O^+ + HSO_4^-$$

Attack on the aromatic ring by the nitronium ion, usually the rate-determining step, is then followed by the rapid loss of a proton to give the nitro derivative.

The ease or difficulty of nitration depends upon the nature of the substituents already present. Electron-donating groups (—OH, —OR, —NHCOR, —CH$_3$) facilitate the nitration, but electron-withdrawing groups (—NO$_2$, —COOH, —COR) retard the reaction. With substituted benzenes, the nitro group may be directed to the *ortho* and *para* positions (by —OH, —NHCOR, —Cl, —Br) or to the *meta* position (by —NO$_2$, —COOR).

EXPERIMENTS

In the experiments in this chapter, we illustrate how nitration can be controlled by the proper choice of conditions with substrates of significantly different reactivities. A variety of methods for the separation and purification of the products are also illustrated in the procedures. The first nitration is done on a macroscale while the last three are carried out on a microscale.

SAFETY NOTE

Concentrated sulfuric acid and concentrated nitric acid can cause burns on contact with the skin. Particular care must be taken in measuring and handling these reagents. Clean up any spills immediately (consult your instructor). Should any acid get on you, wash quickly and thoroughly with cold water and then with soap and water. Wear safety goggles! Aromatic nitro compounds are generally toxic, and some may cause contact dermatitis; avoid contact with any of the products. Nitroanilines also badly stain the skin.

A.   PREPARATION AND SEPARATION OF *o*- AND *p*-NITROANILINE

(Macroscale)

The —NH_2 group activates the benzene ring so strongly for electrophilic substitution that direct nitration of aniline cannot be controlled. For the preparation of mononitroanilines, therefore, the reactivity is moderated by acetylation of the amine; the acetyl group is subsequently removed by acid hydrolysis. Even with the acetyl "protecting group" the ring is strongly activated, and it is difficult to avoid the formation of some dinitro products.

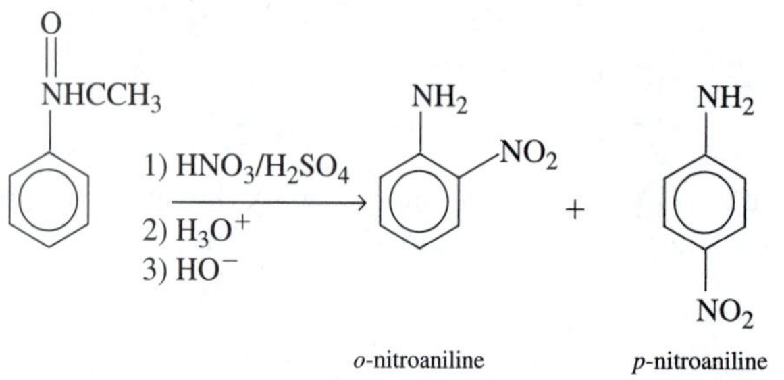

o-nitroaniline
mp 72°C

p-nitroaniline
mp 147°C

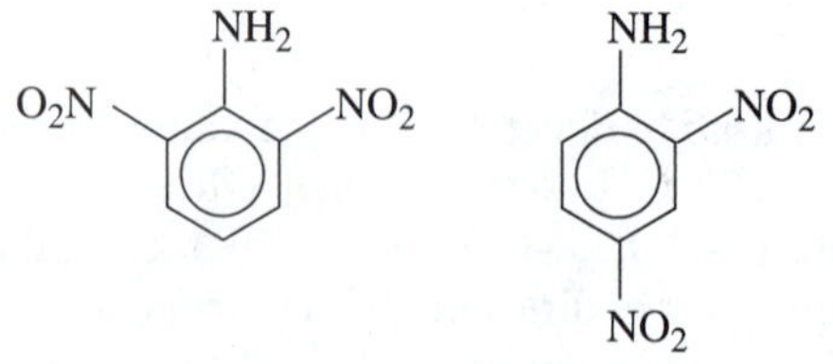

2,6-dinitroaniline
mp 147°C

2,4-dinitroaniline
mp 188°C

The *para* isomer is the major product, and it can be purified readily by recrystallization. A small amount of *o*-nitroaniline and traces of the 2,4- and 2,6-dinitro compounds can be isolated or detected by chromatography.

PROCEDURE

In a 125-mL Erlenmeyer flask, place 2.7 g of acetanilide. Obtain 9 mL of concentrated sulfuric acid in a graduated cylinder, and pour approximately half of the acid into the flask. Swirl and stir the mixture until all but traces of the acetanilide dissolve (a small amount of remaining solid will subsequently dissolve). Cool the solution in an ice bath. With a buret or graduated pipet, measure out 1.5 mL of concentrated nitric acid and add it to the remaining sulfuric acid. Mix the acids by drawing up samples from the bottom with a transfer pipet and bulb and emptying them on top; complete the mixing by discharging a stream of bubbles from the empty pipet at the bottom of the cylinder.

Using the pipet and bulb, add the mixed acid in small portions (approximately 0.5 mL) to the cooled sulfuric acid solution of acetanilide. Swirl the flask in the ice bath after each portion is added. The flask should not become perceptibly warm to the touch; the addition requires 10 to 15 minutes. After 20 minutes, including addition time, add 25 mL of a mixture of ice and water (loosely pack a 16 × 150-mm test tube with ice and fill with water) to the reaction mixture.

The resulting suspension of nitroacetanilides is now hydrolyzed in the same flask, using the aqueous sulfuric acid. Add a boiling stone and heat the flask on a hot plate or wire gauze with a burner flame. Watch carefully as the color darkens and the solid begins to dissolve, and remove the heat source if necessary to avoid excessively vigorous boiling and foaming. Continue heating at a gentle boil for 15 minutes, and then cool the flask and contents in an ice bath. Place the bath and flask in the hood, and add, in five or six portions, 25 mL of concentrated aqueous ammonium hydroxide; swirl in the ice bath after each addition.

Collect the precipitated nitroaniline in a Büchner funnel by suction filtration. Rinse the flask with 3 mL of water, disconnect the suction tubing, pour the rinsing liquid into the funnel, and then suck the water through the precipitate; press the solid, and allow air to be drawn through it for 2 or 3 minutes.

Transfer the yellow crystals, which can be peeled cleanly from the filter paper, to a 25 × 100-mm test tube and add 4 mL of ethanol. Heat and stir the mixture on the steam bath until the alcohol boils and most of the solid dissolves, and then cool in an ice bath for 10 to 15 minutes to crystallize the *p*-nitroaniline. Collect the crystals by suction filtration, wash with a minimum volume of ethanol, and dry on glassine paper. Save the mother liquor for the examination of minor products. Determine the melting point and weight, and calculate the percent yield of crude *p*-nitroaniline. Run a TLC on samples of the solid and mother liquor, using methylene chloride for development.

For crystallization, dissolve the crude product in a minimum volume of hot ethanol. If the crude nitroaniline is very dark, add a small amount of charcoal and filter through fluted paper. Cool, collect the crystals, and determine the melting point. Dispose of the filtrate in the organic waste container.

For examination of the minor products, evaporate the mother liquor from the *para* isomer to a thick syrup and filter off any additional solid that crystallizes. Prepare a chromatography column with 15 g of alumina, using methylene chloride as the solvent. A 10 × 300-mm chromatography tube or a 25-mL pinch-cock buret can be used for the column.

Dissolve the nitroaniline residue in a minimum volume of methylene chloride; 1 mL should suffice. Apply the sample to the adsorbent, and then develop and elute the column with methylene chloride. **Be careful not to inhale the vapors of methylene chloride.** Collect the eluate in test tubes in 5-mL fractions, beginning with the first colored solution that emerges from the column. After five 5-mL fractions, collect one final large fraction until either all of the colored material is eluted or the volume of this fraction reaches 25 mL. Examine each fraction by TLC (three fractions per plate), and pool fractions on the basis of TLC behavior; if two fractions show the same TLC spot(s), combine them.

Evaporate and weigh any fractions or groups of fractions that contain a single component. Compare the TLC with authentic samples of the possible minor products; determine the melting point and weight of the crystalline residues. Record the yield of *o*-nitroaniline and any other compounds isolated from the chromatogram.

B.   NITRATION OF TOLUENE AND ANALYSIS OF PRODUCTS

(Microscale)

SAFETY NOTE

In addition to care in using concentrated nitric acid and sulfuric acid, caution should be used with diethyl ether, which is highly flammable. There should be no flames in the room; do not work near a hot plate or other heating device.

Place 1 mL of concentrated nitric acid in a 25-mL Erlenmeyer flask and then add 1 mL of concentrated sulfuric acid. The temperature will rise somewhat; cool the flask in a cold water bath for 1 to 2 minutes. Using a Pasteur pipet, add 1 mL of toluene dropwise with swirling over 10 to 15 minutes so that the temperature does not rise above 35 to 45°C (quite warm but not hot to the touch; again cool the flask briefly from time to time in a cold water bath). Swirl the flask at room temperature for 5 minutes and then pour the contents into 10 mL of water in a separatory funnel. Rinse the flask with 10 mL of diethyl ether and add the ether to the separatory funnel. Remove the aqueous layer and wash the ether layer with 10 mL of 10% sodium bicarbonate solution. Finally wash the ether layer with 5 mL of water. The aqueous layers may later be combined, neutralized, and washed down the drain with ample water. Dry the ether layer over anhydrous sodium sulfate and decant off the ether through a small cotton pad or remove it with a pipet. Evaporate the ether to a volume of 1 to 2 mL and analyze by gas chromatography. Consult your instructor about operating conditions; typical conditions include 170°C at 30 psig using a column of 8% Carbowax 1540 on chromosorb WHP. The order of elution is normally *ortho, meta,* then *para;* retention times may be checked versus one or more knowns. Dispose of the remaining nitrotoluenes in the organic waste container.

From your gas chromatogram, determine the areas under each peak and calculate the percent of *ortho-, meta-,* and *para*-nitrotoluene in the mixture. What percentages would you expect if all positions were equally reactive? Calculate the ratio of *ortho* to *para* product.

Ethylbenzene and isopropylbenzene may similarly be nitrated and analyzed.

QUESTIONS

1. Which compound in each of the following pairs is more reactive towards aromatic nitration? Explain your answers.
   a. phenol or nitrobenzene
   b. methyl benzoate or phenol
   c. nitrobenzene or methyl benzoate
   d. benzene or toluene
2. Show all possible mononitration products of toluene. Do you think it would be easy to isolate and identify all of them if each is produced in some amount?

PRELABORATORY QUESTIONS

1. List three different combinations of reagents used for nitration of aromatic compounds.
2. Which is nitrated faster, toluene or nitrobenzene? Explain.
3. In Experiment A, what is the reagent for hydrolysis of the crude nitroacetanilide?
4. In Experiment B, how is the ratio of nitrotoluene isomers determined?
5. In an aromatic nitration reaction, what is the electrophile that is produced by the reaction of sulfuric and nitric acids?

REFERENCES

de la Mare, P. B.; Ridd, J. H. *Aromatic Substitution: Nitration and Halogenation;* Academic Press: New York, 1959.

M. Davis, et al. *J. Chem. Educ.,* 1978, *55*(34).

Aromatic Nucleophilic Substitution

Chlorobenzene and aryl chlorides generally are extremely unreactive toward nucleophilic substitution. Displacement reactions that occur under drastic conditions, as in the preparation of phenol from chlorobenzene with NaOH at 250°C, often occur by elimination to give an aryne (Chapter 36) followed by addition.

With nitro groups *ortho* or *para* to the halogen, however, nucleophilic substitution becomes a facile and useful reaction. The preparation of 2,4-dinitrophenol is an example. This compound cannot be obtained satisfactorily by the nitration of phenol or mononitrophenols because of the susceptibility of the ring to oxidation, but displacement of halogen in 2,4-dinitrochlorobenzene to give the phenol can be carried out in good yield in boiling aqueous base. The reaction occurs by addition of the nucleophile to give an intermediate with an sp^3 carbon, followed by loss of halide anion.

This addition-elimination sequence leads to two characteristic differences between aromatic nucleophilic substitution and S_N2 reactions of alkyl halides. In the latter case, the reactivity sequence within a series of halides is I>Br>Cl>F, corresponding to the progressively greater strength of the C—X bond, which is broken in the rate-determining step of the reaction. In aromatic substitution this reactivity difference is not observed, and the fluoro compound is in fact the most reactive, since the highly electronegative fluorine atom increases the stabilization of the transition state for the addition step.

A second distinctive feature of aromatic nucleophilic substitution as contrasted to aliphatic S_N2 reactions is the **α-effect.** This term denotes a large acceleration of the rate with nucleophiles such as NH_2NH_2, NH_2OH, ClO^-, or HO_2^- in which an electronegative atom with an unshared electron pair is located adjacent (i.e., α) to the nucleophilic atom. This effect is also seen in reactions of esters and other carbonyl compounds with α-nucleophiles. It has been suggested that the α-effect arises from the interaction of the orbitals containing the unshared electron pairs on adjacent atoms, leading to a situation in which both bonding and antibonding molecular orbitals are occupied (Fig. 28.1). The electrons in the antibonding orbital are at a higher energy than the unshared pairs in simple nucleophiles and are more readily donated to an unsaturated electrophile.

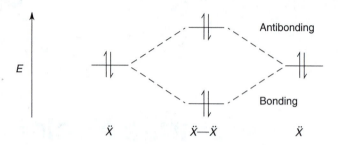

Figure 28.1 Interaction of adjacent electron pairs in α-nucleophiles. The $\ddot{X}$ denotes an atom with one or more unshared pairs of electrons (e.g., N, O, Cl).

The α-effect in reactions of 2,4-dinitrochlorobenzene is a very large one, the rate constant k_2 for reaction with hydroperoxide ion being nearly 10^5 times greater than that for hydroxide ion. The consequences of such an increase in rate can be translated in practice to faster reactions at lower temperatures. In the preparation of 2,4-dinitrophenol, for example, a traditional procedure requires reaction with aqueous base at 130 to 200°C for several hours. As will be demonstrated in the first experiment in this chapter, the phenol can be obtained in the presence of hydroperoxide in a few minutes at 50 to 60°C. In this procedure the product formed initially is probably the aryl hydroperoxide, which rapidly undergoes attack by hydroxide to give the phenoxide anion.

Likewise, differences in reactivity due to the α-effect and also due to the nature of the halogen can be seen in the reactions of dinitrohalobenzene with hydrazine and with ammonia.

2,4-dinitroaniline
mp 188°C

2,4-dinitrophenylhydrazine
(2,4-DNPH) mp 197°C

The compound 2,4-dinitroaniline can be obtained more economically by another route, but the reaction with hydrazine is the only practical process for preparing the very useful reagent 2,4-dinitrophenylhydrazine (2,4-DNPH). The reaction of 2,4-dinitrofluorobenzene (Sanger's reagent) with amines has an important application in determining the sequence of amino acids in peptides.

EXPERIMENTS

A.   PREPARATION OF 2,4-DINITROPHENOL

(Microscale)

SAFETY NOTE

2,4-dinitrochlorobenzene is very irritating to the skin; wear gloves and avoid any contact. Both 2,4-dinitrochlorobenzene and 2,4-dinitrophenol are toxic materials. Wash hands thoroughly at end of experiment.

In a 25-mL round-bottom flask with a ground glass joint, place 0.2 g of 2,4-dinitrochloroben-zene, 4.0 mL of 2 M aqueous KOH solution, and 0.5 mL of methanol. Warm the mixture on the steam bath with swirling until the solid melts. Obtain 1.0 mL of 3% H_2O_2 or 0.5 mL of 6% H_2O_2 (30% solution diluted 1:4), and add about one fourth of the peroxide solution to the flask. Place a ground glass stopper in the flask, hold the stopper in place with your hand, and shake the flask vigorously for 3 to 4 minutes; then remove the stopper. (**CAUTION:** Handle the stop-per with a paper towel and avoid contact of the solution with the skin; rinse thoroughly with water if contact occurs.)

Warm the flask to about 50°C (as warm as can be comfortably held in the hand), add one fourth to one third of the peroxide solution, restopper, and shake again for several minutes. When solid begins to appear in the dark solution, remove the stopper and warm the flask on the steam bath until the solid redissolves. Add the remaining peroxide, and swirl the flask on the steam bath until no more oily droplets can be seen on the bottom of the flask (approximately 5 minutes' heating at 60 to 65°C should suffice).

Pour the solution into an Erlenmeyer flask, rinse with a few tenths mL of water, and cool the solution to 5 to 10°C in an ice bath. Mix 0.8 mL of concentrated HCl with approximately 1 mL of crushed ice, and add this dilute acid to the solution until the mixture is a light yellow color. Stir the mixture for 2 minutes, and rub the solid against the wall to insure that all of the phe-noxide is acidified. Collect the solid on a small Büchner funnel, wash with a few tenths mL of dilute HCl, followed by water, and dry thoroughly (60°C oven for 30 minutes or longer at room temperature). Determine the melting point (lit. mp 114–115°C) and calculate the percent yield. Submit the product to your instructor as directed or save to use in the Chemiluminescence ex-periment in Chapter 30. Neutralize the filtrate and discard as directed by your instructor.

B. REACTION OF 2,4-DINITROHALOBENZENES WITH AMMONIA AND HYDRAZINE

(Microscale)

SAFETY NOTE

2,4-dinitrochlorobenzene, 2,4-dinitrophenol, hydrazine, and 2,4-dinitrofluorobenzene are toxic materials; the latter two are suspected carcinogens. All reactions should be carried out in a hood and gloves should be worn.

In a 10 × 75-mm test tube (Tube 1) dissolve 0.06 g of 2,4-dinitrochlorobenzene in 1.0 mL of ethanol. Transfer half of the solution to a second test tube (Tube 2). In a third tube (Tube 3) dissolve 0.04 mL of 2,4-dinitrofluorobenzene (d = 1.47 g/mL) in 1.0 mL of ethanol and transfer half of this solution to a fourth tube (Tube 4).

To the first tube, add 0.02 mL of 85% hydrazine hydrate (suspected carcinogen) ($N_2H_4 \cdot H_2O$; 0.55 g N_2H_4/mL), and shake the tube briefly to mix the solutions. Note any changes and the time required. In the same way, add 0.02 mL of hydrazine solution to Tube 3, and note changes.

To Tubes 2 and 4, add 0.02 mL of concentrated ammonium hydroxide (0.26 g NH_3/mL), mix well, and record any changes.

After 30 minutes, collect any crystalline products that were obtained, and wash with a few mL of alcohol. Dry the products, weigh, and calculate percent yields. Place the filtrates and any unreacted solutions in the halogenated waste solvent receptacle. Label the products and retain for later use or submit them as directed.

Compare the results in the four tubes and explain the reactivity order observed.

QUESTIONS

1. The reaction of a polypeptide with 2,4-dinitrofluorobenzene is used in an analytical procedure for identifying the N-terminal amino acid in the chain. After reaction, the amide bonds are hydrolyzed, liberating all but the N-terminal amino acid (containing R_1).

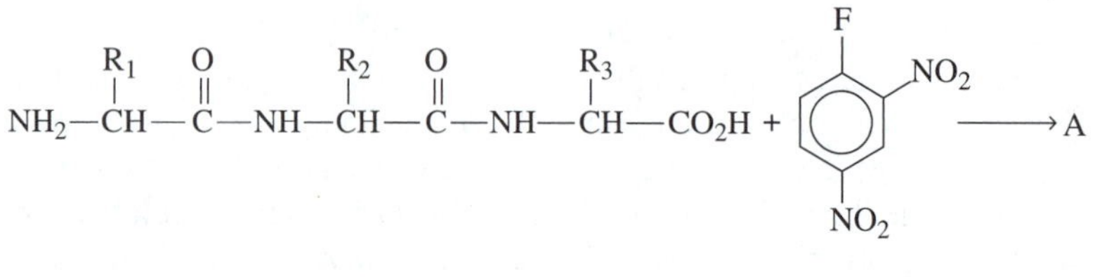

$$A \xrightarrow{H_3O^+} B + NH_2\text{—}\underset{\displaystyle R_2}{CH}\text{—}CO_2H + NH_2\text{—}\underset{\displaystyle R_3}{CH}\text{—}CO_2H$$

Write structures for compounds A and B. Suggest how this reaction sequence permits the identification of the amino acid containing R_1.

2. Groups that are deactivating in electrophilic substitution reactions (e.g., $—NO_2$) become activating groups in nucleophilic substitution reactions. Explain.
3. Smiles rearrangements are those that follow an intramolecular nucleophilic substitution course, as exemplified by the following reaction.

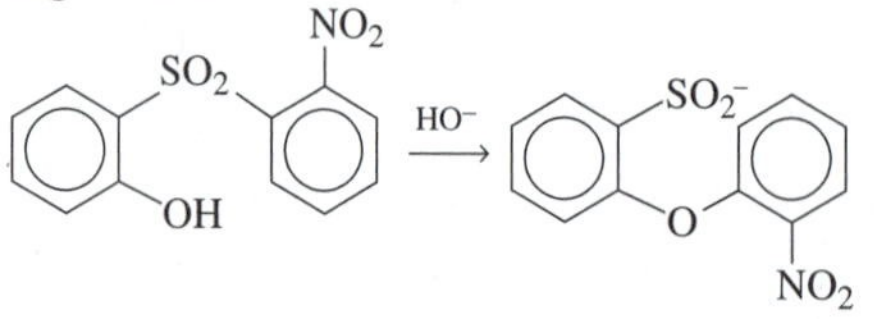

Write the mechanism by which this rearrangement takes place.

PRELABORATORY QUESTIONS

1. 2,4,6-trinitrochlorobenzene reacts with aqueous NaOH at 35 to 45°C. Write an equation to illustrate this reaction.
2. Draw the structural formula for the product of the nucleophilic substitution reaction of 2,4-dinitrochlorobenzene with methylamine.
3. Show the reaction of Sanger's reagent with glycylalanine, $H_2NCH_2CONHCH(CH_3)COOH$.
4. What is the order of reactivity of 2,4-dinitrohalobenzenes in the addition-elimination reaction of aromatic nucleophilic substitution as a function of the particular halogen? Explain.

REFERENCES

Bernasconi, C. F. *Accounts of Chemical Research,* 1978, *11*, 147.

Bunnet, J. F. *J. Chem. Education,* 1974, *51*, 312.

Miller, J. *Aromatic Nucleophilic Substitution;* American Elsevier: New York, 1968.

Properties of Amines

A mines have an unshared pair of electrons on nitrogen and, therefore, act as bases, nucleophiles, and compounds that may be oxidized. As Lewis bases, they form salts with acids and form coordination complexes with metal cations. As nucleophiles, they displace halogen from alkyl halides and acid chlorides to give more highly alkylated amines and amides, respectively. They may be oxidized by a variety of oxidizing agents including oxygen, permanganate ion, hydrogen peroxide, and nitrous acid.

The basicity of an amine is influenced by the number and type of groups attached to the nitrogen atom. Aliphatic amines are stronger bases than ammonia because the alkyl groups are electron donors relative to hydrogen. Aromatic amines are weaker bases than ammonia because delocalization of the unshared electron pair on nitrogen into the ring lowers the electron density on nitrogen.

Amines up to approximately five carbon atoms are soluble in water. Higher molecular weight amines that are water insoluble dissolve in aqueous acid through the formation of salts. This provides a convenient method for separating such amines from water insoluble neutral and acidic compounds (see Chapter 4, Experiment B).

Many amines, as well as other nitrogen compounds, are physiologically active and should be handled with caution. Amines that occur naturally in plants are called alkaloids; the potency of many of these such as morphine, cocaine, and nicotine is widely known.

EXPERIMENTS

SAFETY NOTE

Many amines are toxic substances; inhalation of their vapors or contact with the skin should be avoided. If any should contact the skin, wash with water, then with 5% acetic acid solution, and again with water. Benzenesulfonyl chloride is corrosive and gives off irritating vapors. 2-naphthol is a toxic irritant.

A. SOLUBILITY IN WATER

Place 1 mL of water in a 13 × 100-mm test tube, add 1 drop of cyclohexylamine, and shake the tube to mix the two materials. Note whether the amine is soluble; if it is, determine the solubility of a second and third drop. Save the mixture for use in Experiment B.

Repeat this solubility test with aniline, pyridine, and triethylamine. Also save the mixtures for Experiment B.

B.   BASICITY

1.   Determine the pH of each of the mixtures from Experiment A by placing a drop of the aqueous phase on a piece of universal pH paper.
2.   To those mixtures from Experiment A in which the amine is insoluble, add 10% HCl dropwise with stirring. Does the amine dissolve?

C.   HINSBERG TEST

Amines react as nucleophiles toward acid chlorides and acid anhydrides to yield amides. Many of these amides, such as those of acetic, benzoic, and benzenesulfonic acid, are crystalline solids that may be used as derivatives to identify the amine.

   The reaction of an amine with benzenesulfonyl chloride can be useful in determining whether the amine is primary, secondary, or tertiary. The sulfonamide from a primary amine is acidic and dissolves in sodium hydroxide solution, but the sulfonamide of a secondary amine lacks the amide hydrogen and is insoluble. Tertiary amines do not form amides. This overall process is known as the **Hinsberg test.**

   Place two drops of amine, 4 drops of benzenesulfonyl chloride, and 4 mL of 10% sodium hydroxide in a 13 × 100-mm test tube. Stopper the tube and shake it for approximately 5 minutes. If a clear or nearly clear solution is obtained, a primary amine is indicated; acidification with hydrochloric acid should precipitate the sulfonamide. The presence of a solid or liquid residue in the original reaction mixture indicates a secondary or tertiary amine. Separate the residue and determine its solubility in 10% HCl. A tertiary amine will dissolve; if the residue does not dissolve, then it is a sulfonamide of a secondary amine. Perform the Hinsberg test on aniline, *N*-methylaniline, and triethylamine, and then determine whether an unknown is a primary, secondary, or tertiary amine.

D.   REACTIONS OF DIAZONIUM SALTS

(Microscale)

Amines undergo a variety of reactions with nitrous acid depending on whether the amine is a primary, secondary, or tertiary amine and whether it is aliphatic or aromatic. The reaction of primary aromatic amines with nitrous acid is unique and provides a very versatile synthetic intermediate. The process is called **diazotization,** and the intermediates formed are called diazonium salts. Aqueous solutions of diazonium salts are stable at 0 to 5°C for several hours; however, they lose nitrogen on warming and form phenols. Solid diazonium salts sometimes crystallize from solution but should *never* be separated by filtration as they can decompose explosively.

$$\text{Ph}-NH_2 \xrightarrow[\text{HCl}]{\text{HONO}} \text{Ph}-\overset{+}{N}\equiv N\ Cl^-$$
$$0\text{--}5°C$$

Diazonium salts react with nucleophiles to liberate N_2 and form phenols, aryl halides, and nitriles with such nucleophiles as water, halide ion, and cyanide ion, respectively. They react with phenols and tertiary aromatic amines (aromatic compounds particularly reactive towards electrophilic aromatic substitution) to give azo compounds by a reaction known as **azo coupling.** In the following experiment, *p*-nitroaniline will be reacted with nitrous acid to form the diazonium salt, *p*-nitrobenzenediazonium chloride, which will then be used to prepare *p*-iodonitrobenzene and/or the dye Para Red.

1.   Preparation of the Diazonium Salt

$$O_2N-\text{C}_6H_4-NH_2 \xrightarrow[\text{HCl}]{\text{HONO}} O_2N-\text{C}_6H_4-N_2^+\ Cl^-$$
$$0\text{--}5°C$$

In a 50-mL Erlenmeyer flask, place 16 mL of water, 2 mL of 6M hydrochloric acid, and 0.552 g (0.004 mole) of *p*-nitroaniline. Warm the mixture, stir until the amine has dissolved, and then cool the solution to 0 to 5 °C. Add approximately 2 g of chipped ice and keep the mixture at 0 to 5°C in an ice bath.

Next prepare a solution of 0.276 g (0.004 mole) of sodium nitrite in 2 mL of cold water. Add the sodium nitrite solution to the solution of amine hydrochloride while swirling the latter vigorously. Continue swirling for 3 to 5 minutes, and then keep the solution at ice temperature until ready for use in Part 2 and Part 3 below. If appreciable solid remains, filter the cold solution through a small cotton plug.

The diazonium solution may be tested for nitrous acid with potassium iodide–starch paper (a purple to black coloration indicates HNO_2; the test paper must darken immediately) and for the actual formation of a diazonium salt (streak a solution of 2-naphthol in aqueous NaOH across a piece of filter paper, and then cross it with a streak of the reaction solution; an orange to red coloration at the junction indicates the presence of a diazonium salt).

2.   Preparation of *p*-Iodonitrobenzene

$$O_2N-\text{C}_6H_4-N_2^+ + I^- \longrightarrow O_2N-\text{C}_6H_4-I + N_2 \uparrow$$

Dissolve 0.6 g of potassium iodide in 2 mL of water and cool the solution to 0 to 5°C. Add it, with swirling, to three-fourths of the diazonium salt solution at such a rate that the temperature of the latter is maintained below 5°C. When the addition is complete, allow the mixture to stand in an ice bath for 3 to 5 minutes with occasional swirling. Collect the product by vacuum filtration, dry it, and recrystallize it from ethanol. Determine the weight, melting point, and percent yield.

3.   Preparation of Para Red

In a 50-mL Erlenmeyer flask, dissolve 0.288 g of 2-naphthol (0.002 mole) in a mixture of 3 mL of 10% sodium hydroxide and 20 mL of water. Chill this solution in an ice bath until the temperature is below 5°C (a few chips of ice may be added to lower the temperature). Pour the remaining diazonium solution all at once into the cold solution of 2-naphthol, and swirl the mixture for 5 to 10 minutes. Add 1 mL of concentrated hydrochloric acid, heat the mixture with stirring for 20 to 30 minutes on a steam bath, and filter the product. Allow the dye to dry and weigh the crude material. Do not determine a melting point or percent yield since the product contains inorganic salts.

QUESTIONS

1. Write an equation for the reaction of butylamine with one mole of HCl.
2. Write equations for the reaction of aniline with
    a.   acetyl chloride
    b.   benzoyl chloride
    c.   acetic anhydride
3. How might one distinguish between the following pairs of substances using reactions described in this chapter?
    a.   butylamine and dibutylamine
    b.   cyclohexylamine and aniline
4. Write equations to illustrate the diazotization of aniline and subsequent reaction with phenol; with dimethylaniline.

PRELABORATORY QUESTIONS

1. Write an equation for the reaction of butylamine with hydrochloric acid.
2. Is the benzenesulfonamide of aniline soluble in aqueous NaOH? Explain your answer.
3. Write the structure of the product of the reaction between benzenediazonium chloride and potassium iodide.
4. Write an equation for the redox reaction between nitrous acid and potassium iodide. Why does the KI/starch paper turn black?
5. Which is a stronger base, aniline or butylamine?

Chemiluminescence

Chemiluminescence is the phenomenon in which visible light is emitted from a chemical reaction that occurs at or near room temperature. The most familiar examples of chemiluminescence are those that occur in fireflies and several marine organisms. These bioluminescent processes are enzymatic reactions of molecular oxygen with rather complex heterocyclic substrates, as exemplified by firefly luciferin.

Chemiluminescence occurs by emission of energy in the visible region of the electromagnetic spectrum from a molecule in an electronically excited state. The process generally involves three steps: (1) formation of an energy-rich intermediate, I, from the reactants; (2) fragmentation of the intermediate to form products, of which one, P^*, is in an electronically excited state; and (3) emission of a quantum of light from P^* as it decays to its ground state.

$$R \xrightarrow{\;1\;} I \xrightarrow{\;2\;} P^* \xrightarrow{\;3\;} P + h\nu$$

The overall efficiency or quantum yield in chemiluminescence, as in any chemical process, is the product of the yields in Steps 1 through 3. The limiting factor is usually Step 2, in which a molecule in an excited state is produced. This reaction must be sufficiently exothermic (50 to 80 kcal/mole) to provide the energy for excitation. Even when this degree of exothermicity is available, significant chemiluminescence is only seldom observed because most of the energy is dissipated as vibrational, rather than electronic, excitation of the products.

A few systems have been discovered, however, in which the number of vibrational modes is reduced by a rigid structure, and the overall efficiency of chemiluminescence is nearly 25%. This value can be compared with an efficiency of approximately 3% for the conversion of energy to light by an incandescent lamp. The most efficient nonenzymatic chemiluminescence is produced by the reaction of diaryl oxalate esters with the highly reactive nucleophile hydrogen peroxide. Devices based on this oxalate-peroxide reaction are used in the familiar, commercially available Cyalume lightsticks, in which a glass capsule containing one of the reactants is broken in a tube containing the other to activate the light.

$$\underset{}{\text{ArO}-\overset{\displaystyle O}{\underset{}{C}}-\overset{\displaystyle O}{\underset{}{C}}-\text{OAr}} + H_2O_2 \longrightarrow \text{ArO}-\overset{\displaystyle O}{\underset{}{C}}-\overset{\displaystyle O}{\underset{}{C}}-\text{OOH} + \text{ArOH}$$

aryl hydroperoxy oxalate

$$\text{ArO}-\overset{\displaystyle O}{\underset{}{C}}-\overset{\displaystyle O}{\underset{}{C}}-\text{O}-\text{OH} \longrightarrow \text{ArOH} + \underset{\displaystyle O-O}{\overset{\displaystyle O \;\;\;\; O}{C-C}}$$

1,2-dioxetanedione

$$\underset{\displaystyle O-O}{\overset{\displaystyle O \;\;\;\; O}{C-C}} + \text{acceptor} \rightarrow \text{acceptor}^* + 2CO_2$$

$$\text{acceptor}^* \rightarrow \text{acceptor} + h\nu$$

The energy-rich intermediate in this reaction has been suggested to be the strained peroxide, dioxetanedione, produced by cyclization of the hydroperoxy oxalate. Decomposition of dioxetanedione to two molecules of the very stable compound carbon dioxide can be estimated to be exothermic by at least 100 kcal/mole. The energy liberated in this reaction is not released directly as light, but it is first transferred to an acceptor molecule. The excited state of the acceptor is the species that emits light.

The final emission step is identical to that observed in fluorescence, in which radiation absorbed at one energy is reemitted at a lower energy. In chemiluminescence, the energy of a chemical reaction is used instead of radiation to pump the acceptor to an excited state. The acceptors used in peroxyoxalate chemiluminescence can be any compounds that undergo efficient visible fluorescence. The characteristic feature of most fluorescent compounds is a rigid polycyclic aromatic structure. As the conjugated system becomes more extended, the wavelength of the fluorescent emission becomes longer.

One group of compounds that are efficient acceptors for chemiluminescence is linear polycyclic hydrocarbons and phenyl derivatives of these structures, such as 9,10-diphenylanthracene, naphthacene, and rubrene.

C$_6$H$_5$

C$_6$H$_5$

8,9-diphenylanthracene

naphthacene

C$_6$H$_5$ C$_6$H$_5$

C$_6$H$_5$ C$_6$H$_5$

rubrene (5,6,11,12-tetra-phenylnaphthacene)

Another type of acceptor is based on the highly fluorescent xanthene dyes, such as fluorescein and the rhodamines. The latter dyes are used as pigments in fluorescent signs, tape, and fabric. Minor changes in structure—for example, the addition of alkyl groups or the substitution of an ester for an acid substituent—cause shifts in the wavelength of the fluorescent emission and a corresponding change in the color of the chemiluminescence, from blue through green and yellow to red.

HO   O   O

CO$_2$H

fluoroscein

(C$_2$H$_5$)$_2$N   O $\overset{+}{N}$(C$_2$H$_5$)$_2$

CO$_2$H

rhodamine B

(C$_2$H$_5$)$_2$N   O $\overset{+}{N}$(C$_2$H$_5$)$_2$

CH$_3$ CH$_3$

CO$_2$C$_2$H$_5$

rhodamine 6G

EXPERIMENTS

A. BIS-(2,4-DINITROPHENYL) OXALATE

(Microscale)

The reaction used in this preparation is a common one—esterification with an acid chloride in the presence of a tertiary amine. The ester is quite susceptible to hydrolysis, and contact with water must be avoided. It is possible to separate it from the triethylammonium chloride by washing with chloroform since it is less soluble than is the chloride. However, the washing invariably dissolves appreciable amounts of ester as well. Since the triethylammonium chloride by-product does not interfere with the subsequent chemiluminescence reactions, the ester can be used without further purification.

$$2\ \text{(2,4-dinitrophenol)} + \text{ClCOCOCl} + 2(C_2H_5)_3N \longrightarrow \text{bis-(2,4-dinitrophenyl) oxalate, mp 185°} + 2(C_2H_5)_3\overset{+}{N}HCl^-$$

oxalyl chloride   triethylamine

triethylammonium chloride

SAFETY NOTE

Oxalyl chloride is toxic and hydrolyzes in air to produce HCl. Work with it in a well-ventilated hood. 2,4-dinitrophenol is toxic and can cause skin burns and will stain. Wear gloves or wash hands thoroughly at the end of the experiment.

PROCEDURE

Add 0.12 g of *dry* 2,4-dinitrophenol to a graduated centrifuge tube and dissolve the phenol in 0.6 mL (estimated in the centrifuge tube) of reagent-grade *dry* acetone. Add 0.1 mL of triethylamine, and cool the solution to below room temperature. In the hood, add 0.1 mL of a 30% (v/v) solution of oxalyl chloride in methyl *tert*-butyl ether (0.03 mL or 0.045 g of oxalyl chloride) *dropwise* to the acetone solution, and constantly stir the mixture until it is a thick cream-colored suspension. Back-titrate by adding 1 drop of triethylamine. If a yellow/orange color develops, add an additional 0.05 mL (1 or 2 drops) of oxalyl chloride until the loss of the yellow/orange color. (It is imperative that the mixture be kept cold during these operations.)

After 10 minutes, warm the mixture gently on a steam or hot water bath and stir to evaporate most of the acetone—creamy yellow crystals give the best light. When the mixture is a thick pasty consistency, cool in an ice bath, and remove the remaining solvent under reduced pressure, shaking the receiver by hand or by using a rotary evaporator. Have the system close to vertical because some bumping and spattering may occur. After most of the remaining acetone has been evaporated, warm the evacuated tube to room temperature, and release the vacuum.

Remove the product from the centrifuge tube with a microspatula and let dry. Use this material without further purification in Experiment B.

B.   CHEMILUMINESCENCE

(Microscale)

PROCEDURE

The chemiluminescent reaction can be carried out in several ways. A simple and effective procedure is to add the solid bis-(2,4-dinitrophenyl) oxalate to a solution of the fluorescent acceptor and hydrogen peroxide. One or more of the hydrocarbon or xanthene emitters will be available as 0.002 M solutions in dimethyl phthalate.

Label test tubes with the fluorescent emitters to be used, and obtain 1 mL of each solution. To each tube, add 1 mL of a 0.2 M solution of hydrogen peroxide in dimethyl phthalate/t-butyl alcohol (2 mL of 30% H_2O_2 in 20 mL t-butyl alcohol plus 80 mL of dimethyl phthalate). Mix the solutions well and take the tubes, your notebook, and your vial of bis-(dinitrophenyl) oxalate to a dark room or a shielded box in the laboratory. Add a few crystals of the oxalate to each tube and record your observations.

To observe the effect of added base on the reaction, mix 2 mL each of 0.002 M 9, 10-diphenylanthracene solution and 0.2 M hydrogen peroxide and divide the mixture between two tubes. Dissolve a few mg of your oxalate in 2 mL of dimethyl phthalate and obtain a few drops of a 10% solution of triethylamine in dimethyl phthalate. In the dark area, add 1 mL of the oxalate solution to each of the tubes of fluorescer, shake the tubes, and then add a drop of the triethylamine solution to one tube and compare the two reactions.

To observe the effect of temperature, place the other tube in an ice bath and then rewarm it.

QUESTIONS

1. Bis-(2,4-dinitrophenyl) oxalate and the 2,4,5-trichlorophenyl ester give much higher light yields than diphenyl oxalate in peroxy–oxalate chemiluminescence. Suggest a reason in terms of the mechanism of the process.
2. It has been reported that when hydrogen peroxide is added to a solution of bis-(dinitrophenyl) oxalate, and a stream of nitrogen is bubbled through the solution and into a solution of rubrene, light is emitted from the latter. Suggest an explanation for this observation.
3. Firefly luminescence requires "high-energy" phosphate in the form of ATP, and the process has been suggested to involve formation of structure A followed by further steps. Suggest the subsequent steps and a plausible high-energy intermediate I.

PRELABORATORY QUESTIONS

1. How does chemiluminescence compare with fluorescence?
2. How is the efficiency of a chemiluminescent reaction measured?
3. What is the oxidizing agent in the bis-(2,4-dinitrophenyl) oxalate reaction?
4. Explain how "acceptor" molecules function.
5. If you have one solution giving off yellow light and one giving off blue light, what color will the light be if you mix the two solutions together?

REFERENCES

Adam, W. *J. Chem. Educ.,* 1975, *52,* 138.

Burr, J. G., Ed.; *Chemi- and Bioluminescence;* Marcel Dekker: New York, 1985.

Mohan, A. G.; Turro, N. J. *J. Chem. Educ.,* 1974, *51,* 528.

Rauhut, M. M. *Accounts of Chemical Research,* 1969, *2,* 80.

Smith, P. E.; Johnston, K.; Reason, D. M.; Bodner, G. M.; Parker, G. A. *J. Chem. Educ.,* 1992, *69,* 827.

Photodimerization of Benzophenone

Photochemical reactions, involving molecules in high-energy electronically excited states, provide synthetic pathways to otherwise inaccessible compounds. A simple illustration is the photodimerization of benzophenone to benzopinacol. In this experiment, the benzophenone **1** absorbs a photon of ultraviolet (UV) light to give the triplet excited state, which can be drawn as the diradical **2**. The oxygen radical abstracts a hydrogen atom from the isopropyl alcohol, to give the carbon radicals **3** and **4**. Radical **4** donates a hydrogen atom to another molecule of **1** to give acetone and another equivalent of **3**. Two of the **3** radicals then couple to give the product benzopinacol **4**.

Other types of photochemical reactions include 2+2 cycloadditions and remote C-H activation.

mp = 185°C

EXPERIMENTS

A. PHOTOCHEMICAL PREPARATION OF BENZOPINACOL

(Microscale)

Combine 400 mg of benzophenone, 2.8 mL of isopropyl alcohol, and 1 drop of acetic acid in a small test tube. Stopper and label the test tube, then set it in a 50-mL Erlenmeyer flask. Irradiate with UV light (a "Spectroline Shortwave UV display lamp" from Fisher works well, as does direct strong sunlight) until heavy colorless crystals appear. Filter the product and record the yield and the melting point. Save a small amount for a TLC sample.

The conversion of benzopinacol **4** to benzopinacolone **5** proceeds by the familiar pinacol rearrangement. The melting points of **4** and **5** are similar, but the TLC R_f's are quite different from each other.

B. PREPARATION OF BENZOPINACOLONE

(Microscale)

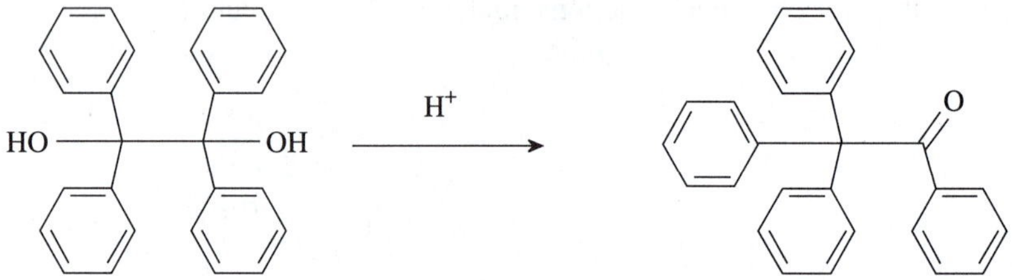

4

5

mp = 183°C

PROCEDURE

Combine 100 mg of benzopinacol and 1 mL of acetic acid in a test tube. Add a small crystal of iodine (the solution should be just brown, not black), and warm the test tube gently. Follow the reaction by TLC. When it is complete (just a few minutes), chill the reaction mixture and filter the resulting crystals. You may have to add some cold ethanol to have enough liquid to filter. Wash the crystals with a little cold ethanol, then record a melting point and a yield.

QUESTIONS

1. Would you expect the dimerization of benzophenone to work as well in methanol? Why or why not?
2. Draw an arrow-pushing mechanism for the conversion of benzopinacol to benzopinacolone.

PRELABORATORY QUESTIONS

1. Why might benzopinacol be less soluble in 2-propanol than is benzophenone?
2. How might you tell that the solid that crystallizes from solution on irradiation is not just the starting benzophenone?
3. What would happen if you were to look directly at the UV light source when it was turned on?

REFERENCE

Coyle, J. D. *Introduction to Organic Photochemistry;* Wiley: New York, 1986.

The Diels–Alder Reaction

The Diels–Alder reaction is an important synthetic tool for building cyclic systems. The reaction comprises the cycloaddition of a conjugated diene and another unsaturated compound, the dienophile. The dienophile usually contains a double or triple bond conjugated with an electron-withdrawing group or groups. Occasionally an acid catalyst is found to increase the rate of reaction, presumably by making the dienophile more electrophilic (dienophilic). A typical synthetic application of the Diels–Alder reaction is the condensation of butadiene and 4-methoxy-2,5-toluquinone; this is the first step in a classic synthesis of the steroid hormones.

The Diels–Alder reaction occurs by a concerted cycloaddition process in which bonds are broken and formed in a continuous change from reactants to products, without ionic or free-radical intermediates. The transition state for such a process requires a well-defined orientation of the reactants, and this in turn leads to high stereospecificity. Thus, if groups in the dienophile are *cis,* they remain *cis* in the product.

A second "rule" governing the stereochemistry of the cycloaddition is that in the formation of a bridged bicyclic product from a cyclic diene, the reaction proceeds with the orientation that places the new double bond closer to the substituents in the dienophile. This rule is illustrated by the reaction to be carried out in this chapter, in which cyclopentadiene and maleic anhydride react to give *endo*-norbornene-5,6-*cis*-dicarboxylic anhydride. The two possibilities for reaction give *endo* (inside) and *exo* (outside) products. In the *endo* product, the substituent is *trans* to the shorter bridge.

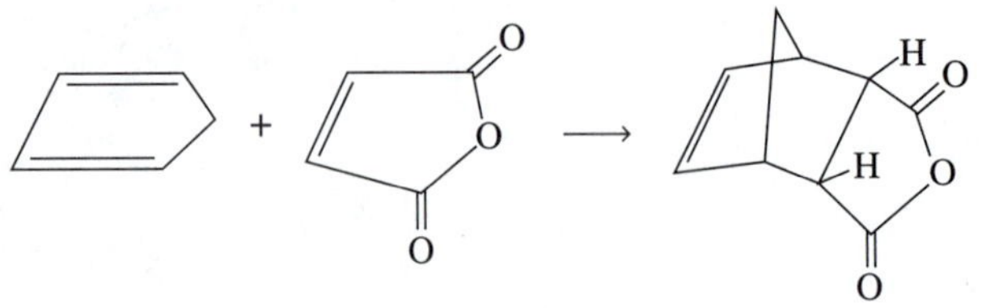

cyclopentadiene
bp 41°C, d 0.80

maleic anhydride
mp 60°C

endo-norbornene-5,6-*cis*-
dicarboxylic anhydride
mp 165°C

The preference for the *endo* orientation is suggested to arise from the more complete overlap of *p* orbitals in the transition state, compared with that for the *exo* product. This stereoselectivity is noteworthy, since in this case the *exo* anhydride is sterically less hindered and is the lower-energy product.

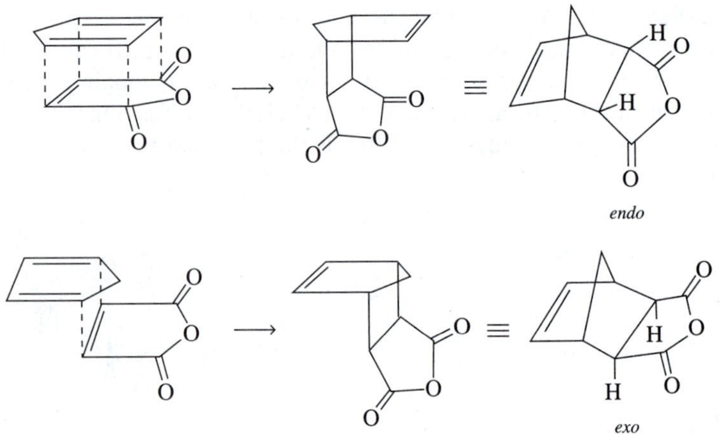

endo

exo

Diels–Alder cycloadditions are reversible reactions, and the adducts can often be converted back to the diene and dienophile by heating. The reverse process is illustrated in this experiment by using the reaction to obtain the diene employed in the synthesis. Cyclopentadiene is extremely reactive as a diene component, and on standing it dimerizes by utilizing one mole as a diene and another as dienophile. The compound is therefore stored and sold as the dimer, called dicyclopentadiene. The reverse reaction is carried out simply by heating the dimer (cracking) and distilling out the monomeric cyclopentadiene as it is formed.

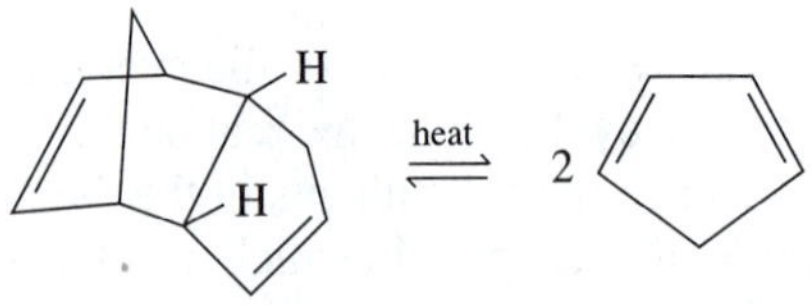

dicyclopentadiene

EXPERIMENTS

A. *ENDO*-NORBORNENE-5,6-*CIS*-DICARBOXYLIC ANHYDRIDE

(M a c r o s c a l e)

SAFETY NOTE

Cyclopentadiene and maleic anhydride are toxic and maleic anhydride dust is irritating. It is best to carry out the macroscale procedure in a hood to reduce exposure to cyclopentadiene fumes. Ethyl acetate is flammable and mildly toxic; avoid breathing the vapors.

PROCEDURE

Place 4 g of maleic anhydride in a 125-mL Erlenmeyer flask and dissolve it in 15 mL of ethyl acetate by warming on the steam bath. Add 15 mL of hexane (or petroleum ether), and then cool the solution in an ice bath.

From your instructor obtain 4 mL of cold cyclopentadiene in a 10-mL graduated cylinder, add it to the maleic anhydride solution, and swirl the solution to effect mixing. Wait until the product crystallizes from solution, then heat it on the steam bath to redissolve, and allow it to recrystallize. Collect the product by suction filtration, and record the melting point, weight, and percentage yield. Dispose of the filtrate in the organic solvent waste container.

B. *ENDO*-NORBORNENE-5,6-*CIS*-DICARBOXYLIC ANHYDRIDE

(M i c r o s c a l e)

PROCEDURE

Place 0.4 g maleic anhydride in a 25-mL Erlenmeyer flask and dissolve it in 1.5 mL of ethyl acetate by warming it on a steam bath. Add 1.5 mL of hexane, and then cool the solution in an ice bath. Obtain 0.4 mL of cyclopentadiene from your instructor and add this to the cold maleic anhydride solution. Swirl the solution to cause mixing, replace it in the ice bath, and scratch the inside of the flask with a stirring rod to induce crystallization. Once the product crystallizes, heat the mixture on the steam bath to redissolve the solid and recool it to induce recrystallization in the same container. Collect the product by suction filtration on a Hirsch funnel, record the melting point of the dry crystals, and calculate the percent yield. Dispose of the filtrate in the organic solvent waste container.

QUESTIONS

1. Write the structures of the products that would be obtained in the following Diels–Alder reactions:

    a.

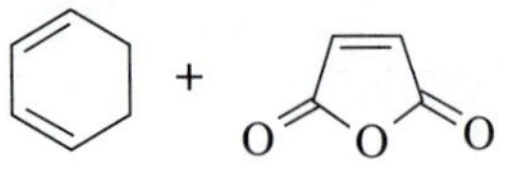

    b. 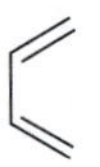   + CH_2=$CHCO_2Me$

    c.    + $MeOCC$≡$CCOMe$

2. Write the products that would be obtained by the thermal reverse Diels–Alder reaction of the following:

    a.

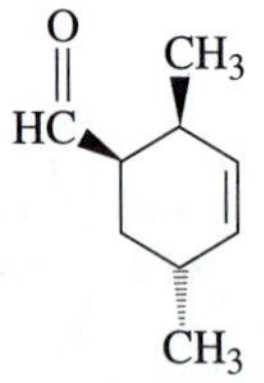

    b.

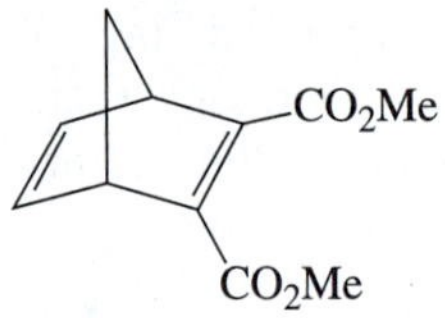

    c.

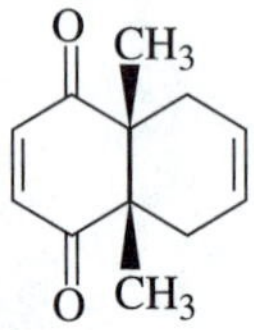

3. After collecting a first crop of norbornene-5,6-dicarboxylic anhydride, TLC indicated the presence of further product in the mother liquor. The solution was concentrated on the steam bath to isolate a second crop of product, but the residue proved to be an oily liquid, and the desired product could not be crystallized. Suggest the reason for the problem. What was the oily liquid?

PRELABORATORY QUESTIONS

1. What is the product of the following Diels–Alder reaction? (Show stereochemistry where applicable.)

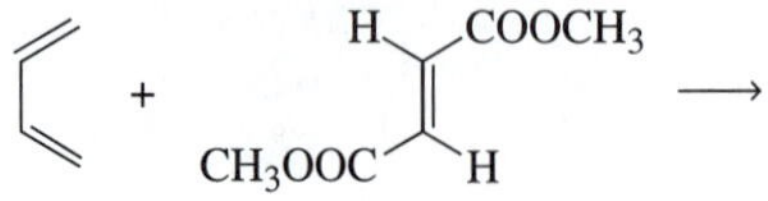

2. The Diels–Alder reaction is said to be "stereospecific." What does that mean in terms of the product shown in Question 1?
3. Why cannot cyclopentadiene be purchased directly from a commercial supplier?
4. Draw the structure for the "exo" product from the reaction of cyclopentadiene with maleic anhydride.
5. How is the "exo" form of a bridged, bicyclic compound defined?

REFERENCE

Fringuell, F.; Tatuchi, A. *Dienes in the Diels-Alder Reaction;* Wiley: New York, 1990.

Phase Transfer Catalysis

he traditional approach to bringing about a chemical reaction between two substances of different types, such as an organic compound and an inorganic salt, is to use a solvent in which both reactants are at least partially soluble. In some cases, alcohols can be used; however, dipolar protic solvents such as dimethylsulfoxide (DMSO) or dimethylformamide (DMF) are often more suitable. A different and very general way to effect such reactions is to use a **phase transfer catalyst.** In this method the reactants may be in two different liquid phases. The catalyst is a quaternary ammonium salt, $R_4N^+ X^-$, in which long alkyl groups provide solubility in organic solvents and the cation is the hydrophilic portion. Crown ethers may also be used.

Many nucleophilic reactions can be enhanced by the use of phase transfer catalysts. Typically, an S_N2 reaction of an alkyl halide (RX_{org}, soluble in organic phases) with a nucleophilic anion (Nuc_{aq}^-, soluble in polar phases) can be carried out using a catalyst such as benzyltributylammonium chloride.

$$RX_{org} + Nuc_{aq}^- \longrightarrow RNuc_{org} + X_{aq}^-$$

The catalyst transfers the nucleophile to the organic phase, thus setting up an equilibrium cycle between the two phases as illustrated in the following diagram:

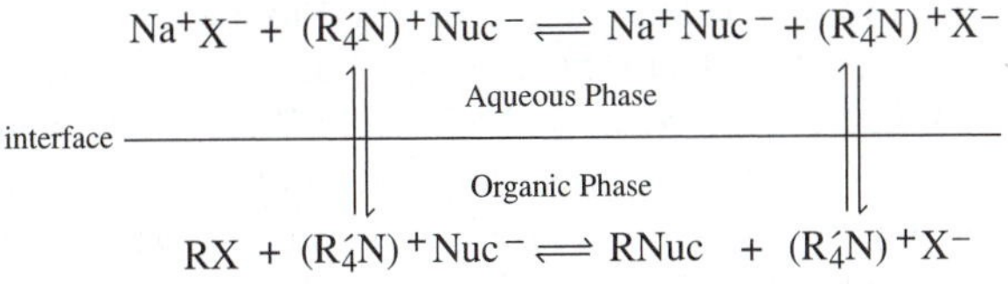

$$Na^+X^- + (R_4'N)^+Nuc^- \rightleftharpoons Na^+Nuc^- + (R_4'N)^+X^-$$

interface ———————————— Aqueous Phase / Organic Phase

$$RX + (R_4'N)^+Nuc^- \rightleftharpoons RNuc + (R_4'N)^+X^-$$

The reaction to be studied involves the preparation of Nerolin from an alkyl halide and a naphtholate ion that have very different solubility characteristics. The phase transfer catalyst is tetrabutylammonium chloride, an organic ion that is soluble in organic solvents because of the alkyl groups and, at the same time, is slightly soluble in aqueous media because of its ionic nature. The overall reaction is as follows:

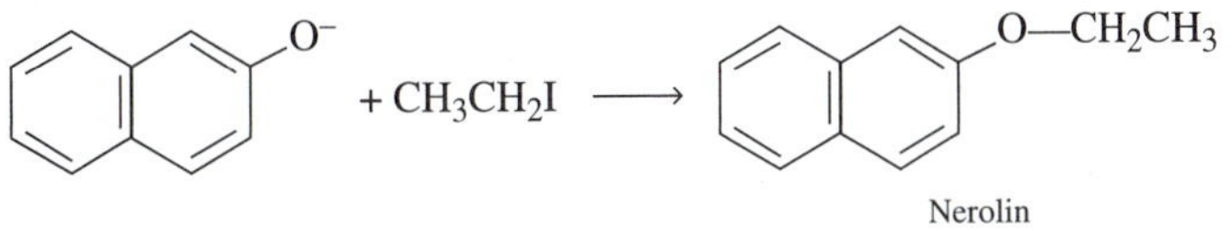

Nerolin
mp 37–38°C

Nerolin [2-ethoxynaphthalene; ethyl β-naphthyl ether] is used as a perfume in some soaps and as a fixative in soaps and perfumes. The purpose of a fixative is to bind other volatile, fragrant ingredients so that they will not evaporate as rapidly, which then allows the product to retain its pleasant odor for a longer time.

Mixed ethers like Nerolin may be made by the Williamson synthesis, a nucleophilic substitution reaction. In this case the conjugate base of 2-naphthol is reacted with ethyl iodide. Normally this reaction is carried out in an anhydrous, one-phase system involving methanol as the solvent. It may also be carried out as a two-phase reaction if a phase transfer catalyst is used. The phase transfer catalyst used facilitates the dissolution of the naphthoate ion in the alkyl halide layer.

EXPERIMENT

Safety Note

Ethyl iodide and 2-naphthol are both toxic irritants. Handle these materials with care.

PREPARATION OF NEROLIN

(Microscale)

PROCEDURE

In a large test tube, combine 1.44 g of 2-naphthol (TOXIC!) and 4.0 mL of 10% aqueous NaOH, rinsing any naphthol from the walls of the test tube. Add 4.0 mL of distilled water and swirl the tube until most of the brown solid has dissolved. The remainder of the procedure must be carried out in the fume hood because of the toxicity of the reagents! Add 10 drops of 10% aqueous tetrabutylammonium chloride, 0.88 mL of ethyl iodide (TOXIC!), and a boiling chip and, heating the very bottom of the test tube in a sand bath, bring the mixture to a gentle boil, stirring with a glass rod. Follow the reaction by TLC (3:7 ethyl acetate/hexane works well), visualizing by UV. When the starting material has almost all been consumed (~ 30 minutes), cool the mixture to room temperature, then rinse it into a small separatory funnel with ether and extract the aqueous layer with 2×20 mL of ether. Wash the ether extracts with saturated aqueous $NaHCO_3$ solution, dry them over Na_2SO_4, and concentrate.

Take up the product in 3 mL of hot methanol, add hot water dropwise until cloudiness just appears, then one more drop of methanol. Add activated carbon to decolorize the solution and filter through Celite into a small test tube. Cool the test tube in an ice-water bath to effect crystallization. Filter the crystals and determine the melting point and the yield. Consider the ^{1}H-NMR spectrum of the product—where might you expect to find each proton (chemical shift), and what would you expect the splitting pattern to be?

QUESTIONS

1. Write a reaction to show the Williamson synthesis of benzyl phenyl ether.
2. Sketch the transition state for the S_N2 reaction of ethyl iodide with the conjugate base of 2-naphthol.
3. Give an example of a compound other than tetrabutylammonium chloride or benzylammonium chloride that might be used as a phase-transfer catalyst. (Consult your textbook if necessary.)
4. What effect would doubling the concentration of the nucleophile in the alkyl halide layer have on the reaction rate of an S_N2 reaction?

PRELABORATORY QUESTIONS

1. Write an equation to show the Williamson synthesis of anisole from sodium phenoxide and methyl iodide.
2. Draw the structure of Nerolin. What is it used for?
3. Draw the structural formula for the phase transfer catalyst used in this experiment.
4. Give examples of two phenols other than 2-naphthol that could be used as nucleophiles in the Williamson synthesis.
5. What is meant by the term "phase transfer catalyst"?

REFERENCES

McIntosh, J. M. *J. Chem. Ed.,* 1978, *55*, 235.

Rowe, J. E. *J. Chem. Ed.,* 1980, *57*, 162.

34

Derivatives of Carboxylic Acids: Synthesis of a Plant Hormone

Plant hormones are substances synthesized in the cells of a plant that regulate growth and physiological function. One of the major plant hormones is indoleacetic acid, found in several parts of the plant. This compound and other plant hormones that promote and control growth are called **auxins.** In addition, several structural analogs of indoleacetic acid elicit the same growth response in plants. Among these synthetic auxins are indolebutyric acid, 1-naphthaleneacetic acid, and 2,4-dichlorophenoxyacetic acid (2,4-D). The application of auxins at very low concentrations promotes stem growth and stimulates rooting. High concentrations can cause destruction of the plant, as in the use of 2,4-D as a weed killer.

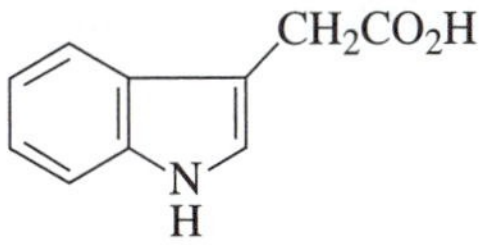

indoleacetic acid

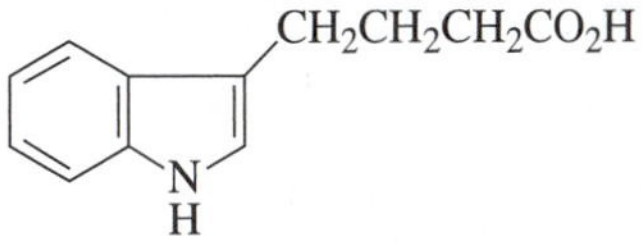

indolebutyric acid

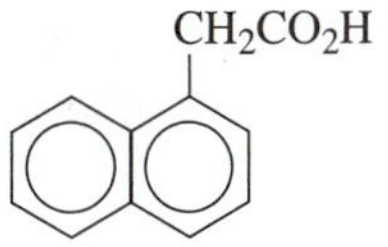

naphthaleneacetic acid

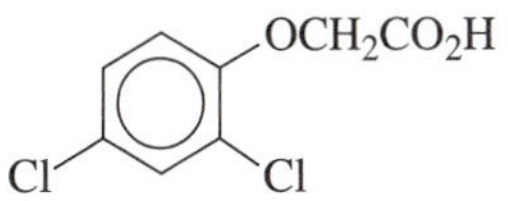

2,4-dichlorophenoxyacetic acid

One of the important uses of auxins is in the propagation of plant cuttings. For this and several other horticultural purposes, the preferred auxin is the amide of 1-naphthaleneacetic acid. This compound probably acts by the same mechanism as the acid but is more effective. Commercial preparations such as Transplantone® or Rootone® contain indolebutyric acid as the principal active ingredient (less than 0.1%) together with fertilizer or inert carrier. Fungicides are often added to protect the developing roots from fungal attack. In this experiment, 1-naphthaleneacetamide will be synthesized by two different methods.

SYNTHESES OF 1-NAPHTHALENEACETAMIDE

One method of synthesis illustrates a general approach to several types of carboxylic acid derivatives, namely the preparation and reaction of an acid chloride with a nucleophile, in this case ammonia. The second method involves the conversion of a nitrile to an amide by hydration.

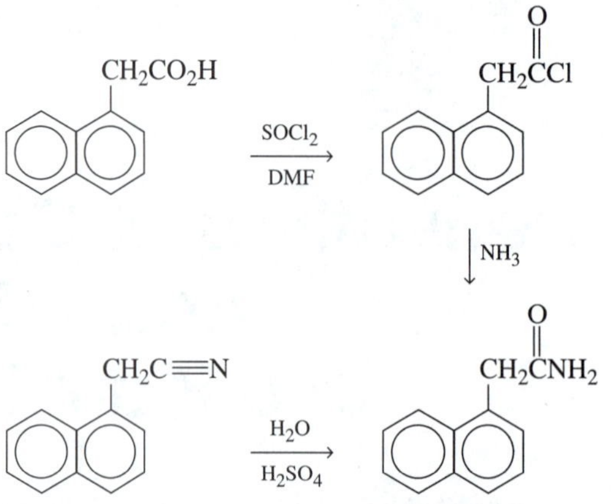

1-naphthaleneacetamide
mp 183–184°C

The most commonly used reagent for converting an acid to the acid chloride is thionyl chloride, $SOCl_2$; the reaction is convenient because the by-products SO_2 and HCl are gases and are readily removed. A complication, however, is the possibility of chlorination at the α-position of the acid on heating with thionyl chloride, particularly in the case of an acid of the type $ArCH_2CO_2H$, in which the α-hydrogens are readily enolized. An effective way to circumvent this problem is to use dimethylformamide (DMF) as a catalyst. The DMF is converted to dimethylformamide chloride, which then reacts with the acid at room temperature to give the acid chloride:

$$HC(=O)N(CH_3)_2 + SOCl_2 \longrightarrow H{-}C(Cl){=}\overset{+}{N}(CH_3)_2Cl^- + SO_2$$

$$RCO_2H + HC(Cl){=}\overset{+}{N}(CH_3)_2Cl^- \longrightarrow RC(=O)Cl + HCl + HC(=O)N(CH_3)_2$$

In the second synthesis, hydration of the nitrile is carried out in 67% (v/v) sulfuric acid solution. Strong acid protonates the nitrile, and water then attacks the nitrilium ion to give the imidic acid, which isomerizes to the amide:

$$RC{\equiv}N \xrightarrow{H^+} \underset{\text{nitrilium ion}}{RC{\equiv}\overset{+}{N}H} \xrightarrow[-H^+]{H_2O} \underset{\text{imidic acid}}{RC(OH){=}NH} \longrightarrow RC(=O)NH_2$$

Since hydrolysis of the amide to the carboxylic acid can occur, the hydration must be carried out under conditions that are as mild as possible.

EXPERIMENTS

A. ACID CHLORIDE METHOD

(Microscale)

SAFETY NOTE

Thionyl chloride is a fuming, corrosive liquid and reacts violently with water. Concentrated aqueous ammonia has choking fumes. Handle all chemicals in the hood and avoid breathing vapors; if materials are spilled on the skin, wash thoroughly with water.

PROCEDURE

In a dry, heavy-wall 10-mL Erlenmeyer flask or 13 × 100-mm test tube, place 0.47 g of 1-naphthaleneacetic acid (1-naphthylacetic acid). In the hood, obtain 0.25 mL of thionyl chloride (graduated pipet) and add it all at once to the acid. Cool the mixture of acid and thionyl chloride in an ice bath, and then remove from the bath. Add 0.1 mL of dimethylformamide, and swirl the contents (note the temperature of the container with your hand). Loosely cork the flask or test tube, label it with your name, and allow it to stand in the hood. After 1 hour, connect the flask or test tube to an aspirator and evaporate the excess thionyl chloride by keeping it under aspirator vacuum for about 2 minutes and warming to 30°C.

Mix 1.3 mL of concentrated aqueous ammonia and approximately 3 mL of crushed ice in a graduated cylinder, and pour the mixture into the flask containing the acid chloride. Stir with a rod or spatula until all of the oil is converted to solid. Collect the solid by vacuum filtration, wash with 1 mL of cold water, and press out excess water. Recrystallize the solid from 1 to 2 mL of hot ethanol. Dry the product and determine its weight, melting point, and the yield. The remaining filtrate may be flushed down the drain with ample water.

B. NITRILE METHOD

(Microscale)

SAFETY NOTE

Concentrated sulfuric acid causes burns on contact with skin. If any is spilled on the skin, wash immediately and thoroughly with water, followed by soap and water. Naphthalene-1-acetonitrile is an irritant.

PROCEDURE

In a 10-mL Erlenmeyer flask, place 1 mL of water and then slowly add 2 mL of concentrated sulfuric acid. (**CAUTION:** The flask will become very hot.) Add 0.5 g of naphthalene-1-acetonitrile (1-naphthylacetonitrile) and place the flask on a steam bath. Using a strip of paper towel in a loop to protect your hand, swirl the mixture intermittently to keep the oily layer suspended. When all of the oil dissolves (approximately 15 to 20 minutes), allow the hot solution to stand for an additional 10 minutes and then cool. Add approximately 4 g of ice, and stir the oil until it solidifies. Collect the solid by vacuum filtration, press out excess water, and recrystallize from 1 to 2 mL of hot ethanol. Save a sample of the mother liquor for TLC. Dry the product and determine the weight, melting point, and yield. The remaining filtrate may be flushed down the drain with ample water.

QUESTIONS

1. A student omitted the aqueous ammonia in preparation of the amide; ice was added to the acid chloride, then the oil was rubbed and stirred to give a white solid that was collected, washed with water, and recrystallized. What was this product?
2. The 1999 prices of 1-naphthaleneacetic acid and 1-naphthylacetamide were $27.35/100 g and $48.80/100 g, respectively. Ignoring the cost of solvents, other reagents, your time, etc., calculate the cost of 100 g of amide based on the cost of the acid and your yield in Experiment A of the amide preparation.
3. The 1999 prices of 1-naphthylacetonitrile and 1-naphthylacetamide were $263/100 g and $48.80/100 g, respectively. Ignoring the cost of solvents, other reagents, your time, etc., calculate the cost of 100 g of amide based on the cost of the nitrile and your yield in Experiment B of the amide preparation.
4. A 2.0-oz bottle (56.8 g) of a commercial rooting hormone containing 0.10% of indolebutyric acid costs $5.49. The cost of 98% indolebutyric acid is $56.10 for 25 grams. Calculate the nominal cost of the indolebutyric acid in the commercial product.

PRELABORATORY QUESTIONS

1. Write an equation to show the reaction of 1-naphthaleneacetic acid with thionyl chloride to give the acid chloride.
2. Write an equation to show the reaction of the product formed in Question 1 above with concentrated aqueous ammonia.
3. What properties of thionyl chloride make it a reagent to be used with caution? Why is it prudent to use it in a hood?
4. Show an "arrow pushing" mechanism for the hydrolysis of benzonitrile to the corresponding amide by aqueous acid.

35

Synthesis of Coumarin

major process in synthetic organic chemistry is the condensation of an enolate with a carbonyl group to form a carbon–carbon bond. The **Knoevenagel condensation** takes advantage of the greater acidity of the α-hydrogens in a compound with two activating groups such as RCO, CO_2R, or CN. A weak base such as an amine can be used, and the reactions usually proceed rapidly under mild conditions to give the unsaturated product in high yield.

$$CH_2 \Big\langle {}^{COOR'}_{COOR'} \xrightarrow{\text{B:}} {}^-CH \Big\langle {}^{COOR'}_{COOR'}$$

$$\underset{\text{RCH}}{\overset{O}{\parallel}} + {}^-CH \Big\langle {}^{COOR'}_{COOR'} \longrightarrow \underset{\text{RCHCH}}{\overset{O^-}{\mid}} \Big\langle {}^{COOR'}_{COOR'}$$

$$+\ H^+ \downarrow$$

$$RCH = C \Big\langle {}^{COOR'}_{COOR'} \xleftarrow{-H_2O} \underset{\text{RCHCH}}{\overset{OH}{\mid}} \Big\langle {}^{COOR'}_{COOR'}$$

To obtain a monocarboxylic acid, the condensation can be carried out with malonic acid, as one of the CO_2H groups subsequently is removed by decarboxylation. Alternatively, the less expensive malonic ester can be used, followed by hydrolyses of the ester groups and decarboxylation in separate steps:

$$RCH = C \Big\langle {}^{CO_2R'}_{CO_2R'} \xrightarrow[\text{2) H}^+]{\text{1) 2 KOH}} RCH = C \Big\langle {}^{CO_2H}_{CO_2H} \xrightarrow{\Delta} RCH = CHCO_2H + CO_2$$

SYNTHESIS OF COUMARIN

In the sequence of experiments in this chapter, the Knoevenagel condensation, ester hydrolysis, and decarboxylation are illustrated by a classic synthesis of the aromatic lactone, coumarin. This compound occurs in nature in many plants, including sweet clover, the herb woodruff, and the tonka bean, from a South American plant that was for many years the principal source. Coumarin has a strong, sweet odor of newmown hay and is widely used in perfumery. It was formerly used with vanilla as a flavoring agent, but because of its toxicity, coumarin is now prohibited for use in human food. It has been classified as a suspected carcinogen, but the hazards of coumarin to humans are low.

The synthesis begins with the Knoevenagel condensation of salicylaldehyde and diethyl malonate using piperidine and acetic acid as the catalyst. In this step, the C=C bond is formed and the lactone ring is closed by transesterification involving the *o*-hydroxy group and one of the carboxylate esters. The second step is a simple hydrolysis of the other ester group with KOH, and the final step involves the removal of the CO_2H groups by decarboxylation. The first two steps can be carried out in one 3-hour laboratory period and the last reaction in a second period. The coumarin is isolated by short-path distillation in a setup in which the distillate is collected as a solid on a cold-finger condenser.

salicylaldehyde   diethyl malonate   ethyl coumarin-3-carboxylate
                                                    mp 92–93°C

1) KOH
2) HCl

coumarin   −CO₂ Δ   coumarin-3-carboxylic acid
mp 69–70°C   mp 191–192°C

EXPERIMENTS

SAFETY NOTE

None of the starting materials in this sequence is particularly toxic or hazardous. Piperidine has a very unpleasant odor and should be transferred in a hood. Both KOH and HCl are corrosive and should be handled with care.

A. ETHYL COUMARIN-3-CARBOXYLATE

(Macroscale)

PROCEDURE

In a 50-mL round-bottom flask, place 2.7 mL (3.0 g) of salicylaldehyde, 4.8 mL (5.0 g) diethyl malonate, and 12 mL of absolute ethanol. With a graduated pipet, add 3.0 mL of piperidine, and then add 5 drops of glacial acetic acid. Add a boiling stone, fit the flask with a reflux condenser, and heat the solution at reflux for approximately 1 hour. If desired, samples (1 to 2 drops) of the solution can be removed before starting and at 30-minute intervals to follow the progress of the reaction by TLC (use fluorescent silica gel, develop with CH_2Cl_2, and visualize with a UV lamp).

While the solution is being heated, prepare a setup for suction filtration with a small Büchner funnel. For the next step, prepare a solution of 4.0 g of KOH in 20.0 mL of water and 8.0 mL of ethanol.

When the condensation reaction is complete—confirmed by TLC analysis—cool the solution to approximately 60°C, add 15 mL of warm water, and then chill the solution in an ice bath. When the mixture is cold, collect the crystals and wash with a small amount of chilled 50% ethanol–water. Press the crystals thoroughly on the funnel and then spread them out on paper. Remove a small sample, allow it to dry completely, and then record the melting point. Meanwhile, continue directly to Experiment B.

B. COUMARIN-3-CARBOXYLIC ACID

Since this reaction is carried out in aqueous ethanol solution, there is no need to completely dry the ester from the preceding step. Place the damp crystals in a 50-mL round-bottom flask, and add 28.0 mL of the aqueous alcoholic KOH solution prepared above and a boiling stone. Heat the solution to reflux for 30 minutes. During this time, obtain 12 mL of concentrated HCl, mix it in a 100 mL Erlenmeyer flask with 12 mL of water, and cool the solution in the icebath.

After the alcoholic KOH solution has been heated for 30 minutes, cool in an ice bath and add the cold HCl solution all at once. Collect the crystals of the acid on a small Büchner funnel, and wash them thoroughly with water. Press the solid on the funnel, and spread it out to dry until the next laboratory period. Record the melting point and weight, and calculate the yield for the two-step sequence.

C. COUMARIN

SAFETY NOTE

In this part, you will be carrying out a distillation at reduced pressure. Be sure to keep safety goggles on at all times.

In a 125-mL heavy-wall side arm filter flask, place 2.0 g of the dried carboxylic acid from Experiment B. Clamp the flask about 1 cm deep in a sand bath with a thermometer in the bath. Place a rubber stopper in the neck of the flask. Connect a piece of rubber tubing to the side arm, attach a short length of glass tubing or a pipet to the other end of the rubber tubing, and place the glass tip in a test tube containing 2 to 3 mL of water so that bubbling can be observed. (**CAUTION:** Arrange the tube with the tip *just* below the surface of the water so that water cannot be sucked back into the side arm flask if the system is cooled.) Heat the sand bath to 220°C, and then maintain the temperature between 220 and 240°C until CO_2 evolution (bubbling) ceases; this usually requires 20 to 25 minutes. Tap the rubber stopper occasionally to dislodge droplets of distillate from the walls.

When gas evolution has stopped, remove the heat source, raise the flask out of the bath, disconnect the tubing, and allow the flask to cool, first in air and then in cold water. (Be careful not to use the cold water too quickly, to avoid cracking the flask.) Remove the stopper, scrape off any oil or solid, and return it to the flask. Place a cold-finger condenser (a 16 × 150-mm test

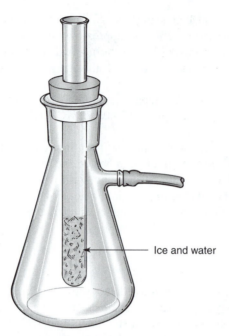

Figure 35.1 Simple sublimation apparatus.

tube fitted through a rubber cone or a rubber stopper containing a wide hole; see Fig. 35.1) in the flask with the test tube approximately 2 cm above the bottom of the flask.

Connect the side arm of the flask to the aspirator, evacuate the flask, and pack the cold finger with small pieces of ice. (Ice can be prepared for use in the cold finger by crushing it finely in a towel.) Be sure that you have and maintain vacuum! Clamp the flask in the sand bath and heat to 200°C. Maintain the temperature at 210 to 220°C for 25 minutes. Every few minutes, remove water from the cold finger with a pipet and bulb and replace it with more ice. Take care not to get water inside the flask. Some splattering usually occurs during the distillation, and it may be difficult to see inside the flask. Most of the coumarin distills in the first 20 minutes; a total time of 25 to 30 minutes at 220°C is usually sufficient to complete the distillation.

Raise the flask out of the bath and allow it to cool while still under aspirator vacuum, first in air and then in water. Pipet out the ice water in the cold-finger test tube (to avoid condensation of moisture), and then remove the aspirator tube. Carefully lift out the cold finger, and scrape the distilled coumarin onto a piece of weighing paper. The material is usually a mixture of brown and white solid at this point. With the spatula, scrape the white nodules of distillate from the walls of the flask and add them to the material from the cold finger. Record the weight of the solid and take a melting point of the crude sample.

Clean out the flask (use a few mL of acetone to loosen the residue) and dry it with aspirator suction. Place the solid back in the flask, replace the cold finger, fill with ice, connect the flask to the aspirator, evacuate, and distill the crude coumarin again at a bath temperature of approximately 140 to 160°C. Remove the product as described above, weigh, and report the yield and melting point. Compare the melting point with that of the crude coumarin.

QUESTIONS

1. Show the mechanistic steps involved in the base-catalyzed aldol condensation of propanal with itself.
2. Write a series of reactions to illustrate the preparation of cinnamaldehyde from benzaldehyde and ethyl acetate, using a base catalyst.
3. The first synthesis of coumarin in the nineteenth century was carried out by Perkin using salicylaldehyde, acetic anhydride, and sodium acetate. Write a series of reactions to illustrate this synthesis.

PRELABORATORY QUESTIONS

1. Write an equation to show the base-catalyzed reaction of benzaldehyde and dimethyl malonate.
2. Write two resonance forms for the intermediate enolate ion formed in the reaction above.
3. Draw the structure of coumarin.
4. In Experiment B of synthesis of coumarin, what is the purpose of adding KOH, boiling, and then adding HCl? What is the product after the KOH has boiled?

REFERENCE

Jones, G. *Organic Reactions: The Knoevenagel Reaction;* John Wiley & Sons: New York, 1967; Vol. 15, p. 204.

36

Benzyne

Benzynes, or dehydrobenzenes, were first recognized in 1953 as reactive intermediates in certain nucleophilic substitution reactions of aryl halides. Since their discovery, a number of different approaches have been devised for generating these intermediates. Several of the methods that have been used to prepare benzyne are shown in the following diagram:

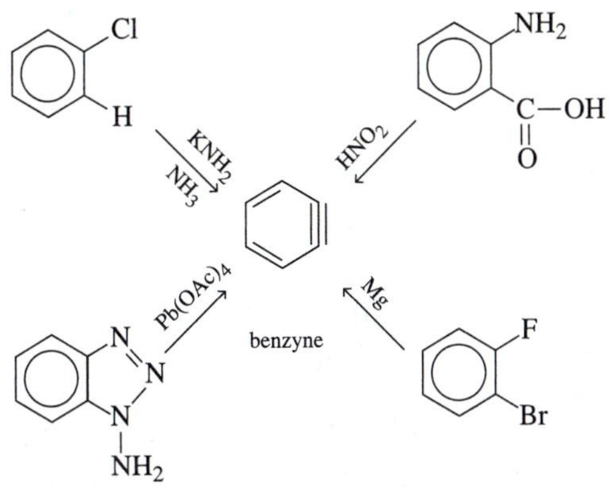

The main synthetic utility of benzynes arises from their great reactivity as electrophiles and dienophiles. As examples of the former property, when generated in water or liquid ammonia, benzyne rapidly reacts to form phenol or aniline, respectively.

The Diels–Alder reactivity of benzyne is illustrated in the experiment in this chapter, in which furan is used as the diene. The benzyne is generated in the presence of the furan and reacts in situ. The product is 1,4-dihydronaphthalene-1,4-endoxide, which can be isomerized very readily in the presence of acid to 1-naphthol.

Benzyne in this experiment is generated by the diazotization of anthranilic acid with a nitrite ester, followed by loss of carbon dioxide and nitrogen. To minimize side reactions of benzyne with the starting materials, the anthranilic acid and isoamyl nitrite are added at the same time so that neither is present in large excess.

EXPERIMENTS

SAFETY NOTE

Isoamyl nitrite is toxic and is a coronary vasodilator and 1-naphthol is a toxic irritant. Handle both compounds with care and immediately wash off any that get on the skin. Isoamyl nitrite and 1,2-dimethoxyethane are flammable liquids as are diethyl ether and methyl t-butyl ether.

anthranilic
acid

isoamyl nitrite
d 0.87

1,4-dihydronaphthalene-
1,4-endoxide, mp 56°C

furan
d 0.94

1-naphthol, mp 96°C

A.   1,4-DIHYDRONAPHTHALENE-1,4-ENDOXIDE

(Microscale)

In a 25-mL round-bottom flask place 2 mL of furan, 2 mL of 1,2-dimethoxyethane (glyme), and a boiling stone, and attach the flask to a reflux condenser. In separate 10×13-mm test tubes, dissolve 1 mL of *iso*-amyl nitrite in 2 mL of glyme and 0.55 g of anthranilic acid in 2 mL of glyme. (If the anthranilic acid does not dissolve easily, heat gently on a steam bath.) Heat the furan solution to reflux on a steam bath and at 3- to 4-minute intervals add 0.5 mL of each of the two solutions through the condenser. Use separate transfer pipets for each of the reagents, since they react with one another. One-mL graduated syringes are convenient for these transfers. After the additions are complete, continue at reflux for 3 minutes.

Cool the reaction mixture and add 5 mL of 2% NaOH solution. Transfer the mixture to a separatory funnel with a Pasteur pipet and extract with 5 mL of pentane. Discard the aqueous layer and rinse the pentane solution with four 3-mL portions of water. Add charcoal to partially decolorize the pentane solution, filter through a pipet containing $MgSO_4$, and concentrate to approximately 1 mL on a steam bath. If an oil separates at this point, decant the warm solution into a clean test tube and rinse the oil with 1 to 2 mL of pentane. Let the solution cool and scratch to induce crystallization of the cyclic ether (endoxide). If the solution does not crystallize, add pentane, place on a steam bath, reduce the volume, and chill to induce crystallization. Collect the product and recrystallize it from approximately 5 mL of pentane. Weigh, determine the melting point, and calculate the percentage yield of the 1,4-dihydronaphthalene-1,4-endoxide. If time permits, a purer product can be obtained by sublimation at 100°C under aspirator vacuum rather than by recrystallization. The aqueous filtrate may be neutralized and flushed down the drain with ample water. The pentane mother liquor should be placed in the solvent waste container.

B.   1-NAPHTHOL

(Microscale)

Place 0.1 g of 1,4-dihydronaphthalene-1,4-endoxide and 2 mL of ethanol in a centrifuge tube, and add 1 mL of concentrated hydrochloric acid. Stir and let stand for 10 minutes. Add 4 mL of diethyl ether or methyl *t*-butyl ether and 3 mL of water, stopper and shake the tube, and draw off the lower aqueous layer with a pipet. Wash the ether layer with two 1-mL portions of water and dry the ether layer by filtering it into a 15-mL Erlenmeyer flask through a column of Na_2SO_4 contained in a Pasteur pipet. Rinse the pipet with a small amount of solvent and evaporate the solution to an oil on a steam bath. Add 3 mL of hexane, heat to dissolve the product, and let cool to crystallize. Filter the slightly pink crystals and air dry them in the funnel. Weigh and determine the percentage yield and the melting point of the 1-naphthol. The filtrate should be placed in the solvent waste container.

QUESTIONS

1. Write out the complete reaction for the preparation of benzyne from chlorobenzene and potassium amide/ammonia; show intermediates and by-products.
2. Write out the complete reaction for the preparation of benzyne from *o*-bromofluorobenzene and magnesium; show intermediates and by-products. (*Note:* aryl fluorides do not form Grignard reagents.)
3. Write equations involving benzyne intermediates for the following reactions:
   a. *p*-bromotoluene + KNH_2/liq NH_3 → *m*- and *p*-toluidine
   b. *o*-fluorobromobenzene + Mg + anthracene → triptycene
   c. 3-(*m*-chlorophenyl)propionitrile + KNH_2 → 1-cyanobenzocyclobutane
4. Careful analysis of the gas evolved in the reaction of anthranilic acid and isoamyl nitrite shows that nitrogen is lost first, followed by carbon dioxide. Suggest a possible uncharged intermediate that might be formed by the loss of nitrogen.
5. Suggest why four water rinses of the pentane layer are required in the isolation of 1,4-dihydronaphthalene-1,4-endoxide.
6. Suggest the mechanism for the acid-mediated isomerization of 1,4-dihydronaphthalene-1,4-endoxide to 1-naphthol.

PRELABORATORY QUESTIONS

1. Draw the structure of benzyne.
2. Draw the structure of the product that forms when benzyne reacts with ammonia.
3. Draw the structure of anthranilic acid.
4. Draw the structure of the product that forms when furan is reacted with benzyne.
5. What properties of isoamyl nitrite require that it be handled with caution?

REFERENCES

Fieser, L.F.; Haddadin, M.J. *Can. J. Chem.*, 1965, *43*, 1599.

Gilchrist, T.L. In *The Chemistry of Functional Groups,* Supplement C; Patai, S.; Rappoport, Z., Eds. Wiley: New York, 1983; Part 1, pp. 383–419.

Levin, R.H. *React. Intermed.* (Wiley) 1978, *1*, 1–26; 1981, *2*, 1–14; 1984, *3*, 1–18.

Heterocyclic Syntheses: Pyrimidines, Pyridones, Isoxazoles, Quinolines, & Indoles

The most common heterocyclic compounds are those containing nitrogen and other heteroatoms in a fully unsaturated five- or six-membered ring. Such ring systems are isoelectronic with cyclopentadiene anion and benzene, respectively, and are aromatic compounds; typical examples are pyrrole, pyrazole, isoxazole, pyridine, and pyrimidine. These and other rings are present in important natural substances such as hemoglobin, vitamins, and nucleic acids; they also form the basic structure of many synthetic drugs and dyes. Rings with a number of combinations of heteroatoms are readily prepared, and heterocyclic compounds constitute the bulk of all known organic molecules.

pyrrole   pyrazole   isoxazole

pyridine   pyrimidine

Many heterocyclic compounds are prepared by simple reactions such as the condensation of carbonyl groups with amino and activated methylene groups in bifunctional molecules. The experiments in this chapter demonstrate how a number of the fundamental heterocyclic rings can be built from a single starting material, acetylacetone. The β-diketone system in acetylacetone is largely enolized, and the compound and its reactions should be represented with this tautomeric form. This enolic ketone contains two electrophilic centers and a highly reactive nucleophilic central carbon. Condensations can occur between any two of these functional groups and two complementary groups in another molecule or molecules. This is indicated schematically below:

Type A   Type B

TYPE A CONDENSATIONS

The first group of syntheses to be carried out are condensations of type A, in which two nucleophilic centers in the second component condense with the dicarbonyl system, eliminating two moles of water. Typical examples are the formation of the pyrazole (1) and the pyrimidine (2), from the enolic dibenzoylmethane (eq. 37.1)

$$\text{(37.1)}$$

Condensations of this type are very simple to carry out, and with the very reactive acetylacetone, these preparations involve little more than mixing the reagents in an appropriate medium. When the solubilities of starting materials and products permit, water is a very effective solvent. The rate of these carbonyl condensations in aqueous solution is pH-dependent; a slightly basic solution is optimal for the reactions studied here.

EXPERIMENTS

A.   REACTIONS OF ACETYLACETONE WITH NUCLEOPHILES

SAFETY NOTE

In this experiment and the other parts of this chapter, a number of different reagents are used in small amounts. None of these are particularly toxic, but the usual precautions should be taken to avoid inhalation of any vapors or contact with the skin. Wash your hands thoroughly with soap and water if any spills occur. Hydrazine is classified as a suspected carcinogen. Ethers are flammable.

Procedures are indicated in the following paragraphs for reactions of acetylacetone with the four substances listed alphabetically in Table 37.1. The names of the products are given, also in alphabetical order, in Table 37.2. Before proceeding with the experiment, determine the product that will be obtained from each starting material (see Question 2 at end of chapter). Then carry out one or more of the reactions as directed by your instructor. Only minimum directions are given; use your own judgment and experience in isolating the products. Submit products, labeled with yield, melting point, and structural formula, to your instructor. Dispose of all filtrates in the solvent waste container.

Table 37.1 Starting Materials Used

STARTING MATERIALS	FORMULA
Cyanoacetamide	$NCCH_2CONH_2$
Guanidine	$(NH_2)_2C{=}NH$
Hydrazine	NH_2NH_2
Hydroxylamine	NH_2OH

Table 37.2 Products

2-Amino-4,6-dimethylpyrimidine, mp 156°C (from MeOH)
3-Cyano-4,6-dimethyl-2-pyridone, mp 290°C (from H_2O)
3,5-Dimethylisoxazole, bp 142°C
3,5-Dimethylpyrazole, mp 107°C (from H_2O)

1. REACTION WITH CYANOACETAMIDE

Dissolve 0.84 g cyanoacetamide and 0.5 g of sodium carbonate in 15 mL of water, add 0.01 mole (1 mL) of acetylacetone.

2. REACTION WITH GUANIDINE

Dissolve 0.9 g of guanidine carbonate and 0.5 g of sodium acetate in 2.5 mL of water, and add 1 mL of acetylacetone. After partial air drying, recrystallize the product from a minimum volume of methanol; filter the hot solution to remove traces of inorganic solid.

3. REACTION WITH HYDRAZINE

Mix 0.5 mL of 85% hydrazine hydrate and 2.5 mL of water, and add 0.5 mL of acetylacetone.

4. REACTION WITH HYDROXYLAMINE

Dissolve 2.8 g of hydroxylamine hydrochloride and 3.3 g of anhydrous sodium acetate in 5 mL of water and add 4 mL of acetylacetone. Withdraw the oily product with a pipet, filter through granulated K_2CO_3 into a 10-mL distilling flask and distill.

VARIATIONS OF TYPE A CONDENSATIONS

A variation of type A condensation discussed earlier is illustrated in the preparation of 2,4,6-trimethylquinoline. The intermediate enamine $\left(\!-\overset{|}{C}{=}\overset{|}{C}-\overset{|}{N}-\!\right)$ from an aromatic amine and the β-diketone is prepared and isolated in a first step and then cyclized. The second step is a typical electrophilic aromatic substitution reaction involving the protonated carbonyl group of the enamine, as illustrated in Eq. 37.2.

(37.2)

2,4,6-trimethylquinoline • $2H_2O$
(mp 62°C)

EXPERIMENTS

B.   4-(*p*-TOLUIDINO)-3-PENTEN-2-ONE

(Microscale)

In a 10-mL distilling flask, place 0.42 g of *p*-toluidine, 0.6 mL of acetylacetone, and a boiling stone. Insert a thermometer in the neck so that the bulb is immersed in the liquid as far as possible without touching the bottom. Use an iced 10×75-mm test tube as a receiver. Heat the flask over a wire gauze until the temperature reaches 140°C and boiling commences. Adjust the heating to maintain gentle boiling with a minimum of distillation into the side arm for 5 minutes, then increase the heating rate and collect water and excess acetylacetone until the temperature reaches 215°C. Cool the flask and pour the contents into another 10×75-mm test tube and rinse with 0.5 mL of pentane. Chill the pentane solution in an ice bath; after crystallization, collect the product and wash on the filter with a small volume of chilled pentane. Determine the weight of the crude enamino ketone. Dispose of the filtrate in the organic solvent waste container.

C.   2,4,6-TRIMETHYLQUINOLINE

(Microscale)

In a 25-mL Erlenmeyer flask, place 2.5 mL of concentrated sulfuric acid and add 0.4 g of enamino ketone. (Reduce the amount of H_2SO_4 proportionately if less enamine is available.) Warm the solution for 20 to 25 minutes on the steam bath, then cool and add approximately 10 g of ice. Chill the solution in an ice bath; after the salt crystallizes, collect it on a Hirsch funnel and wash with a few mL of water. Suspend the moist salt in 2 mL of water and add concentrated ammonium hydroxide dropwise (see Question 5). Chill and crystallize the resulting oil, and collect the solid. Dissolve the moist solid in 1 mL of ethanol, add water until the solution is not quite turbid (approximately 1 mL of water), seed, stopper loosely, and allow the mixture to stand. The dihydrate of the quinoline crystallizes in long needles. Record the melting point and the yield (calculated on the basis of trimethylquinoline • $2H_2O$) and submit the product to your instructor.

TYPE B CONDENSATIONS

Type B condensations again involve the reaction of ammonia or an amine with the enolic

diketone to give an enamino ketone, $-NH{=}C-C-C{=}O$. These compounds are stabi-

lized by conjugation of the amino —NH and carbonyl —CO groups, and are better regarded as "vinylogous" amides than amino ketones. The enamine system can provide one or two nucleophilic centers for further reactions.

One important application of this type of condensation is the **Knorr pyrrole synthesis,** shown in Eq. 37.3. An α-amino ketone is generated in the presence of the enol, and the intermediate enamine cyclizes to a pyrrole. In the following procedure, the amino ketone is formed by the reduction of isonitrosoacetophenone.

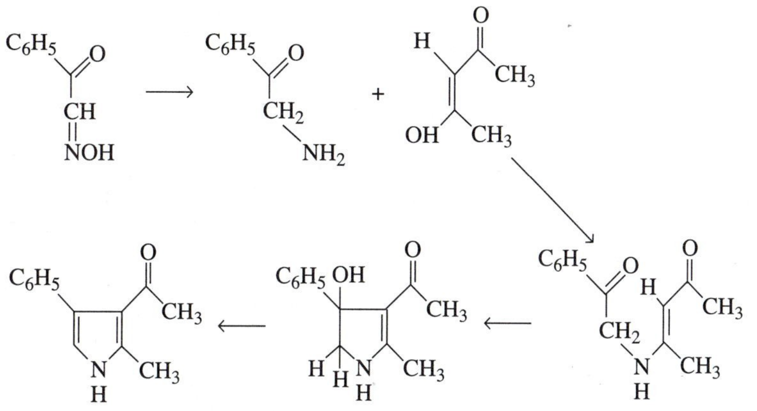

(37.3)

mp 151°C

EXPERIMENTS

D. 3-ACETYL-2-METHYL-4-PHENYLPYRROLE

(Macroscale)

Place 6 mL of acetic acid and 1.5 mL of water in a 50-mL Erlenmeyer flask and add 1.8 g of isonitrosoacetophenone and 1.3 mL of acetylacetone. To this solution add, in four portions, a total of 2.5 g of zinc dust. Swirl vigorously between each addition, cooling in an ice bath if the flask becomes too hot to hold in the hand. After all of the zinc has been added, heat the reaction mixture to gentle boiling for 2 to 3 minutes. Decant the hot solution from residual zinc into 50 g of ice. Add alkali until $Zn(OH)_2$ just begins to precipitate and then extract the mixture twice with diethyl ether or methyl t-butyl ether. Wash the ether solution with dilute base and then water, dry over $MgSO_4$, and evaporate to about 5-mL volume. Cool, crystallize and collect the product, and recrystallize it from methanol. The pyrrole tends to crystallize in extremely fine needles that occlude the oily mother liquor, and thorough washing with cold solvent on the funnel (with the aspirator disconnected) is important. Determine the yield and melting point and submit the product, labeled with the compound name, structure, and melting point.

VARIATIONS OF TYPE B CONDENSATIONS

Two other condensations of type B are performed under very similar conditions, with an additional component present in one case. In these reactions, ammonia, conveniently added in the form of ammonium acetate to buffer the solution, gives the enamino ketone (Eq. 37.4). This intermediate can supply two nucleophilic centers for condensation with another molecule of acetylacetone (or a second molecule of enamine) (Eq. 37.5). If a more reactive carbonyl group is available, however, the reaction of two molecules of enamine can occur at this center. In the reaction with acetaldehyde described below, the acetaldehyde can very conveniently be added in the form of its ammonia addition product, called "acetaldehyde ammonia" (Eq. 37.6).

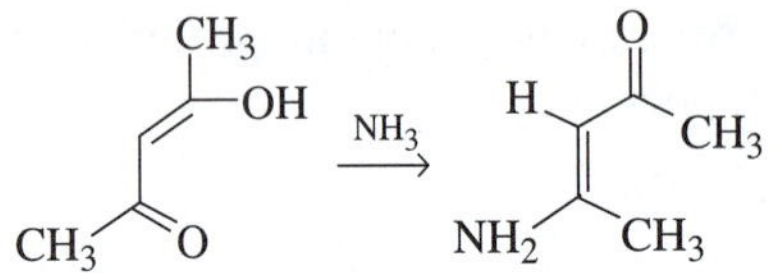

(37.4)

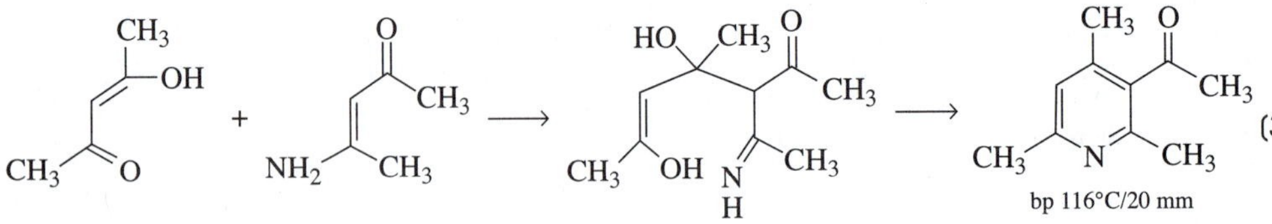

(37.5)

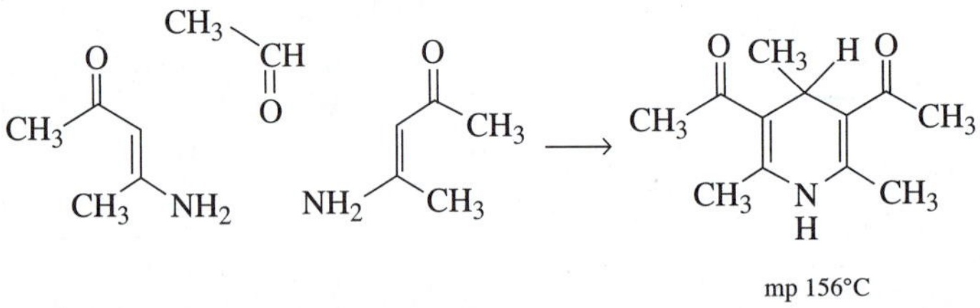

bp 116°C/20 mm

(37.6)

mp 156°C

EXPERIMENTS

E.   3-ACETYL-2,4,6-TRIMETHYLPYRIDINE

(Macroscale)

In a 125-mL Erlenmeyer flask mix 8 g of ammonium acetate and 10 mL of acetylacetone. Warm on the steam bath for 30 minutes, cool, and add saturated sodium carbonate solution in portions until CO_2 evolution stops. Extract the mixture with 25 mL of diethyl ether or methyl t-butyl ether and then with 15 mL of ether. Dry the combined ether solutions (do not backwash with water) with $MgSO_4$ and evaporate to an oil. Transfer the oil with a pipet to a 50-mL round-bottom flask and set up for vacuum distillation using only a distilling head with thermometer, a vacuum take-off adapter, and a tared 50-mL round-bottom flask as a receiver; a condenser is unnecessary. Add several boiling stones, connect to aspirator suction, and distill with a heating mantle or a low flame. Some splashing will occur, but a nearly colorless distillate can be obtained. Report the yield and boiling range of the product.

F.   3,5-DIACETYL-2,4,6-TRIMETHYL-1,4-DIHYDROPYRIDINE

(Macroscale)

Dissolve 0.6 g of acetaldehyde ammonia and 0.8 g of ammonium acetate in 10 mL of water and add 2 mL of acetylacetone. Warm for approximately 2 minutes at 50°C until all liquid droplets dissolve; allow to cool. Remove a few drops of the solution and chill to obtain a seed; then seed the main solution and allow it to crystallize for several days at room temperature if time permits. Collect the product and report the yield and melting point. (This preparation illustrates the point of view, held by some individuals, that organic chemistry is one of the more rewarding art forms.)

QUESTIONS

1. Explain how pyrrole qualifies as an aromatic compound according to the $(4n + 2)\pi$ electron rule. From these considerations, predict whether pyrrole or pyridine would be the more strongly basic compound.
2. Write equations with structural formulas for reactions of the four starting materials in Table 37.1 with acetylacetone.
3. A β-keto ester, $RCOCH_2CO_2R'$, is often used instead of a β-diketone in condensations of type A; the product then contains a carbonyl group adjacent to the heteroatom. Write the structural formulas for the products that would be obtained in the reaction of hydrazine and hydroxylamine with ethyl acetoacetate.
4. In the Knorr condensation (Eq. 37.3), no pyrrole is obtained if the acetylacetone is added after the addition of the zinc. A product with the formula $C_{16}H_{14}N_2$ can be isolated instead. Suggest the structure of this product.
5. Heterocyclic bases, and amines in general, can form two types of salts with sulfuric acid: sulfate $(R_3NH^+)_2SO_4^=$ and bisulfate $(R_3NH)^+HSO_4^-$. Suggest the formula of the salt that precipitates in the quinoline preparation. Describe the appearance of the reaction mixture as ammonia is added, and account for the observations in terms of the ionic species present.

PRELABORATORY QUESTIONS

1. Draw structural formulas for pyrrole and pyridine.
2. The common starting material utilized in all the syntheses in this chapter is what diketone?
3. Show an enol form of acetylacetone.
4. Draw the structure of the enamine that would form from propionaldehyde and dimethylamine.

REFERENCES

Texts

Acheson, R.M. *An Introduction to the Chemistry of Heterocyclic Compounds,* 3rd ed.; Interscience: New York, 1976.

Gilchrist, T.L. *Heterocyclic Chemistry,* 2nd ed.; John Wiley and Sons: New York, 1992.

Katritzky, A.R. *Handbook of Heterocyclic Chemistry;* Pergamon: Elmsford, New York, 1985.

Katritzky, A.R.; Rees, C.W.; Scriven, E.F. *Comprehensive Heterocyclic Chemistry II;* Elsevier Science: New York, 1999.

Newkome, G.R.; Paudler, W.W. *Contemporary Heterocyclic Chemistry;* Wiley-Interscience: New York, 1982.

Experimental Procedures

Fitton, A.O.; Smalley, R.K. *Practical Heterocyclic Chemistry;* Academic: New York, 1968.

Heterocyclic Synthesis: Synthesis of Phenytoin

Epilepsy is a neurological disorder that has been known since antiquity. Not until 1857, however, was the first effective anticonvulsant drug, potassium bromide, introduced for its treatment. Early in the twentieth century, the barbiturate phenobarbital began to be widely used, and further development of anticonvulsant drugs often involved barbiturates or compounds that are structurally similar to them. By 1940, phenytoin (5,5-diphenylhydantoin) had been developed and had become the leading drug for convulsive seizure disorders. It has been the drug of choice for grand mal epilepsy during the intervening years and is marketed under such trade names is Dilabid, Dilantin, and Divulsan.

In the United States, epilepsy is the second most frequently encountered neurological disorder, exceeded only by stroke. It has been estimated that there are over 1 million people who have recurrent seizures and that 2 million persons have had two or more seizures. In addition to phenytoin, which has been widely used in the treatment of epilepsy, other compounds such as phenobarbital, primidone, carbamazepine, and valproic acid have been found to be effective.

phenobarbital

primidone

carbamazepine

valproic acid

One preparation of phenytoin (III) involves the reaction of benzil (I) with urea (II) under basic conditions. The course of the reaction sequence is illustrated on the following page.

EXPERIMENTS

PREPARATION OF PHENYTOIN

(Microscale)

SAFETY NOTE

Concentrated sodium hydroxide solution is quite caustic and can cause damage to eyesight within seconds. Wear safety goggles at all times and clean up any spills immediately. If any contacts the skin, wash with copious amounts of water. Avoid exposure to phenytoin; it is a suspected carcinogen.

PROCEDURE

In a 10-mL round-bottom flask dissolve 0.525 g of benzil (0.0025 mole) and 0.300 g of urea (0.0050 mole) in 6 mL of ethanol. Then, add 1.2 mL of 6 N sodium hydroxide solution, add a boiling stone and heat the mixture to reflux for 1.5 hours. (**CAUTION:** Ground joints may fuse when heated with caustic liquid; be sure to grease well the joint between flask and condenser, and **disconnect the condenser after heating** but before the apparatus cools.) Allow the mixture to cool, and filter it through a filter pipet to remove any solid material. Chill the filtrate in an ice-water bath and carefully neutralize it using 10% hydrochloric acid until it is acid to litmus paper. The 5,5-diphenylhydantoin that precipitates during the neutralization is collected by vacuum filtration and dried. The product is reasonably pure as isolated from the neutralization procedure, but may be recrystallized from 95% ethanol, if required. The melting point is 295 to

298°C (decomposition); the melting point should not be determined using an oil bath because of the high temperature.

Disposal: Neutralize any excess acid or base before disposal. Dispose of organic waste as directed by your instructor.

Determine the weight and percentage yield of the product, record the IR and/or the NMR spectra, and submit the product to your instructor in an appropriately labeled container.

QUESTIONS

1. The infrared spectrum of phenytoin contains the following absorptions. Make assignments for these.

   3275, 3205 cm^{-1}
   3064 cm^{-1}
   1740, 1719 cm^{-1}
   1599, 1496 cm^{-1}
   747, 690 cm^{-1}

2. The NMR spectrum of phenytoin shows peaks at 7.3, 9.2, and 10.9 ppm with relative areas 10:1:1. Make assignments for these peaks.
3. The pK_a of phenytoin is 8.3. Could the equivalent weight of this compound be determined by titration with standard NaOH in water or in water-acetone? Explain.

PRELABORATORY QUESTIONS

1. List three substances that have been used as anticonvulsants.
2. Draw the structure of benzil. Give the IUPAC name for it.
3. Draw another resonance form for the anion formed in the next-to-last step in the synthesis of phenytoin.
4. Suggest a possible analogue of phenytoin that might have similar physiological properties. What reagents would you use to synthesize it?

REFERENCES

Biltz, H. *Ber.,* 1908, *41,* 1391; 1911, *44,* 411.

Pankaskie, M.C.; Small, L. *J. Chem. Educ.,* 1986, *63,* 650.

Philip, J., et al.; In *Analytical Profiles of Drug Substances;* Florey, K., Ed.; American Pharmaceutical Association: Washington, DC, 1984; Vol. 13, p. 417.

Synthesis of Sulfanilamide

S ulfa drugs, like some other medicinal agents, comprise a group of compounds having a key structural feature that imparts a specific pharmacological property. The common feature of sulfa drugs is the *p*-aminobenzenesulfonamido group, and the compounds are useful as antibacterial agents. In fact, the first discovered sulfa drug, Prontosil, is not a *p*-aminobenzenesulfonamide but is converted to the pharmacologically active *p*-aminobenzenesulfonamide in the body. When this fact was discovered, the metabolite known as sulfanilamide became the preferred agent for treatment. Attempts to increase its effectiveness toward specific infections led to the synthesis and testing of numerous derivatives. Sulfapyridine and later sulfadiazine and sulfamethazine became the agents of choice for streptococcal (throat) and pneumococcal (lung) infections such as pneumonia. Sulfathiazole is the drug of choice for staphylococcal (skin) infections, and sulfaguanidine for intestinal infections.

Prontosil
2′,4′-diaminoazobenzene-4-sulfonamide

sulfonilamide
p-aminobenzenesulfonamide

sulfapyridine

sulfadiazine

sulfamethazine

sulfathiazole

sulfaguanidine

The compounds act by competitively inhibiting the incorporation of *p*-aminobenzoic acid, an essential component for cell growth of the microorganisms. The structurally comparable sulfonamides, with a pK_a similar to that of the carboxylic acid, block the metabolic pathway. Thousands of analogs of sulfanilamide have been prepared and tested for antibacterial properties. A number of these compounds are used clinically, although they have been replaced to a considerable extent by naturally occurring antibiotics, such as penicillin and streptomycin.

The preparation of the parent compound, sulfanilamide, involves the introduction and removal of an acetyl blocking group to control the course of the synthesis (see Chapter 27). The three-step synthesis, starting with acetanilide (see Chapter 3), can be carried out without purification of intermediates and can be completed in a single 3-hour laboratory period.

EXPERIMENTS

SAFETY NOTE

Chlorosulfonic acid is a very hazardous reagent. *It reacts violently with water to produce HCl gas, and causes severe burns* even on brief contact with the skin. Do not pour excess reagent into the sink; dispose of it carefully in the fume hood. Be certain any glassware coming in contact with chlorosulfonic acid is absolutely dry. Concentrated ammonium hydroxide fumes are irritating to the nose and eyes. Avoid breathing vapors.

Instructor's Note. The chlorosulfonic acid (use a fresh bottle) and the ammonia may be dispensed by a buret. This results in better safety practice and less waste.

PROCEDURE

CAUTION: Be sure all glassware is dry before starting to use chlorosulfonic acid. When using chlorosulfonic acid wear protective rubber gloves.

Disposal: Combine all filtrates, neutralize them with dilute hydrochloric acid or sodium carbonate solution as required, and wash them down the drain with copious amounts of water.

A. SYNTHESIS OF *para*-ACETAMIDOBENZENESULFONYL CHLORIDE

For ease of handling, connect a clamp to a 25-mL round-bottomed flask and add 2.70 g (0.02 mole) of acetanilide to the flask. In a fume hood, add carefully, with swirling, and in small portions, 4.0 mL (7.0 g) of chlorosulfonic acid. Fumes will be evolved and the mixture will become hot. When all of the chlorosulfonic acid has been added, connect the flask to a reflux condenser fitted with a trap for HCl fumes and immerse the flask in a warm water bath (60 to 70°C) for 1.5 hours. The mixture should turn to a liquid during this period.

In a hood, carefully and slowly pour the contents of the flask onto 60 grams of crushed ice in a 100-mL beaker. Break up the lumps with a stirring rod and filter with suction using a Büchner funnel. Wash the precipitate with a few mL of ice water. (The crude *p*-acetamidobenzenesulfonyl chloride does not have to be dried at this point.)

B. AMINATION OF *p*-ACETAMIDOBENZENESULFONYL CHLORIDE

Place the crude *p*-acetamidobenzenesulfonyl chloride in a 50-mL beaker or flask and add 5 mL of concentrated ammonium hydroxide (be careful; this is an exothermic reaction). Stir the mixture well for 15 minutes. Complete the conversion to the sulfonamide by heating the mixture in a hot (80°C) water bath until the odor of ammonia has disappeared. Cool the beaker in an ice bath, filter the solid with suction, and wash with a few mL of ice water. Use the crude product directly in the next step.

C. SULFANILAMIDE

Place the crude amide in a 50-mL round-bottomed flask, add 4 mL of concentrated HCl and 4 mL of water, and heat the mixture to reflux for 30 minutes. (If the solution is colored, cool it and add a pinch of activated carbon, then heat to boiling and filter hot through a conical funnel fitted with fluted filter paper.) Cool the solution and add concentrated ammonium hydroxide dropwise until the mixture is neutral to litmus. Cool in an ice bath and allow the crystals to form. Filter the sulfanilamide with suction and allow it to dry. Record the weight of the crude sulfanilamide and recrystallize it from a minimum amount of water. Record the weight, percentage yield, and melting point range of the purified product.

QUESTIONS

1. What product would be expected if aniline rather than acetanilide were treated with chlorosulfonic acid?
2. What is the compound in solution after boiling the acetamidosulfonamide in hydrochloric acid?
3. What would be the result if excess sodium hydroxide solution were used for the final neutralization after acid hydrolysis? (Hint: see Chapter 29.)
4. Write a balanced equation for the overall reaction of acetanilide and chlorosulfonic acid and indicate the mechanism of the reaction(s).
5. Which enzyme system of a bacterium is affected by sulfanilamide (i.e., what is the mechanism of its biological activity)?
6. Prontosil is actually an orange-red dye that was found to be active in the treatment of patients with infections. What structural features of Prontosil impart its color?

PRELABORATORY QUESTIONS

1. Draw the structure of Prontosil. What part of the structure is the "biologically active" part?
2. Name three "sulfa" drugs.
3. Draw the structure of chlorosulfonic acid. Write an equation for its reaction with water.
4. Why does the final reaction mixture need to be neutralized before sulfanilamide crystals can form?

Resolution of
α–Phenylethylamine

Chiral molecules exist as two nonsuperimposable mirror-image isomers, or **enantiomers;** that is, they exhibit handedness. Enantiomers are identical with respect to most physical and chemical properties such as melting and boiling points, solubilities, and reactivities with symmetrical reagents. An important exception is the behavior of enantiomers when placed in a beam of polarized light. When passing through samples of the individual enantiomers, the plane of polarization is rotated in equal but opposite directions. This phenomenon is termed **optical activity,** and enantiomers are sometimes called **optical isomers.**

The compound α-phenylethylamine contains one stereogenic center and is chiral. The material obtained by reductive amination of acetophenone is optically inactive. The rotation is zero because the synthetic product is a **racemic mixture;** that is, both enantiomers are present in exactly equal amounts. Optical activity can be observed, of course, only when one of the enantiomers is present in excess over the other. The separation of a racemic mixture into its individual enantiomers is called **resolution.**

Enantiomers can be separated only in the presence of a chiral reagent or environment. Classically, separation has been effected by resolution, the conversion of the racemic mixture to a diastereomeric pair by bonding (either covalently or by salt formation) to an enantiomerically pure substance. This second substance is then called a "resolving agent." More recently, other strategies have been introduced. It may be possible to find a microorganism or pure enzyme that will selectively transform one of the two enantiomers in the mixture. Alternatively, it may be possible to separate the two enantiomers by chromatography, using a chiral stationary phase. Chromatographic resolution is especially widely used as an analytical tool, to establish precisely the ratio of enantiomers in an unknown mixture. The formation and separation of diastereomeric salts will be used in this experiment to resolve α-phenylethylamine (1-phenylethylamine).

When a racemic amine is treated with one enantiomer of a chiral acid, the resulting salts are diastereomers. If the two enantiomers of the amine are designated $(+)$ and $(-)$, and the acid is designated $(+)$, the salts are $(+)(+)$ and $(-)(+)$; this is seen in Equation 40.1.

$$(+)RNH_2 \qquad\qquad (+)RNH_3^+(+)R'CO_2^-$$

$$+ (+)R'CO_2H \longrightarrow \qquad\qquad\qquad (40.1)$$

$$(-)RNH_2 \qquad\qquad (-)RNH_3^+(+)R'CO_2^-$$

racemic amine $\qquad\qquad$ diastereomeric salts

Since the salts no longer have a mirror-image relationship, they have different physical properties. If a solvent can be found in which one salt is more soluble than the other, they can be separated by fractional crystallization.

In this experiment the diastereomeric salts are obtained using tartaric acid, a chiral compound whose (+) enantiomer is produced as a by-product in wine making. The hydrogen tartrate obtained from the (+) acid and (−) α-phenylethylamine is less soluble in methanol than the (+)(+) salt, and the (+)(−) salt crystallizes in nearly pure stereoisomeric form. After separation of the pure diastereomeric salt, the amine enantiomer is obtained by treating the salt with excess strong base (Eq. 40.2).

(40.2)

THE MEASUREMENT OF OPTICAL ACTIVITY

The rotation of plane polarized light is measured with a relatively simple instrument called a polarimeter, shown schematically in Figure 40.1. Proceeding from left to right, randomly oriented light is polarized by passing through a Nicol prism (calcite) or a sheet of Polaroid film. The light beam then passes through a sample tube containing a solution of the compound being analyzed; if the compound is a liquid and a large enough sample is available, the neat liquid may be used. If the tube contains an optically active compound, the plane of the polarized light is rotated as it passes through the sample. The analyzer, another polarizer whose orientation is adjustable, is next in the path of the light. In using some polarimeters, the analyzer is rotated until it completely blocks the polarized light reaching it. A scale calibrated in degrees of arc is attached to the analyzer, and the amount of rotation is read from this scale. In a properly calibrated polarimeter, the reading will be 0° if there is no optically active sample in the tube. By convention, if it is necessary to rotate the analyzer clockwise (from the user's viewpoint) to cause extinction of the light, the rotation is designated positive, (+); counterclockwise rotation is designated (−).

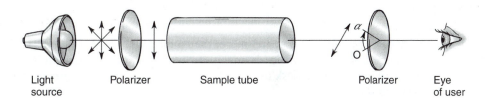

Light   Polarizer   Sample tube   Polarizer   Eye
source   of user

Figure 40.1   Schematic diagram of a polarimeter with a sample giving $\alpha \sim +30°$. Arrows indicate the direction of polarization of the light beam.

The observed rotation of a substance depends on the structure of the compound and the number of molecules in the light path between the polarizers, the later factor being a function of the path length and the concentration. A quantity termed **specific rotation,** $[\alpha]$, is defined as follows:

$$[\alpha] = \frac{\alpha}{lc}$$

where α is the observed rotation, l is the path length in decimeters, and c the concentration in g/mL. In addition, the temperature, solvent, and wavelength of the light used are variables that affect optical rotation and must be specified in reporting the value. Thus the specific rotation of (+)-tartaric acid has been reported as $[\alpha]_D^{20} + 12.0°$ ($c = 20$ g/100 mL in H_2O). The subscript D indicates that the wavelength of light used was that of the yellow D line from a sodium vapor lamp; that is, 589 nm, and the superscript 20 indicates that the measurement was made at 20°C. The concentration (in g/100 mL) is also reported, since the rotation is exactly proportional to concentration only in dilute solutions.

Commercial polarimeters come in a variety of models. The major differences are in the type of polarizers and the accuracy with which the positions of the analyzers can be measured. In models designed to be used with a light source that is not monochromatic, a filter or filters are incorporated in the light path. Provisions may also be made for controlling the temperature of the sample. Some polarimeters are available that automatically adjust the polarizers and provide a readout of the rotation on a meter or send data to a computer; a photocell replaces the user's eye in such an instrument. The accuracy obtained with the most expensive instruments is ±0.001 to 0.002°.

An inexpensive commercial polarimeter that is adequate for this experiment (accuracy ±0.5°) is illustrated in Figure 40.2. Light from a strong flashlight or high-intensity desk lamp is reflected by the mirror up through a sheet of Polaroid film and through the sample, contained in a flat-bottomed tube. The polarized light then passes through a yellow filter to a second polarizer, which is rotated until the observed field reaches maximum darkness.

EXPERIMENT

(Macroscale)

SAFETY NOTE

Methanol is toxic and flammable; avoid inhalation of vapor. α-Phenylethylamine is toxic and corrosive; handle with care. Tartaric acid is an irritant.

PROCEDURE

Dissolve 11.9 g of (+)-tartaric acid in 165 mL of methanol in a 250-mL Erlenmeyer flask by heating the mixture on a steam bath. Slowly add 10.0 mL of racemic α-phenylethylamine to the warm solution while swirling the flask to mix well. Salt formation is exothermic and the solution will become fairly hot.

The desired salt crystallizes slowly as large clear prisms. The first crystals obtained are often fine white needles instead of prisms; these needles do not provide optically pure amine (see Question 2) and must be avoided. To insure that the desired prisms form, seeds of the correct composition are needed. To obtain the seeds, first transfer approximately 1 mL of the hot solution to a test tube and concentrate this solution to approximately half the original volume by heating on the steam bath. As soon as crystals form in the boiling solution, pour the suspension of crystals back into the flask containing the remainder of the solution. Cork and label the flask and set it aside until the next laboratory period. (Check the next day; if the crystallized salt is in the form of needles rather than prisms, reheat the mixture and allow it to recrystallize slowly.)

After allowing it to stand at least 24 hours, decant the solution from the crystals. (The solution can be processed for eventual recovery of the (+)-amine.) Add 10 mL of methanol to the flask containing the crystals and break the mass up into individual crystals with a stirring rod.

Using a small Büchner funnel and suction flask, collect the crystals. Wash any remaining crystals out of the flask with a little of the filtrate and rinse the crystals on the funnel with 10 mL of cold methanol. Spread the crystals on paper to dry, weigh the product, and calculate the percentage yield of (−) amine (+) hydrogen tartrate. Place the filtrate in the organic solvent waste container.

To recover the amine, place the crystalline salt in a 125-mL Erlenmeyer flask. Add 50 mL of 2 M (8%) aqueous NaOH solution (or use 40 mL of water plus 5 mL of 50% aqueous NaOH). Swirl the mixture until all of the crystals have dissolved. Transfer the mixture to a separatory funnel and rinse the flask with 10 mL of methylene chloride. Add the rinse to the separatory funnel, stopper the funnel, and shake well. Remove the methylene chloride layer into a clean, dry, 50-mL Erlenmeyer flask. Extract the aqueous phase twice more with 5-mL portions of CH_2Cl_2 and combine these extracts with the first. Dry the solution of the amine by adding approximately a gram of granular anhydrous K_2CO_3, swirling, and letting it stand for a few minutes. Remove the K_2CO_3 by filtering the solution through a cotton plug into a tared

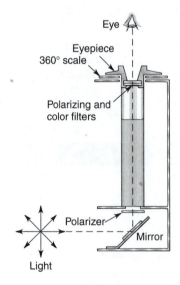

Figure 40.2   Diagram of an I^2R polarimeter.

25 × 100-mm test tube containing a boiling stone. Carefully warm on a steam bath to evaporate most of the solvent. This operation should be carried out in a fume hood, or a tube can be connected to the aspirator vacuum and inserted in the neck of the tube to remove the vapors as they distill. When nearly all of the methylene chloride has evaporated as indicated by cessation of boiling, remove the remainder of the solvent under reduced pressure (Fig. 4.4A) while heating the tube gently on the steam bath. Weigh the test tube and its contents, remove the boiling stone, and calculate the yield of amine.

To determine the optical purity of the amine, transfer it with a pipet to a polarimeter tube. Measure the length of the liquid column to the nearest millimeter (see Fig. 40.2), and measure its rotation, α. The rotation can also be measured in an appropriate solvent such as methylene chloride or methanol. The mechanics of filling the tube and measuring the rotation will depend upon the type of polarimeter available. Consult your instructor for specific directions. Calculate the specific rotation of the amine using the formula $[\alpha] = \alpha/cl$. Note that the path length is specified in dm and that the concentration of a neat liquid is equal to its density (0.97 g/mL in this case). The reported specific rotation of pure (−) α-phenylethylamine is $[\alpha]_D^{25} - 40.4 \pm 0.2°$ (neat). Calculate the optical purity of your sample (see Question 3). Dispose of the remaining amine or amine solution in the appropriate waste container.

QUESTIONS

1. If a pure (+) enantiomer has $[\alpha]_D + 100°$, what will be the rotation of a mixture containing 80% of the (+) enantiomer and 20% of the (−) isomer?
2. The α-phenylethylamine recovered from the hydrogen tartrate salt that crystallizes as needles was found to have a specific rotation of $-20 \pm 1°$. What is the percent composition of this mixture of (+) and (−) enantiomers?
3. Optical purity is defined as the ratio of the measured specific rotation to the specific rotation of the pure enantiomer. Calculate the optical purity of the amine described in Question 2.
4. Suggest how the procedure given in this experiment could be expanded to provide a product of higher optical purity.
5. Describe with equations how the (−)-α-phenylethylamine obtained in this experiment could be used to resolve racemic 2-phenylpropionic acid.

PRELABORATORY QUESTIONS

1. Draw the Fischer projection formula for (R)-α-phenylethylamine.
2. Draw the structure of (−)tartaric acid, which could also be used to resolve the enantiomers of α-phenylethylamine.
3. Write an equation to illustrate the recovery of hexylamine from its hydrochloride salt.
4. What is meant by the term "specific rotation" as opposed to observed rotation?
5. What are the units for the terms l and c, used to calculate specific rotation?

REFERENCE

Ault, A. *J. Chem. Educ.,* 1965, *42,* 269.

Chemistry of Sugars

Sugars make up a specialized but important branch of organic chemistry. Because of their polyfunctional structures, sugars undergo a broad variety of reactions, including some complex rearrangements. In addition to transformations of sugars with acids, bases, and other simple reagents, enzyme-catalyzed reactions play a prominent role in carbohydrate chemistry, and sugars are good substrates for studying enzymatic processes.

Experimental work with sugars involves some special problems and techniques. Sugars are extremely soluble in water and only very slightly soluble in the usual organic solvents. They are complex and sensitive molecules and cannot be distilled or sublimed. Most sugars can be crystallized, although sometimes with great difficulty. One of the major problems in carbohydrate work is the characterization and differentiation of closely related sugars in aqueous solutions. To this end, a number of color reactions have been developed for detection and analysis of specific types of sugars.

SOME REACTIONS OF SUGARS

Monosaccharides (simple sugars) exhibit reactions of both hydroxyl and carbonyl groups plus others that depend upon interactions between these groups. Aldohexoses exist as equilibrium mixtures of cyclic hemiacetals with a negligible amount of the aldehyde tautomer present, but carbonyl reactions occur readily since the interconversion of these forms is rapid.

α-D-glucopyranose   aldehyde form   β-D-glucopyranose
                           of D-glucose

Tests for reducing properties, condensations with amines, and certain TLC color reagents involve the reaction of the carbonyl group of aldose or ketose sugars. When the OH group of the hemiacetal is replaced by OR, as in glycosides, these carbonyl reactions do not occur unless the acetal is hydrolyzed under the conditions of the test. In most disaccharides, such as maltose or lactose, one hexose unit is tied up as an acetal or glycoside linkage but the second unit has an aldehyde ⇌ hemiacetal group. An important exception is sucrose, in which the carbonyl groups of both hexose units are linked together as an "acetal-ketal."

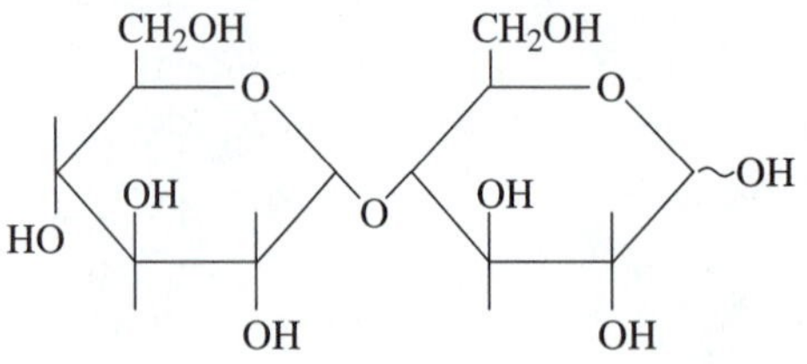

maltose

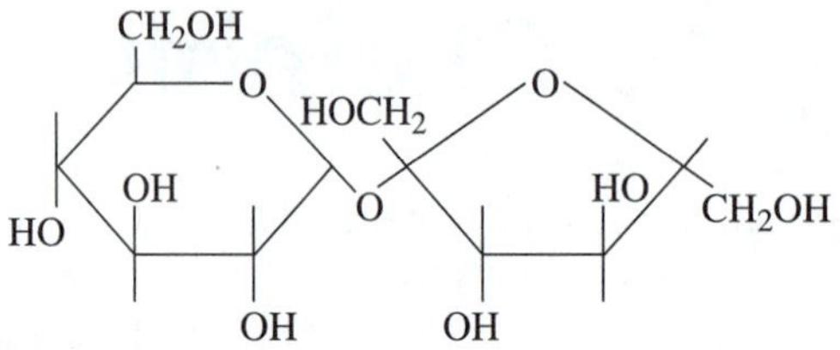

sucrose

Reducing Properties

A characteristic property of glucose and other sugars with a "free aldehyde" or hemiacetal group is oxidation by alkaline copper (II) ion. The blue cupric ion is reduced to red Cu_2O, and this reaction, with various modifications, is a general method for the detection and quantitative clinical analysis of reducing sugars. The standard reagent for qualitative work is Benedict's solution, which contains Na_2CO_3 and citrate ion to complex the cupric ion.

$$RCHO + 2Cu^{2+} + 5OH^- \rightarrow RCO_2^- + Cu_2O + 3H_2O$$

Reactions in Strong Acid

Disaccharides such as maltose and sucrose are hydrolyzed by warming with aqueous acid. Monosaccharides are relatively stable in acidic solution, but a series of dehydration reactions occurs when they are heated in strong acid, ultimately leading to furan derivatives. The process is carried out on an industrial scale to obtain furfural from pentose sugars found in grain chaff and corn cobs. These reactions occur most readily with sugars that exist in the furanose form, such as pentoses and 2-ketohexoses. Subsequent condensation of the furfural with a phenolic compound gives colored products that are useful for characterization. A typical example is the reaction of fructose with hydrochloric acid and resorcinol, called Selivanoff's reagent. A deep red color develops, which can be used to estimate the amount of fructose in a solution.

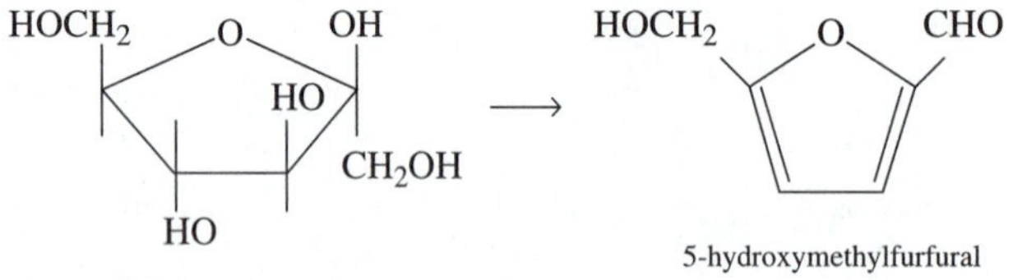

β-D-fructofuranose   5-hydroxymethylfurfural

Reactions in Base (The Lobry de Bruyn–van Ekenstein Transformation)

In basic solution an equilibrium is established, by means of an enediol, between aldose and ketose sugars. When glucose is warmed briefly with dilute base, a mixture of glucose, fructose, and a small amount of mannose results. Further reactions then occur to give dark, extremely complex mixtures of other sugars together with saccharinic acids. Several types of acid have been isolated; these products arise from further enolization and rearrangement steps.

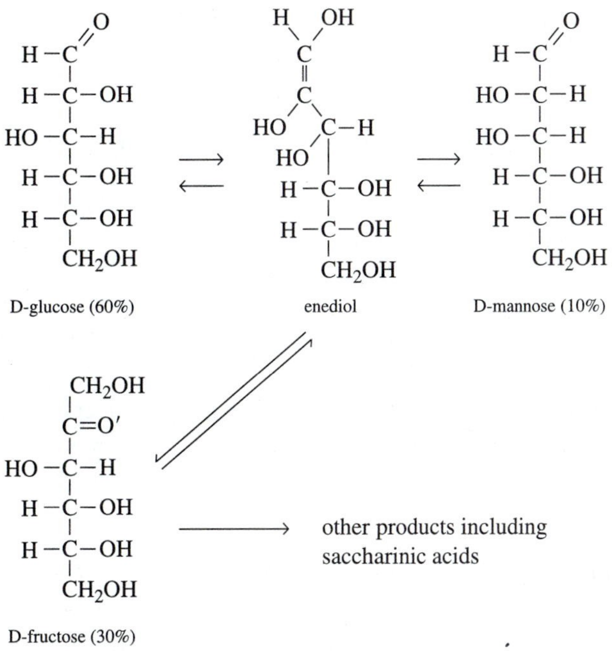

D-glucose (60%)   enediol   D-mannose (10%)

D-fructose (30%)

other products including
saccharinic acids

(41.1)

Reactions with Amines

The reaction of sugars with ammonia or amines gives *N*-glycosyl compounds in which the hemiacetal—OH group is replaced by nitrogen. Derivatives of this type are intermediates in the biosynthesis of nucleosides and nucleic acids. Enolization of these *N*-glycosides occurs with a trace of acid, and a 1-amino-1-deoxyketose results; the overall reaction is called the **Amadori rearrangement.**

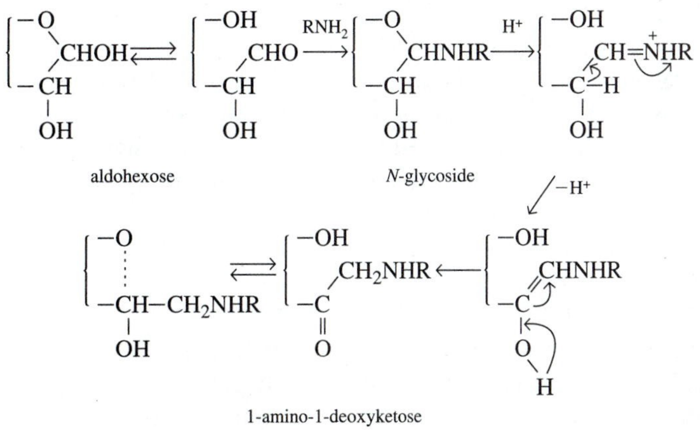

aldohexose   *N*-glycoside

1-amino-1-deoxyketose

A traditional reagent for the characterization of sugars is phenylhydrazine. Excess reagent acts as a selective oxidizing agent for the adjacent OH group and the product is the yellow osazone, which is obtained from either of two epimeric aldoses or from the 2-keto sugar. Due to the presence of two aromatic rings, osazones of monosaccharides are only slightly soluble in water and provide derivatives that can be readily isolated from dilute aqueous solutions.

$$
\begin{array}{c}
\text{CHO} \\
| \\
\text{H}-\text{C}-\text{OH} \\
| \\
\text{R}
\end{array}
$$

$$
\begin{array}{c}
\text{CHO} \\
| \\
\text{HO}-\text{C}-\text{H} \\
| \\
\text{R}
\end{array}
\xrightarrow{3\text{C}_6\text{H}_5\text{NHNH}_2}
\begin{array}{c}
\text{CH}=\text{N}-\text{NHC}_6\text{H}_5 + \text{C}_6\text{H}_5\text{NH}_2 \\
| \\
\text{C}=\text{N}-\text{NHC}_6\text{H}_5 \qquad + \text{NH}_3 \\
| \\
\text{R} \qquad\qquad + 2\text{H}_2\text{O} \\
\text{osazone}
\end{array}
$$

$$
\begin{array}{c}
\text{CH}_2\text{OH} \\
| \\
\text{C}=\text{O} \\
| \\
\text{R}
\end{array}
$$

EXPERIMENTS

Safety Note

Some of the reagents in this experiment are strongly acidic or basic and should be handled with care. Phenylhydrazine is toxic and a suspected carcinogen; use it with particular caution. If you get any on your skin, wash it off immediately.

A.　THIN LAYER CHROMATOGRAPHY OF SUGARS

Chromatographic methods are widely used in the separation and isolation of sugars. Qualitative examination of mixtures by chromatography on paper strips or TLC plates is an indispensable tool. Thin layer chromatography provides somewhat less resolving power than paper, but it is faster and more convenient. Since sugars are extremely polar compounds, mixtures of solvents such as butanol, acetic acid, and water are needed to obtain migration. Visualization of compounds on the developed chromatogram is accomplished by spraying with color-forming reagents. Combinations of aromatic amines and acids result in condensation products that have colors more or less characteristic of certain types of sugars. Nonreducing sugars such as sucrose usually yield much weaker color reactions and are more difficult to visualize.

PROCEDURE

Commercial silica gel G strips on plastic backing should be used. For applying samples, use the micro capillary tubes and the same general technique described in Chapter 8.

In 10×75-mm test tubes, obtain 10- to 20-mg samples of arabinose, fructose, or glucose, maltose, and a commercial syrup such as Karo® or honey. Dissolve the sugars in 1 drop of water, add 0.2 mL of methanol, and shake or stir to mix the heavy aqueous syrup and solvent before spotting. Apply spots no larger than 1 mm in diameter on a 1-inch wide strip; four samples can be applied to each strip. Allow the strip to dry for 2 to 3 minutes after the samples have been applied. Develop the chromatogram in a solvent mixture of 1-butanol:acetic acid:ether:water in a volume ratio of 9:6:3:1. Approximately 20 minutes are required for development.

Dry the strips in air, and spray with a solution of *p*-anisidine and phthalic acid in ethanol, or a packaged aniline/phthalic acid aerosol spray. Attach the strip by one corner with plastic tape to a large piece of cardboard and place it in the hood for spraying. Hold the sprayer tip about 6 inches from the strip and move it back and forth. Apply a fine mist; the strip should not be soaked. Place the strip in an 80° oven or warm briefly over a hot plate to permit visualization of the spots. Dispose of excess solvent mixture in the organic waste container.

B. CHARACTERIZATION REACTIONS

Procedures are given for three reactions that are useful in characterizing sugars. Arabinose, fructose, glucose, maltose, and sucrose are representative sugars that can be used to illustrate these tests; samples of corn syrup (Karo®) and honey can also be examined if desired. Obtain samples of a few sugars and/or unknowns as directed by your instructor, apply the tests, and account for your observations.

Benedict's Reagent for Reducing Properties

(The reagent solution contains 100 g of Na_2CO_3, 175 g of sodium citrate, and 17.3 g of $CuSO_4 \cdot 5H_2O$ per liter.) In a 13×100- or 18×150-mm test tube, dissolve approximately 10 mg of the sugar in a few drops of water (or use 5 drops of a 2% solution). For the corn syrup or honey, dilute 1 mL of the syrup to 50 mL and use 5 drops.

To each tube, add 3 mL of Benedict's solution and stir or shake to mix the solutions. Label the tubes with a marker or self-stick label that will not come off in hot water. Place all the tubes in a boiling water bath at the same time and remove them after exactly 2 minutes. The amount of reducing sugar in the various samples is best compared before the red cuprous oxide settles.

Filter the copper (I) oxide and dispose of it in the nonhazardous solid waste container. Dilute the filtrate and flush it down the drain.

Selivanoff's Reagent (Furfural Formation)

(The solution contains 50 mg of resorcinol in 33 mL of hydrochloric acid diluted to 100 mL with water.) Empty the tubes from the Benedict test, rinse out the Cu_2O residue using a brush, and add samples of sugars as above. To each tube, add 2 mL of Selivanoff's reagent. Mix and heat in the boiling water bath for 2 minutes and compare the appearance of the tubes. The intensity of the red color is proportional to the fructose content. Neutralize the waste products and dispose of them in the organic waste container.

Phenylhydrazine

(The reagent contains 60 g of $CH_3CO_2Na \cdot 3H_2O$ and 40 g of phenylhydrazine hydrochloride in 400 mL of water.) In 18×150-mm test tubes with hot water–resistant labels, dissolve 50 mg of the sugars to be tested in 2 mL of water (or use 2 mL of a 2% solution). Add 5 mL of the phenylhydrazine solution and place the tubes in a boiling water bath. Keep the tubes in the bath for 15 minutes or until a precipitate appears. The osazone that precipitates can be collected and dried, and the melting point determined. Dispose of the filtrate in the organic waste container.

C. ISOMERIZATION OF GLUCOSE

The objective of this experiment is to follow the reaction of glucose in base by use of the Selivanoff and Benedict reagents. In an experiment of this type, the course of reactions can be monitored by sampling the reaction mixture at intervals and observing changes in tests on successive samples. Relative amounts of substances present can be estimated by the intensity of a color reaction without determining absolute concentrations. In this approach, it is important that all samples are the same size and are handled in the same way.

Procedure

Arrange two series of test tubes labeled 0, 5, 10, 20, and 30. To 10 mL of a 2% solution of glucose in water, add 1 mL of 2 M NaOH solution. Swirl to mix the solution, draw up a sample with a pipet and bulb, and place 5 drops of the solution in each of the tubes labeled 0. Place the glucose solution in a water bath at 70°C and, after 5-, 10-, 20-, and 30-minute intervals, transfer 5-drop samples to each tube as labeled. After taking the last samples, note the odor of the solution.

To one series of samples, add 2 mL of Benedict's reagent to each tube; to the second series add 2 mL of Selivanoff's reagent to each tube. Shake the tubes to mix the contents, and immerse all the tubes in each series at the same time in a boiling-water bath. After 2 minutes, remove the tubes and line them up according to reaction time. Record the intensities of the color or the amount of precipitate in the successive samples in each series and interpret the results. You should observe a rapid buildup and then a slower decrease in one series, but not in the other (Question 3).

Dispose of the waste products as directed in the respective sections above.

D.   *N-p*-TOLYL-D-GLUCOSYLAMINE AND *N-p*-TOLYL-1-AMINO-1-DEOXY-D-FRUCTOSE

The preparation of these derivatives illustrates the effect of minor changes in reaction conditions and provides an exercise in the art of crystallization. Both compounds form very readily; isolation is the challenge.

Procedure

Weigh out 2 g of D-glucose and 1.6 g of colorless *p*-toluidine, and mix the compounds in an 18 × 150-mm test tube. Add 0.2 mL of water and 0.2 mL of ethanol, and heat the mixture in a steam or hot-water bath. Shake the mixture intermittently while heating until a single liquid phase is present, and then transfer half of the solution to another tube.

To one tube, add 0.1 mL (2 to 3 drops) of 2 M (10%) acetic acid. Heat the tube in steam or boiling water for 20 minutes and add 10 mL of ethanol. To the other tube, add 10 mL of ethanol. Place a drop or two of each solution in a small test tube and cork the main solutions. Use the smaller samples to initiate crystallization. Evaporate the sample to a syrup, scratch with a rod, and add a drop of ethanol. Repetition of these steps and a few days' standing may be needed before crystals form. When crystals are obtained, seed the solution by rubbing a seed crystal against the wall at the surface of the solution. Collect the crystalline products, wash with a little ethanol, dry, and determine the melting points.

The melting point reported for *N-p*-tolyl-D-glucosamine is 109 to 113°C; the melting point reported for *N-p*-toly-1-amino-1-deoxy-D-fructose is 150 to 152°C. Dispose of the filtrates in the organic waste container.

QUESTIONS

1. Raffinose (a trisaccharide occurring in cotton seeds), α,α-trehalose (occurring in yeast and various insects), and gentiobiose (present in several glycosides) have the structures shown on the next page.
    a.   Which of these sugars would reduce Benedict's solution?
    b.   Which would give a positive test for fructose with Selivanoff's reagent?
    c.   How many different osazones would be formed after acid hydrolysis of each compound followed by reaction with phenylhydrazine?

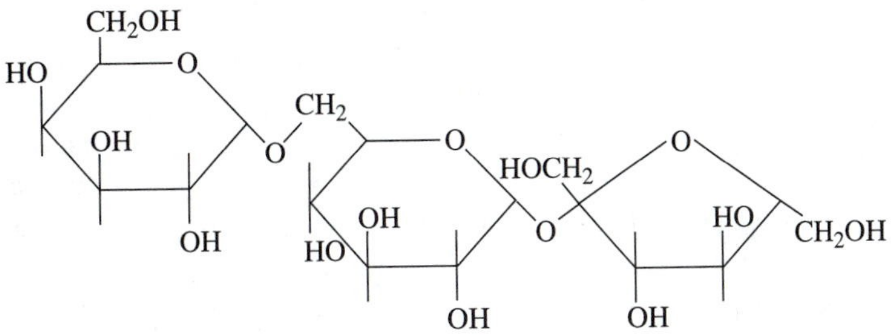

raffinose

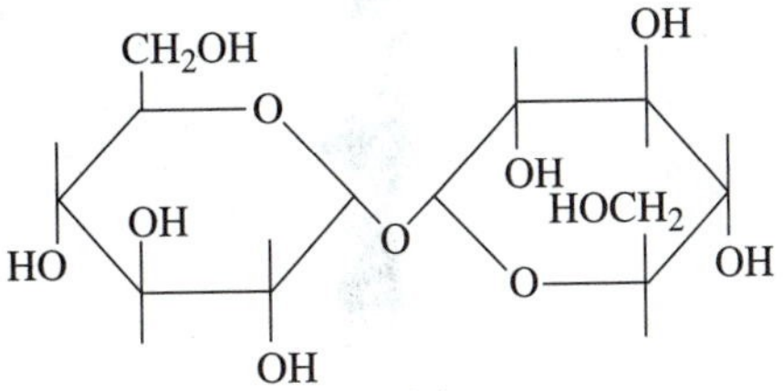

α,α-trehalose

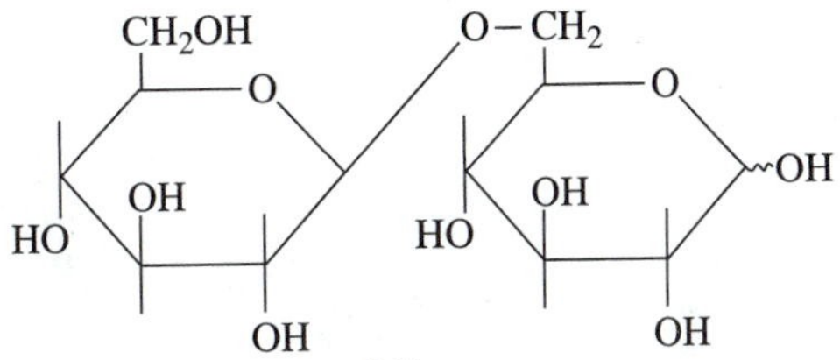

gentiobiose

2. Honey contains sucrose, fructose, and glucose. The Benedict test can be adapted for the quantitative analysis of reducing sugars. With suitable modifications and instruments, the Selivanoff test can be used for the determination of fructose plus sucrose in the presence of glucose. Outline a strategy by which the three sugars in honey could be determined using these two methods.

3. Account for the changes observed on treatment of D-glucose with base, and outline the reactions that occur.

PRELABORATORY QUESTIONS

1. Draw the structures of an aldohexose and a ketohexose.
2. What is the color of complexed copper (II) ion in aqueous solution?
3. Draw the structure of a nonreducing sugar.
4. List the components of Selivanoff's reagent.
5. List the components of honey.

REFERENCE

Robyt, J.F. *Essentials of Carbohydrate Chemistry;* Springer-Verlag: New York, 1997.

Acid-Catalyzed and Enzyme-Catalyzed Hydrolysis of Polysaccharides

Starch and cellulose are polymers formed by linking glucose units through $1 \rightarrow 4'$ glycoside bonds. These polysaccharides make up 60 to 80% of the dry weight of most plants. Because of the abundance of these polymers, their detailed structure and breakdown have been extensively studied.

The basic structure difference between starch and cellulose is the configuration of the glycoside bond, which is α in starch and β in cellulose. Cellulose has a regular structure and very high molecular weight. Naturally occurring cellulose is partially crystalline and forms strong fibers.

Starch, the main storage polysaccharide in grains and other food plants, is a mixture of two types of polymers. One fraction, amylose, usually approximately 25% of the total, is made up of unbranched chains of 250 to 300 D-glucose units linked by $1\alpha \rightarrow 4'$ glycoside bonds. The amylose chains are coiled into a helical structure. The cavity of the helix is the proper size to permit the entrapment of iodine molecules, and addition of iodine to starch gives an intense deep blue color. Amylopectin, the major component of most starch, consists of similar chains with side branches extending from the C—6 CH_2OH group at approximately every twenty-fifth glucose unit, giving a bushlike structure.

CH₂OH

OH

$1\alpha \rightarrow 6'$

CH₂OH   CH₂OH   CH₂OH OH   CH₂

OH $1\alpha \rightarrow 4'$ OH   OH   OH

OH   OH   OH   OH

nonreducing end   amylopectin (partial structure)

In digestion or in fermentation to produce alcohol, starch must be broken down to glucose units by hydrolysis of the α-glycoside bonds. This is accomplished by amylase enzymes present in saliva, digestive fluids, and yeast. There are two main types of amylase enzymes, designated α and β; both are specific for α-glycoside bonds, but they attack the starch chain in different ways. β-Amylase brings about cleavage of two-unit fragments, that is, molecules of the disaccharide maltose, starting from the nonreducing end of the chain. The action of β-amylase is interrupted by the presence of a branch in the chain.

323

α-Amylase cleaves the chain into fragments of six to seven glucose units very rapidly, causing loss of the iodine-complexing properties. Depending upon the type of α-amylase, further degradation to maltose and maltotriose (the three-glucose trisaccharide), and ultimately to glucose, occurs more slowly. α-Amylase of saliva has very little activity for cleavage of the disaccharide or trisaccharide fragments. Maltose and maltotriose therefore accumulate, but very little glucose is produced.

The breakdown of starch by acid-catalyzed hydrolysis occurs in a more random way, and the course of the reaction is quite different from that of the enzymatic hydrolysis. In this experiment the enzyme and acid-catalyzed reactions will be compared by reducing power and TLC.

EXPERIMENTS

SAFETY NOTE

Although no particularly hazardous chemicals are used in these experiments, normal care should be observed. Caution in the use of boiling water baths should be taken to avoid spills and/or burns.

Place 0.6 g of soluble starch in a 125-mL Erlenmeyer flask, make a thin paste with a few mL of water, and then add 40 to 45 mL of boiling water. Boil the solution gently until all starch particles disappear, and allow the opalescent solution to cool.

A. ENZYMATIC HYDROLYSIS OF STARCH

Place 20 mL of the starch solution in an 18×150-mm test tube and label this tube "A." Prepare four 13×100-mm test tubes labeled A15, A30, A45, and A60. Noting the time, add approximately 0.2 mL of saliva to the starch solution, mix the solution with a pipet, and let it stand at room temperature. After 15 minutes, remove 2 mL of the starch solution with a pipet, transfer it to tube A15, and heat the sample briefly in a boiling water bath to stop the reaction by denaturation of the enzyme. Remove additional 2-mL samples after 30, 45, and 60 minutes. Place each sample in an appropriately labeled test tube and heat immediately to denature the enzyme.

For TLC examination, use 1×3-inch silica gel strips and apply spots from the samples taken after 15, 30, and 60 minutes. Also spot a reference sample of a mixture of glucose and maltose (1% of each in water). The hydrolysis samples are quite dilute, and three or four applications should be made. Place a 1-mm spot of each solution on the strip, and then warm and blow gently over the spots until the white zone (due to water) has almost disappeared; then make another application exactly coinciding with the first. Develop and spray the strips as described in Chapter 41.

B. ACID HYDROLYSIS OF STARCH

Place 20 mL of the starch solution in an 18×150-mm test tube and label this tube "B." Prepare four 13×100-mm test tubes labeled B15, B30, B45, and B60. Noting the time, add 0.5 mL of concentrated hydrochloric acid to the starch solution, stir well, and place it in a boiling water bath. After 15 minutes, remove a 2-mL sample of the starch solution with a pipet, transfer it to tube B15, and add a few drops of 2 M NaOH (until just neutral to indicator paper). Remove additional 2-mL samples after 30, 45, and 60 minutes. Place each sample in an appropriately labeled test tube and neutralize each immediately with 2 M NaOH.

Run a TLC on the neutralized samples, together with a glucose–maltose reference, as described for the enzymatic hydrolysis. (The final sample will be somewhat more concentrated because of heating, and only one or two applications are needed.)

After examining the TLC plates from the two hydrolysis reactions, run another TLC with selected samples from both reactions together on the same strip to make a direct comparison and to confirm conclusions.

To each of the eight tubes (four from each hydrolysis) remaining from the TLC samples, add 5 mL of Benedict's solution. Heat the tubes in a boiling water bath for 2 minutes and compare the amount of reducing sugar in the samples. To dispose of the Benedict's test samples, boil them briefly with a pinch of glucose, filter to remove the solid, which may be placed in the non-hazardous waste container, and wash the filtrate down the sink with ample water.

QUESTIONS

1. On partial acid hydrolysis, cellulose gives one disaccharide, whereas the same treatment of starch gives two disaccharides, one major and one minor. Write structural formulas for each of the three disaccharides.
2. If temperature were the only factor, which hydrolysis would you expect to occur more rapidly, the room temperature enzymatic one or the acid-catalyzed one done in a boiling water bath? Explain.
3. Why could the enzymatic hydrolysis not be carried out in a boiling water bath?
4. How would you have expected the rates of enzyme-catalyzed and acid-catalyzed reactions to compare if both had been carried out at room temperature?
5. From examination of your TLC plates, how much glucose was produced in the enzyme-catalyzed reaction compared to the acid-catalyzed reaction?

PRELABORATORY QUESTIONS

1. Cellulose and amylose are polymers of D-glucose. What structural feature makes them different?
2. Distinguish between the terms "amylose" and "amylase."
3. What reagent is used to visualize the TLC plates in the hydrolysis of starch?
4. What types of compounds give a positive Benedict's test?

REFERENCES

Aspinall, G.O. *Polysaccharides;* Pergamon: Elmsford, New York, 1970.

Whistler, R.L. (Ed.). *Methods in Carbohydrate Chemistry;* Academic: New York, 1964; Vol 4.

Analysis of Fats and Oils

The bulk of plant and animal tissue is composed of three main classes of compounds: carbohydrates, proteins, and lipids. The term **lipid** includes all substances in living tissues that are soluble in ether, methylene chloride, or similar organic solvents. The most abundant lipids are triesters of glycerol with long-chain "fatty" acids. The triesters, or **triglycerides,** from animal sources are usually low-melting solids and are called *fats,* while those from plants are viscous *oils* that solidify below 0°C. The general structures are the same, and the difference in properties arises from the fact that the vegetable oils contain a larger proportion of unsaturated acid groups, with one, two, or more double bonds in the chain. Triglycerides such as beef tallow and other depot fats of animals contain mostly acid chains with one or no double bonds. The degree of unsaturation is important in nutrition. Unsaturated fats are essential components of the human diet. However, there is evidence that a high proportion of saturated fats in the diet leads to the deposition of cholesterol in blood vessels (atherosclerosis).

Fatty acids invariably have chains with an *even* number of carbon atoms, most commonly 16 or 18. A few of the more *abundant* fatty acids are listed below; these and several others can occur in any combination in a given triglyceride.

	TYPICAL FATTY ACIDS	TRIGLYCERIDE STRUCTURE
Lauric	$CH_3(CH_2)_{10}CO_2H$	RCO_2—CH_2
Myristic	$CH_3(CH_2)_{12}CO_2H$	$\quad\quad\quad\mid$
Palmitic	$CH_3(CH_2)_{14}CO_2H$	$R'CO_2$—CH
Stearic	$CH_3(CH_2)_{16}CO_2H$	$\quad\quad\quad\mid$
Palmitoleic	$CH_3(CH_2)_5CH{=}CH(CH_2)_7CO_2H$	$R''CO_2$—CH_2
Oleic	$CH_3(CH_2)_7CH{=}CH(CH_2)_7CO_2H$	
Linoleic	$CH_3(CH_2)_3(CH_2CH{=}CH)_2(CH_2)_7CO_2H$	
Linolenic	$CH_3(CH_2CH{=}CH)_3(CH_2)_7CO_2H$	

Butterfat and coconut oil are atypical, since the mixtures of acids in these triglycerides contain a significant number of 8-, 10-, and 12-carbon chains. The compositions of a few fats and oils are given in Table 43.1. These values vary over a fairly wide range, depending on the source of the oil, and there is no one "correct" figure.

Table 43.1 Typical Fatty Acid Content of Selected Fats and Oils

	Constituent Fatty Acids (%)*									
NO. CARBON ATOMS:	<12	12	14	16	16	18	18	18	18	>18
NO. DOUBLE BONDS:		0	0	0	1	0	1	2	3	
Human fat	—	—	3	24	5	8	47	10	—	2
Butterfat	4†	4	12	33	2	12	29	2	—	2
Lard	—	—	1	28	1	16	43	9	—	2
Coconut oil	15‡	46	18	10	—	4	6	—	—	—
Corn oil	—	—	—	10	1	2	25	62	—	—
Cottonseed oil	—	—	1	20	2	1	20	55	—	1
Linseed oil	—	—	—	10	—	4	23	57	6	—
Olive oil	—	—	—	9	—	3	78	10	—	—
Palm oil	—	—	1	45	—	4	40	10	—	—
Peanut oil	—	—	—	10	—	3	48	34	—	5**
Safflower oil	—	—	—	7	—	2	13	75	3	—
Soybean oil	—	—	—	9	—	4	43	40	4	—

* Blank spaces indicate that less than 1% of the corresponding acid is usually present.

† C_6 saturated (caproic acid) and C_{10} saturated (capric acid).

‡ C_8 saturated (caprylic acid) and C_{10} saturated.

** C_{20} saturated (arachidic acid), C_{22} saturated (behenic acid), and C_{24} saturated (lignoceric acid).

The classical procedure for obtaining the acids from a fat is hydrolysis (saponification) with alkali to give the salt (a soap), and then acidification. This hydrolysis is rather slow at room temperature, and at high temperatures, when a glycol is used as solvent, isomerization of the polyunsaturated acids occurs. For analytical purposes, where only a small amount of material is required, this difficulty is avoided by preparing the methyl esters directly from the triglyceride by transesterification (Eq. 43.1). By using a large excess of methanol, the equilibrium is shifted essentially completely to the right.

$$R-\overset{\overset{\textstyle O}{\|}}{C}-OR' + MeOH \underset{}{\overset{acid}{\rightleftharpoons}} R-\overset{\overset{\textstyle O}{\|}}{C}-OMe + R'OH \qquad (43.1)$$

In this experiment, you will be given a sample of a commercially available fat or oil (e.g., a cooking oil) to determine its fatty acid composition by GC of the methyl esters. From this information you can then identify the oil from among a list of possibilities provided by your instructor, using the data in Table 43.1. Again, the percentages reported in the table are typical, but actual compositions may vary by ±5 to 10% among samples of the same oil from different sources.

EXPERIMENTS

SAFETY NOTE

Caution should be observed in the use of concentrated acids and bases. Avoid exposure to solvent fumes by use of a hood.

A. TRANSESTERIFICATION OF TRIGLYCERIDES

(Microscale)

Weigh 0.10 g (approximately 8 drops) of the fat or oil into a 1-dram (4-mL) screw-cap vial. Add 2 mL of toluene to dissolve, and then add 1 mL of a solution of BF_3 in methanol (12 g BF_3/100 mL MeOH). Screw on the cap tightly, shake to mix, and place in a 50-mL beaker, half full of boiling water. Check to see that no vapor escapes; if hissing or bubbling is observed, cool and retighten the cap. After 30 minutes, remove the vial, cool, and add 1 mL of water. Shake well, and let the mixture stand to separate (10 to 15 minutes). The toluene (upper) layer contains the methyl esters.

B. GC ANALYSIS OF METHYL ESTERS

Analyze the toluene solution by gas chromatography on a polar column as directed by your instructor. (Suitable conditions involve use of a 10% diethyleneglycol succinate column at 185 to 200°C. Significantly better results can be obtained using a gas chromatograph with a flame ionization detector. A 10% Apiezon + 2% KOH column at 225°C works well.) By comparison with a GC of a standard mixture of fatty acid methyl esters provided, identify the components and calculate their approximate percent composition in the mixture. Relative peak areas may be measured by cutting out the peaks and weighing them (± 0.001 g) on a balance, calculating peak areas (see Figure 6.4), or using an integrator on the gas chromatograph. Compare the composition with those in Table 43.1 and identify the original oil.

C. ISOLATION OF TRIMYRISTIN

(Microscale)

Place 0.8 g of crushed or ground nutmeg in a 10-mL round-bottom flask and add 4 mL of methylene chloride. Attach a condenser and heat the mixture to reflux gently for 30 minutes. After allowing the mixture to cool somewhat, remove the solvent with a Pasteur pipet with cotton in the tip and place it in a 25-mL Erlenmeyer flask. Repeat the extraction procedure using a fresh 4-mL portion of methylene chloride. Evaporate the combined extracts to dryness (in a hood or use a rotary evaporator) to obtain the crude trimyristin. Weigh the crude trimyristin and calculate the percentage yield based on the weight of nutmeg used. Recrystallize the product using ethanol and submit it to your instructor or reserve the crude product for hydrolysis in Experiment D.

D. HYDROLYSIS OF TRIMYRISTIN

$$CH_2-O-\overset{\overset{\displaystyle O}{\|}}{C}-C_{13}H_{27}$$
$$CH-O-\overset{\overset{\displaystyle O}{\|}}{C}-C_{13}H_{27} + HO^- \longrightarrow$$
$$CH_2-O-\overset{\overset{\displaystyle O}{\|}}{C}-C_{13}H_{27}$$

$$CH_2-OH$$
$$CH-OH + 3\, C_{13}H_{27}\overset{\overset{\displaystyle O}{\|}}{C}-O^-$$
$$CH_2-OH$$

The crude trimyristin obtained is hydrolyzed by heating to reflux in 1 mL of 6 M NaOH and 2 mL of water for 1 hour. At the end of this time, the mixture should have largely clarified, indicating complete hydrolysis. Pour the basic solution with stirring into a mixture of 10 mL of water and 1 mL of 12 M hydrochloric acid. Collect the solid by vacuum filtration using a Hirsch funnel, wash with cold water, and allow it to dry. Determine the weight and melting point of the dry myristic acid, calculate the percentage yield, and submit the sample, in an appropriately labeled container, to your instructor.

QUESTIONS

1. What is the purpose of the BF_3 in the transesterification reaction? Give equations.
2. What happened to the glycerol from the triglyceride; that is, why was it not observed in the GC?
3. Write an equation for the base-catalyzed hydrolysis of glyceryl trimyristate.
4. Why are compounds such as CCl_4, $C_2H_3Cl_3$, etc., used as "dry cleaning" solvents?
5. How could the equivalent weight of an acid obtained from the hydrolysis of a fat be determined?

PRELABORATORY QUESTIONS

1. What is the structural difference between a fat and an oil?
2. Draw condensed structural formulas for palmitic acid and oleic acid.
3. Write an equation to show the transesterification of glyceryl tripalmitate with methanol.
4. Write an equation to show the saponification of glyceryl trimyristate using NaOH.
5. Draw the structure of a triacylglycerol (triglyceride) that could be optically active.

REFERENCES

Akoh, M., Ed. *Food Lipids;* Marcel Dekker: New York, 1998.

Gunstone, F.D. *An Introduction to the Chemistry of Fats and Fatty Acids;* Chapman & Hall: London, 1958.

Morrison, W.R.; Smith, L.M. *J. Lipid Res.,* 1964, *5,* 600.

Nichols, P.L., Herb, S.F.; Riemenschneider, R.W. *J. Am. Chem. Soc.,* 1951, *73,* 247.

Synthetic Organic Polymers

Polymers are large molecules that are built up from many small units called **monomers.** Several well-known polymers that occur naturally are cellulose, rubber, and proteins. In the past 50 years a wide variety of synthetic polymers have been developed for use as fibers, adhesives, protective coatings, and structural materials. Synthetic polymers are now produced in larger volume than any other chemical products.

Polymers can be prepared by two types of reactions: condensation and addition. **Condensation polymers** are obtained from monomers with two or more functional groups that can react to link the monomer units together by splitting out a small molecule such as water or alcohol. Typical examples of condensation polymers are polyesters, polyamides, and epoxy resins; the latter are one of the components of epoxy glues:

$$n \text{ HOCH}_2\text{CH}_2\text{OH} + n \text{ [diacid as diester]} \longrightarrow \text{[polyester]} + 2n \text{ CH}_3\text{OH}$$

| ethylene glycol | diacid as diester | polyester (Dacron) |

$$n \text{ NH}_2(\text{CH}_2)_x\text{NH}_2 + n \text{ HOC(CH}_2)_y\text{COH} \longrightarrow$$

| diamine | diacid |

$$\text{—[NH(CH}_2)_x\text{NHCO(CH}_2)_y\text{CO]}_n + 2n \text{ H}_2\text{O}$$

polyamide (nylon)

Addition polymers are formed by end-to-end combinations of double bonds or rings, each monomer being added to the end of a growing chain. The most important examples are the simple vinyl polymers, including polyethylene, polyvinyl chloride (PVC), polyacrylonitrile (Orlon, Acrilan), polypropylene, polymethyl methacrylate (Plexiglass), and polystyrene. Depending on the monomer, the polymerization may involve electrophilic (cationic), nucleophilic (anionic), or free radical attack on the double bond. In each case, a small amount of an initiating reagent is used to start the polymerization by addition to the first monomer. These processes are illustrated below for several commercially important addition polymers.

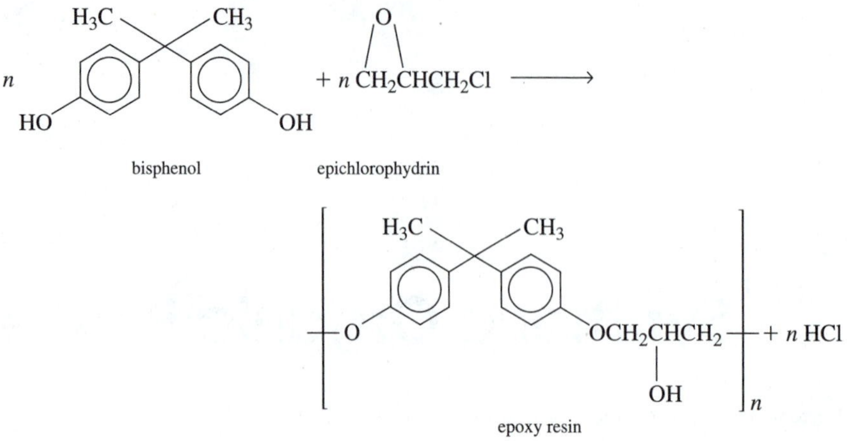

bisphenol   epichlorophydrin

epoxy resin

Cationic

$$n\ CH_2{=}CHCH_3\ \xrightarrow[\text{TiCl}_4]{\text{Et}_3\text{Al}}\ \text{Et}{\left(CH_2CH\right)}_n\ \overset{CH_3}{|}$$

propylene   polypropylene

Free Radical

$$n\ CH_2{=}CHC_6H_5\ \xrightarrow{\text{Z}\bullet}\ \text{Z}{\left(CH_2\overset{C_6H_5}{\underset{|}{CH}}\right)}_n\bullet$$

styrene   polystyrene

Anionic

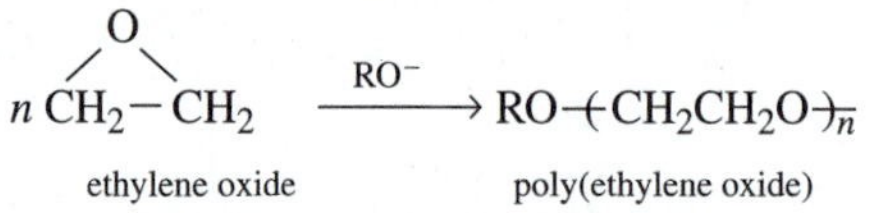

$$n\ \overset{O}{\overset{/\backslash}{CH_2{-}CH_2}}\ \xrightarrow{\text{RO}^-}\ \text{RO}{\left(CH_2CH_2O\right)}_n$$

ethylene oxide   poly(ethylene oxide)

The experimental procedures that follow describe the laboratory-scale syntheses of several polymers. On an industrial scale, where tons of polymer are produced daily, different, more economical methods are generally used. In addition, much greater care must be taken to control the reaction conditions so that a polymer with the desired properties is obtained reproducibly. Some of the properties looked for in specific polymers are plasticity versus rigidity, strength of a drawn fiber, clarity of a film, and stability with respect to elevated temperatures, sunlight, and chemical reagents.

EXPERIMENTS

A. POLYSTYRENE

(Microscale)

The polymerization of styrene can be effected using cationic, anionic, or free-radical initiators. Industrially, polystyrene is generally prepared by heating pure styrene slowly to about 180°C, taking advantage of the fact that above 80°C an added initiator is unnecessary. In the procedure that follows, styrene is polymerized in solution with benzoyl peroxide as the initiator. The O—O bond in benzoyl peroxide is weak and above 60°C breaks homolytically to form two benzoyloxy radicals, which rapidly undergo decarboxylation. The resulting phenyl radicals are the actual initiators that add to the double bond of styrene.

$$C_6H_5\overset{O}{\overset{\|}{C}}-O-O-\overset{O}{\overset{\|}{C}}C_6H_5 \xrightarrow{\Delta} 2C_6H_5\overset{O}{\overset{\|}{C}}-O\cdot \longrightarrow 2C_6H_5\cdot + 2\,CO_2$$

$$C_6H_5\cdot + CH_2\!\!=\!\!CHC_6H_5 \longrightarrow C_6H_5-CH_2-\underset{\underset{C_6H_5}{|}}{CH}\cdot$$

$$C_6H_5-CH_2-\underset{\underset{C_6H_5}{|}}{CH}\cdot \ + CH_2\!\!=\!\!CHC_6H_5 \longrightarrow$$

$$C_6H_5-CH_2-\underset{\underset{C_6H_5}{|}}{CH}-CH_2-\underset{\underset{C_6H_5}{|}}{CH}\cdot$$

$$C_6H_5\!\!-\!\!(CH_2-\underset{\underset{C_6H_5}{|}}{CH})\cdot_{n+2} \xleftarrow{\quad\quad} \overset{n\,CH_2=CHC_6H_5}{\big|}$$

$$C_6H_5\!\!-\!\!(CH_2-\underset{\underset{C_6H_5}{|}}{CH})\cdot_{n+2} + C_6H_5\!\!-\!\!(CH_2-\underset{\underset{C_6H_5}{|}}{CH})\cdot_{m} \longrightarrow$$

$$C_6H_5\!\!-\!\!(CH_2-\underset{\underset{C_6H_5}{|}}{CH})_{n+2}\!\!(CH_2-\underset{\underset{C_6H_5}{|}}{CH})_{m}\!\!-\!\!C_6H_5$$

polystyrene

SAFETY NOTE

Styrene is somewhat toxic and is a flammable liquid; avoid breathing vapor or contact with the skin. Work in the hood. Benzoyl peroxide is unstable to heat and friction and must be handled with care. Transfer by pouring from a paper carton onto a piece of glazed paper, and avoid contact with metal (use a porcelain spatula). Clean up all spills with water, and rinse the glazed paper with water before discarding.

PROCEDURE

Commercial styrene contains a phenol, such as 4-*t*-butylcatechol, which must be removed before it can be used in this experiment. Pass 1.0 to 1.5 mL of styrene through a short column of 1 to 2 g of dry alumina contained in a pipet with a wad of cotton and collect the styrene in a small disposable test tube.

Place 0.5 mL of the purified styrene in a 13 × 100-mm test tube and dissolve it in 2.5 mL of toluene. Obtain approximately 25 mg of benzoyl peroxide (see Safety Note—estimate amount by volume, comparing with a weighed comparison sample). Add the benzoyl peroxide to the styrene solution, and swirl the mixture to dissolve the peroxide. Cork the tube loosely, and heat the solution in a boiling water bath placed on a hot plate. While the solution is being heated, replace water in the bath as it is needed and start one of the other experiments in this or another chapter as directed.

After heating the solution for 60 to 90 minutes, cool it to room temperature and pour it into 5 mL of methanol in an 18 × 150-mm test tube. Stir the mixture well to coagulate the precipitate, which will adhere to the walls. Decant the slightly cloudy liquid from the polymer, and pour it into a waste solvent container. Add an additional 2.5 mL of methanol to the test tube containing the polystyrene, stopper the tube, and shake it vigorously.

Collect the polystyrene in a Hirsch funnel, rinsing the test tube and the polymer in the funnel with another 2.5 mL of methanol. Place the filtrate in the solvent waste container. Air-dry the polymer, weigh it, and calculate the percentage yield.

A polystyrene film may be cast by transferring 50 to 100 mg of the polymer to a microscope slide, adding a few drops of toluene to dissolve the polymer (stir with a matchstick to mix well), and setting the slide aside until the toluene evaporates. Examine the film coating for clarity and hardness.

B. NYLON

(Microscale)

A variety of nylons have been developed to fit different uses, and their production now exceeds 2.5 billion pounds per year. Their properties depend on the degree of flexibility or rigidity of the chain, molecular weight, and other properties. Extremely strong and tough fibers (aramids) are obtained by using aromatic dicarboxylic acids and aromatic diamines.

Nylons or polyamides were the first totally synthetic polymers to come into major use as textile fibers. Different nylons are named by designating the number of *carbon* atoms in each monomer unit. The original nylons, and still the most common type, are obtained by the condensation of diamines and acids. The amine salt obtained by mixing equimolar amounts of the two components is heated with a catalyst to split out water and form the amide bonds. The polymer is then spun from the melt into fibers:

$$n\overset{+}{N}H_3(CH_2)_6\overset{+}{N}H_3 + \overset{-}{O}-\overset{\overset{\displaystyle O}{\|}}{C}(CH_2)_4\overset{\overset{\displaystyle O}{\|}}{C}-\overset{-}{O} \xrightarrow{\Delta}$$

$$-\!\!\left[NH(CH_2)_6NH-\overset{\overset{\displaystyle O}{\|}}{C}(CH_2)_4\overset{\overset{\displaystyle O}{\|}}{C}\right]_{\!n} + 2_n\,H_2O$$

nylon 66

A second type of nylon is obtained by the polymerization of monomer units containing both the amine and acid groups in the same molecule. This type is illustrated by the polymer of 6-aminohexanoic acid, which is actually prepared by the addition polymerization of the cyclic monomer ϵ-caprolactam.

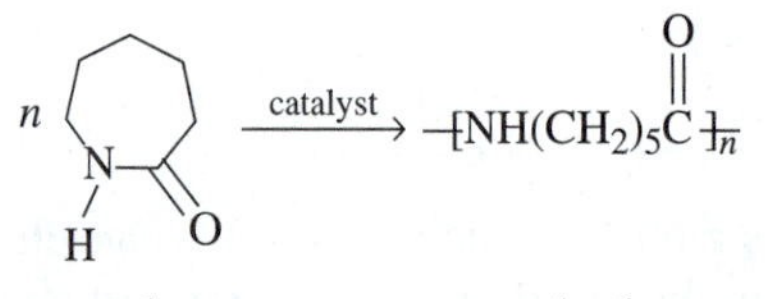

$$n \quad \xrightarrow{\text{catalyst}} \quad -\!\!\left[NH(CH_2)_5\overset{\overset{\displaystyle O}{\|}}{C}\right]_{\!n}$$

ε-caprolactam   nylon 6

Another method of condensation used for some of these nylons involves the use of the diacid chloride and the diamine. This process is called **interfacial polymerization** and is carried out by dissolving the amine in water and the acid chloride in an immiscible solvent. The two solutions mix only at the interface between the layers, and the resulting polymer can be removed as a film. This type of polymerization is illustrated below.

Preparation of Nylon 6,10.

$$n\,NH_2(CH_2)_6NH_2NH_2 + n\,\overset{\overset{\displaystyle O}{\|}}{ClC}(CH_2)_8\overset{\overset{\displaystyle O}{\|}}{CCl} \longrightarrow -\!\!\left[NH(CH_2)_6NH\overset{\overset{\displaystyle O}{\|}}{C}(CH_2)_8\overset{\overset{\displaystyle O}{\|}}{C}\right]_{\!n}$$

hexamethylene   sebacoyl chloride   nylon 6,10
diamine   d 1.12

SAFETY NOTE

Hexamethylene diamine and sebacoyl chloride are corrosive substances. The latter is a lachrymator; avoid breathing vapor and contact with the skin. Avoid contact of the polymer strand with your hands until the strand has been washed thoroughly.

PROCEDURE

Dissolve 0.1 g of hexamethylene diamine and 0.2 g of sodium carbonate in 2 mL of water in a 10-mL Erlenmeyer flask or 13 × 100-mm test tube. Mix separately in a 10-mL beaker, 0.1 mL of sebacoyl chloride and 2 mL of methylene chloride (adipoyl chloride may be substituted and Nylon 6,6 prepared). Carefully and slowly pour the diamine solution down the side of the beaker so that it forms a layer above the methylene chloride solution. The polymer will form at the interface as a white film. Grasp the center of the film with a pair of forceps, and slowly pull the polymer out of the beaker. Note that as the polymer is being removed from the beaker, more forms at the interface, producing a continuous strand or "rope." Wrap the strand around a glass stirring rod and continue to pull the polymer from the beaker in this fashion, wrapping it around the rod until no more polymer forms. (**CAUTION:** The nylon may contain unreacted starting materials, and it should not be handled until the polymer is washed.) Wash the nylon under running water, then squeeze it in a paper towel, allow it to dry in air until the next class period, and record the weight.

When you have finished, **do not** discard the nylon **or** the remaining reaction mixture in the sink; place in designated waste receptacles.

C. POLYURETHANES

(Microscale)

Polyurethanes are generally prepared from a diisocyanate and a dihydroxy compound. The name **polyurethane** is an exception to the standard nomenclature of polymers (e.g., polystyrene), in that polyurethanes are not polymers of urethane ($CH_3CH_2OCONH_2$, ethyl carbamate). Their importance arises from the large variety of R and R' groups that may be used, with a resulting large variation in properties of the polymer.

$$O=C=N-R-N=C=O + HO-R'-OH\longrightarrow$$

$$\overset{O}{\overset{\|}{\text{—}(C}}-NH-R-NH-\overset{O}{\overset{\|}{C}}-O-R'-O\overset{}{)_n}$$

The diisocyanate may be aliphatic, such as hexamethylene diisocyanate, or aromatic, such as the toluene-2,4-diisocyanate (also called tolylene-2,4-diisocyanate). The diol may be as simple as ethylene glycol or 1,6-hexanediol, but generally larger molecular weight diols are used, such as polyethylene glycols, polypropylene glycols, and hydroxyl-terminated polyesters, such as poly(ethylene adipate). Branched (cross-linked) polyurethanes can be produced by using a triol instead of a diol. The popular Lycra fiber is a polyurethane.

Isocyanates

$$OCN \!\!-\!\!(CH_2)_6\!\!-\!\!NCO$$

hexamethylene
diisocyanate

toluene-2,4-
diisocyanate

tolylene-2,6-
diisocyanate

Diols

$$HOCH_2CH_2OH$$

ethylene glycol

$$HO(CH_2)_6OH$$

1,6-hexanediol

$$HO\!\!-\!\!(CH_2CH_2O)_n\!\!-\!\!H$$

poly(ethylene oxide)
"polyethylene glycol"

$$HO\!\!-\!\!(CH\!\!-\!\!CH_2\!\!-\!\!O)_n\!\!-\!\!H \quad (CH_3)$$

"polypropylene glycol"

$$HO\!\!-\!\!(CH_2CH_2O\overset{O}{\overset{\|}{C}}(CH_2)_4\overset{O}{\overset{\|}{C}}\!\!-\!\!O)_n\!\!-\!\!CH_2CH_2OH$$

poly(ethylene adipate)

One of the most important applications of polyurethanes is in polyurethane foams. Other polymers can be produced as foams (e.g., styrofoam, which is foamed polystyrene) by dissolving volatile solvents in the polymer melt and allowing them to vaporize, but polyurethanes are usually foamed during polymerization by adding a small amount of water to the monomer mixture. The water competes with the diol for the isocyanate groups and forms an unstable carbamic acid, which decomposes, liberating the gas, carbon dioxide.

$$R\!\!-\!\!N\!\!=\!\!C\!\!=\!\!O + H_2O \longrightarrow R\!\!-\!\!NH\!\!-\!\!\overset{O}{\overset{\|}{C}}\!\!-\!\!OH \longrightarrow R\!\!-\!\!NH_2 + CO_2 \uparrow$$

In the experiment that follows, a variety of diols and triols can be used to provide foams with different properties. In determining the appropriate ratios of reactants, their equivalent weights must be considered. (Equivalent weight = molecular weight ÷ number of functional groups per molecule.) Several diols and triols that may be used with toluene-2,4-diisocyanate (mol wt 174, equiv wt 87) and their equivalent weights are listed in Table 44.1. Castor oil is the glyceryl triester of a mixture of long-chain carboxylic acids, mostly ricinoleic acid $[CH_3(CH_2)_5CHOHCH_2CH\!\!=\!\!CH(CH_2)_7CO_2H]$. Since 88 to 90% of the acids is ricinoleic, castor oil can be considered as a mixture of 70% triol plus 28% diol and 2% monohydroxy triester, with an average of 2.7 hydroxyl groups per molecule.

Table 44.1   Diols and Triols for Copolymerization

	HYDROXYL COMPOUND	MOLECULAR WEIGHT(g/Mole)	EQUIVALENT WEIGHT (g/Eq)
Ethylene glycol	$HOCH_2CH_2OH$	62.1	31
Diethylene glycol	$(HOCH_2CH_2)_2O$	106	53
Triethylene glycol	$(HOCH_2CH_2\!\!-\!\!O\!\!-\!\!CH_2)_2$	150	75
Glycerol	$HOCH_2CHOHCH_2OH$	92	31
Castor oil		≈930	≈345
Water	H_2O	18	9

In addition to the monomers and water, the procedures call for triethylamine and silicone oil. The tertiary amine functions as a base catalyst for the polymerization reaction and the silicone oil serves to lower the surface tension of the mixture and leads to smaller bubbles. Two specific procedures are given, and if time and materials are available, other combinations of reactants may be tested. In each case the ratio of hydroxyl to isocyanate groups should be kept approximately constant.

SAFETY NOTE

Toluene-2,4-diisocyanate is a toxic substance and a suspected carcinogen; avoid breathing its vapors and contact with the skin. Dispense all reagents and carry out all reactions in the hood. Plastic gloves should be used and hands should be washed thoroughly when the experiment is completed.

PROCEDURE

1. Weigh into a 4-ounce (or smaller) paper cup 2.0 g of castor oil and 0.5 g of glycerol. Add 1 drop each of water, silicone oil, and triethylamine. With a glass stirring rod, mix these components well to form a creamy emulsion. Using a graduated pipet or buret in a fume hood, add 1.5 mL (1.83 g) of toluene-2,4-diisocyanate (or a mixture of the 2,4- and 2,6-isomers). Stir the mixture vigorously until a smooth emulsion is formed and bubbles begin to form. Remove the stirring rod and set the cup aside in the hood for the polymerization to proceed. The foam will remain tacky for some time after the maximum volume is reached and should be left to cure for at least a day before attempting to remove it from the cup.

2. Weigh into a 4-ounce (or smaller) paper cup 0.6 g of glycerol and 0.5 g of triethylene glycol. Add 1 drop each of water, silicone oil, and triethylamine, and continue as described in the preceding procedure, again using 1.5 mL of diisocyanate. The final emulsion is more difficult to obtain owing to the more hydrophilic nature of the glycol mixture. Compare the foam obtained in this experiment with that from the first part of this procedure. Again, allow the mixture to cure at least a day before attempting to remove it from the cup.

QUESTIONS

1. Other than increasing the rate of the polymerization, what would be the effect of using twice as much benzoyl peroxide initiator in the polystyrene preparation?
2. Calculate the percentage yield of the polymer obtained in the nylon 6,10 experiment.
3. What is the role of sodium carbonate in the nylon 6,10 experiment?
4. What other pair of functional groups might be used for the preparation of a polyurethane?
5. Calculate the molar ratio of the two reactants you used in the nylon preparation.

PRELABORATORY QUESTIONS

1. Polymers can be prepared by two types of reaction. What are these two types?
2. The catalyst (initiator) in the polystyrene preparation is what compound? Draw its structure.
3. Draw the structure of nylon 6,6.
4. Name three commercially important addition polymers.
5. Draw the structure for toluene-2,4-diisocyanate.

REFERENCES

Feldman, D.; Barbalata, A. *Synthetic Polymers;* Chapman and Hall: New York, 1996.

Odian, G. *Principles of Polymerization,* 3rd ed.; Wiley: New York, 1990.

Nylon

Morgan, P.W.; Kwolek, S.L. *J. Chem. Ed.* 1959, *36,* 182.

Dyes and Dyeing

Dyestuffs include a diverse group of compounds that have in common the property of producing a permanent color on cloth, leather, or paper. All of the useful dyes are aromatic compounds with highly delocalized electron systems that absorb light at certain wavelengths. The most important structural types are azo compounds, triarymethyl cations, and anthraquinones.

aniline yellow
(azo compound)

malachite green
(triarylmethyl cation)

alizarin
(anthraquinone)

The chemistry of dyeing involves much more than the use of highly colored compounds. The dyeing process is an interaction between a dye and a fiber, and what counts is the final combination. A beautifully colored substance is of no use as a dye if it does not impart its color irreversibly to the fabric. The most important requirements from the standpoint of the user are pleasing shades and fastness or permanence to washing, air oxidation, perspiration, and exposure to strong light.

Another set of considerations comes in the manufacturing process. The dyer must be able to obtain a precisely reproducible shade and depth of color. For economic reasons the manufacturer must also achieve even, level color and at the same time utilize the dye as completely as possible. A major factor in textile dyeing is the large variety of chemical structures present in modern fabrics. These include the natural fibers wool, silk, and cotton and several types of synthetic polymers (Table 45.1).

Table 45.1 Fibers and Dye Types

NAME	TYPE	STRUCTURAL UNIT	TYPE OF DYE
Wool or silk	Protein	R contains CO_2H, NH_2 and OH groups	Cationic (Ar_3C^+) or anionic ($-SO_3^-$ groups)
Cotton or viscose	Cellulose		Substantive (azo) $(-N=N-C_6H_4-C_6H_4-N=N-)$; also vat and mordant
Acetate	Cellulose acetate		Disperse (anthraquinones and others)
Orlon	Acrylic		Cationic
Dacron	Polyester		Disperse
Nylon	Polyamide		Anionic, also disperse

In dyeing and textile practice, dyes have traditionally been classified according to the *dyeing process* or the *fiber* that is dyed. For example, the terms "acid" and "basic" refer to dyes that are applied from an acidic or from an alkaline dye bath. **Direct dyes** are those that can be applied to cotton. **Disperse dyes** are applied from a *suspension* rather than from a solution of the dye. **Vat dyes** are applied by using a soluble, reduced form of the dye in the bath and then oxidizing it to an insoluble, colored form on the fabric. Each of these classes of dyes can include various structural types. Thus, a disperse dye or an acid dye may have any type of chromophoric (color-producing) structure, provided that the dyeing process involves a suspension of the dye or an acid solution, respectively.

Acid and basic dyes are fixed to the fiber by ionic attraction, and the terms *anionic* and *cationic* are more descriptive and meaningful. Typical **"acid" dyes** contain $-SO_3^-Na^+$ groups and combine in acid solution with a fabric containing $-NH_2$ groups. The dyeing process is actually an ion exchange sequence in which the $-NH_3^+$ cation of the fiber replaces Na^+ in the dye. In **"basic" dyes,** the dye molecule has a *cationic* structure and combines with anionic centers in the fiber.

Silk and wool contain both cationic $-NH_3^+$ groups and anionic $-CO_2^-$ groups in the amino acid side chains and can be dyed with either type of ionic dye. Synthetic polyacrylic fibers contain $-SO_3^-$ groups incorporated by design in the polymer to provide dye sites. Nylon can be prepared with either $-NH_3^+$ end groups or $-CO_2^-$ end groups in excess; most nylon fabrics contain excess $-NH_3^+$ groups and can therefore be dyed with anionic (acid) dyes.

Acid dyeing:

$$\text{dye} - SO_3^- Na^+ + \overset{+}{N}H_3 - R \longrightarrow \text{dye} - SO_3^- \overset{+}{N}H_3 - R$$

anionic dye   cationic fiber in   dyed fiber
                 acid solution

Basic dyeing:

$$\text{dye} - \overset{|}{C} = \overset{+}{N}^{/} + {}^-O_3S - R \longrightarrow \text{dye} - \overset{|}{C} = \overset{+}{N}^{/} \ {}^-O_3S - R$$

cationic dye   or

$$\text{dye} - \overset{|}{C} = \overset{+}{N}^{/} + {}^-O - \overset{O}{\overset{||}{C}} - R \longrightarrow \text{dye} - \overset{|}{C} = \overset{+}{N}^{/} \ {}^-O - \overset{O}{\overset{||}{C}} - R$$

anionic fiber in   dyed fiber
neutral or
basic solution

Cotton and modified cellulose (viscose, rayon) do not give good fastness (permanence) with simple anionic or cationic dyes. An important group of dyes that are direct or substantive for cotton are compounds in which azo groups occur at a distance that corresponds to the length of the repeating hexose unit in the cellulose chain. The dye is bound to the fiber by a series of hydrogen-bonding attractions. A single azo group is insufficient for dye fastness; "dis-azo" compounds, with two groups linked to a biphenyl system, provide excellent **substantive dyes.**

Polyester and cellulose acetate fibers have the least affinity for adsorption of dyes from solution. These fibers are dyed by the disperse process. The fabric is heated in a suspension of an insoluble dye and a "carrier" that enhances penetration. The dye becomes dispersed in the hydrophobic fiber. Since there is no bonding to the fiber, **disperse dyes** are often less permanent than other types.

An important development in textile practice is the use of mixed and blended fibers. In dyeing these materials, the differential binding of dyes on the individual fibers provides textured colors. The blended fabric is treated with a mixture of dyes, and the *lack* of affinity of a fiber for certain dyes is just as important as its affinity for others. A blended fiber, or a fabric woven with different fibers, can be **cross-dyed** with a mixture of dye types. In this way, patterned colors can be achieved in a single dyeing operation.

An enormous range of dyes is available to cover the diverse requirements for colors and fabrics. The dyeing process is determined by the type of fabric and the dye, but a number of factors contribute to the final result. The concentration of dye, the presence of electrolytes, and the temperature are important variables in the dye bath. The degree of wetting and swelling of the fiber, the duration of the dyeing, and the addition of surface-active compounds are other factors that affect the overall outcome. Because of the large number of factors that can be varied and the complicated interactions between dye and fiber, textile dyeing is a highly empirical practice, requiring skill and long experience.

CHEMISTRY OF AZO DYES

The preparation of azo dyes involves two reactions—diazotization and coupling. Both reactions are very simple operations that are carried out in aqueous solution. In **diazotization,** an aromatic amine is converted to a diazonium ion with nitrous acid.

$$HONO + ArNH_2 \xrightarrow{H^+} Ar\overset{+}{N}\equiv N + 2H_2O$$
$$\text{diazonium ion}$$

The **coupling reaction** is an electrophilic substitution of a phenol, naphthol, or aromatic amine to give an azo compound. The electrophile is the ArN_2^+ ion, and substitution occurs at positions *ortho* or *para* to the —OH or —NH$_2$ group. To obtain soluble azo dyes, a naphthol containing —SO$_3^-$Na$^+$ groups is used for coupling. A typical reaction is the coupling with the salt of 2-naphthol-3,6-disulfonic acid ("R-salt"). The resulting azo compound can be used as an anionic or acid dye. If a naphthol without solubilizing —SO$_3^-$Na$^+$ groups is employed, the azo compound is insoluble and can be used as a disperse dye.

"R-salt"   anionic azo dye

In certain dyeing applications, particularly textile printing, the azo coupling reaction is carried out *on the fabric*. The cloth is first immersed in an alkaline solution of the naphthol, or a paste of the naphthol is applied in a pattern. The fabric is then immersed in the diazonium solution, and the dye is formed on the surface of the fiber.

EXPERIMENTS

Several aspects of dye chemistry and the dyeability of various fibers are illustrated in the experiments in this chapter. In a brief laboratory experiment it is not possible to obtain optimal dyeing conditions or to evaluate the permanence of a dye. However, the relative affinities of fibers for dye types can be observed simply by exposing samples of various fibers to different dyes under the same conditions. Multifiber cloth containing bands of several different fibers provides a simple means for evaluating and comparing dye affinities. One standard multifiber cloth, Number 10, has six bands in the following sequence: wool, Orlon, Dacron, nylon, cotton, and acetate.

The amines to be used in these experiments are shown in Figure 45.1. In Experiments A and B, you will select one monoamine and one diamine from the list. The amines are available as 0.05 N solutions in 0.1 M HCl—each solution contains 0.05 mmole of monoamine or 0.025 mole of diamine and 0.1 meq of HCl per mL. These solutions are more dilute than those normally used for diazotization, but the rather high dilution is advantageous for an experiment of this type and permits the reactions to be carried out rapidly by a standard procedure.

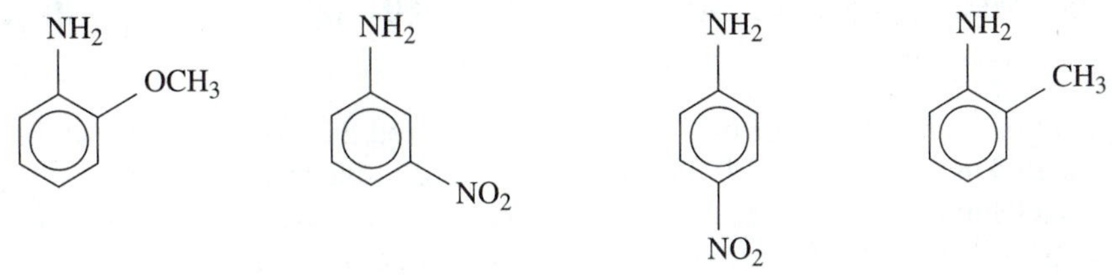

Figure 45.1　Amines to be used in dye syntheses.

General Procedure for Diazotization.

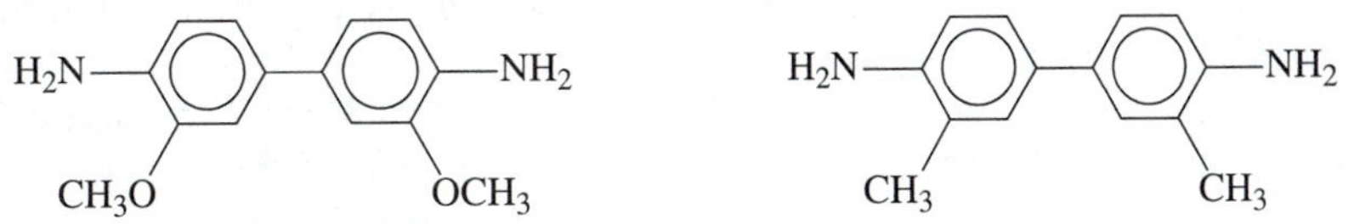

SAFETY NOTE

The amines and diamines used in this experiment are not on the list of suspected carcinogens, but certain compounds of similar structure are carcinogenic. The compounds in this experiment are used in very small amounts and present no significant hazard. Care should be taken to avoid contact of the diazotized solutions, dye solutions, or dyed cloth with the hands or clothing, since skin, fingernails, or fabric can be stained. 2-Naphthol is a toxic irritant.

Place 5 mL (0.25 meq) of a solution of the monoamine or diamine in a graduated centrifuge tube and chill several minutes in an ice bath. Add 0.4 mL of 0.5 M NaNO$_2$ solution to the amine and note any change in appearance. The diazonium solutions must be kept at 0°C in an ice bath until they are used.

A.　DYEING ON THE FIBER

(Microscale)

Obtain one 2 × 4-cm swatch of cotton, lay it flat on a paper towel, and moisten each end with a 1.0 M solution of β-naphthol (the solution contains 0.144 g [1 mmol] of naphthol and 2 meq of NaOH per mL) by placing several drops of the solution on each end of the swatch. Blot with a paper towel and allow the swatch to dry.

To one end of the dried naphthol-treated swatch, add a single drop of a prepared monoamine diazonium solution and to the other end add a single drop of the diazonium solution prepared from a diamine. Rinse the patch thoroughly in running water and blot on a clean paper towel.

B.   COMPARISON OF MONO- AND DIS-AZO DYES IN NEUTRAL AND ACID SOLUTION

(Microscale)

Obtain in a graduated centrifuge tube 3 mL of a diazotized monoamine and diamine different from those used in Experiment A. If the monoamine used is one of the nitroanilines, add 0.4 mL of 1.0 *M* NaOH to the diazonium solution to neutralize the excess acid. To each diazonium solution add 0.3 mL of a 0.5 *M* solution of "R-salt" (this solution contains 0.174 g [0.5 mmole] of R-salt and 1 meq of NaOH per mL).

Place the dye solutions in a water bath at 70 to 80°C, and in each solution immerse a 1-cm-wide strip of number 10 multifiber cloth* and a 1 × 2-cm patch of cotton (this can be cut from a clean, cotton lab towel). Keep the solutions in the bath approximately 5 minutes, and then remove the tubes. Transfer the cloth samples with a glass rod to clean beakers and rinse the samples thoroughly with water and blot on a towel.

To the dye solutions, add 0.4 mL of 2 *M* HCl (or 0.2 mL of 4 *M* acid) and repeat the dyeing in each bath with a cotton patch and a strip of multifiber cloth. After 5 minutes, remove, rinse, and blot the samples. Save the dis-azo dye bath (from the diamine) and discard the other dye solution.

Compare the intensity of color of the cotton strips dyed with the mono- and dis-azo dyes, and also compare the intensity of the color in the various multifiber bands for the two dyes in neutral and acid solutions.

C.   DISPERSE AND CROSS-DYEING

(Microscale)

Place 1 mL of the diazotized *m*-nitroaniline solution in a graduated centrifuge tube. Neutralize with 0.1 mL (approximately 2 drops) of 1 *M* NaOH and then add 0.1 mL (2 drops) of β-naphthol solution. Allow the solution to warm to room temperature and stir and rub, if necessary, to obtain a finely divided suspension of solid. Add approximately 0.5 mg of biphenyl and 1 drop of surfactant solution to the dye mixture and immerse a 1-cm-wide strip of the multifiber cloth. (Alternatively, if available, 1 mL of commercial yellow disperse dye suspension can be used instead of the dye prepared from *m*-nitroaniline.)

Heat the dye bath in a 90 to 100°C water bath for 10 to 15 minutes, and then cool and remove, rinse, and blot the sample. Record the relative intensities of the bands in the multifiber sample.

Combine the remaining dye suspension and the remaining acid solution of dis-azo dye from Experiment B, and repeat the dyeing at 90°C with another multifiber strip. If available, a sample of checked multifiber cloth (number 7403) can be dyed also. Remove, rinse, blot, and record the appearance of the dyed sample.

D.   CATIONIC DYES

(Microscale)

Obtain 1 mL of a 0.1% solution of malachite green (see Chapter 19 for the preparation), and immerse a strip of multifiber cloth. Heat the solution on a steam bath for several minutes and then rinse, blot, and record the relative intensities of color in the bands.

QUESTIONS

1.   Draw the structures of each of the dyes you prepared in this experiment and classify them as cationic, anionic, disperse, and/or direct dyes.

---

* Available from Testfabrics, Inc., P.O. Box O, Middlesex, NJ 08846.

2. Which of the dyes tested had the greatest affinity for cotton? Account for this result.
3. Explain the differences observed from acid and neutral dyeing of the multifiber cloth in Experiment B.
4. Explain the results observed in the cross-dyeing experiment in Experiment C.

PRELABORATORY QUESTIONS

1. Name three important structural types of dyes.
2. Draw the structure of the dye alizarin.
3. What is meant by the term "disperse dye"?
4. Write an equation to illustrate the reaction of phenol with the diazonium salt derived from 4-nitroaniline.
5. Draw the structure of the so-called R-salt. How does its structure affect the water solubility of the resulting dye?

REFERENCE

Wells, K. *Fabric Dyeing and Printing;* Interweave Press, Inc.: New York, 1997.

Qualitative Organic Analyses

Along with synthesis and the examination of reaction mechanisms, an equally important part of organic chemistry has to do with the characterization and identification of compounds, which may be encountered in sources ranging from a laboratory reaction to exotic tropical amphibians. In any case, sufficient information must be accumulated to establish the identity of the compound in question with that of a previously described compound of known structure or else to determine, *ab initio,* the structure of the unknown.

In earlier days, organic chemists relied heavily on chemical behavior in the characterization of compounds and structure elucidation. Various reactions were applied to diagnose the presence of functional groups and structural units, and final evidence for a new structure usually involved systematic degradation to identifiable products. The development of spectroscopic methods has had a revolutionary effect on this area of organic chemistry. Today it is possible to characterize and arrive at the structure of a previously unknown compound entirely by physical and spectroscopic methods, without recourse to any "wet chemistry" at all. Very often, the structure stands revealed as soon as the ultraviolet, infrared, NMR, and mass spectra are in hand. With a single crystal, the entire structure of a highly complex molecule can be determined by x-ray diffraction.

The approach to the problem of identifying or assigning the structure of an unknown organic substance will of course depend on the circumstances and the source of the sample. If the compound has been obtained as a component in a mixture of naturally occurring alkaloids or steroids, it will in all likelihood represent a variation of a known pattern, and the structural problem is relatively restricted, although subtle stereochemical differences, for example, may still present a challenging problem. Similarly, with an unknown arising as a by-product in a synthesis, one can generally assume some relationship to the starting materials, and after a few pieces of information are obtained, a probable structure may be inferred.

Another type of situation, sometimes mentioned in textbook problems but, hopefully, very rarely encountered in practice, is one in which the labels have come off all the bottles in the storeroom. In this case, the only premise that can be made is that most of the unknowns resulting from the disaster are to be found among the 25,000-odd entries in chemical suppliers' catalogs. Although artificial, this is essentially the context of this experiment, in which unknown samples selected from the entire range of simple organic compounds are to be identified.

Chapters 11, 12, 13, and 14 describe the use of spectroscopic methods (^{1}H and ^{13}C NMR, IR, and Mass Spectrometry) to establish the structure of an organic unknown. The focus of this chapter is the use of "wet" methods, including the preparation of crystalline derivatives, to the same end.

The starting point when confronted with an unknown liquid or solid is to establish the empirical formula. This is done by subjecting the compound to combustion analysis: A precisely weighed sample is burned, and the amount of CO_2 and H_2O formed is measured. The result is then expressed in terms of weight percent: C 76.47, H 7.84. Oxygen is done by difference—in this case, O = 15.69. Note that atomic weights based on a *weighted average* of isotopic distribution are used. This is *different* from the precise mass measurement summarized above.

To convert combustion percentages into a molecular formula, it is necessary to divide by the atomic weight of the element:

C: 76.47 gm E ÷ 12 gm/mole = 6.37 mole
H: 7.84 gm E ÷ 1 gm/mole = 7.84 mole
O: 15.69 gm E ÷ 16 gm/mole = 0.98 mole

The formula could then be $C_{61/2} H_8 O$. More plausibly, it would be $C_{13}H_{16}O_2$. Since this technique is beyond the scope of this course, the results of such analysis may be provided for complex cases. From the empirical formula, the likely molecular formula(s) may be calculated, as well as the Index of Hydrogen Deficiency (see Chapter 12) for each proposed molecular formula. This will indicate the number of rings, and/or double or triple bonds. An Index of Hydrogen Deficiency of at least four may indicate that an aromatic ring is present.

GENERAL APPROACH

The systematic approach starts with a determination of **physical properties** such as boiling point, melting point, specific gravity, and refractive index. Determination of **solubility class** will provide much information concerning which functional groups are present and can be confirmed by simple chemical tests (**classification tests**). Knowing the functional group narrows the number of choices as to the compound's identity. Tests for the presence or absence of **halides, sulfur, and nitrogen** provide further narrowing. Consultation with the tables of compounds and derivatives at the end of this chapter or with library sources will lead to a few choices for the proposed structure and selection of possible derivatives. The final step in the approach is the **synthesis of derivatives** that confirm the structure. This problem-solving method can be adapted to many different situations.

The unknowns provided in this experiment are of the purity normally encountered in commercial organic chemicals, which is usually in the range of 95 to 99%. Minor impurities will generally not interfere in the identification procedure, but preliminary recrystallization or distillation of a sample of the unknown may be desirable.

A. PHYSICAL PROPERTIES

The melting point of a solid or the boiling point of a liquid is traditionally one of the first physical constants cited in characterizing an organic compound; in early work these temperatures were among the few measurements that could be made. Several other properties, such as the refractive index and density of liquids and optical rotation of chiral compounds, are often recorded for additional characterization, but they are not particularly useful in locating a compound in the literature.

In using the observed melting point or boiling point of an unknown or a derivative for comparison with literature values, it is necessary to allow sufficient "leeway" and to consider a range of several degrees in literature values on either side of the observed temperature. Literature values as well as the observed ones are subject to inaccuracies; frequently, more than one value can be found because of differences among individual investigators.

The melting point of a compound conveys little structural information in itself, since it depends on such diverse factors as molecular size, symmetry, rigidity, and polarity of functional groups. Although there is a regular progression of melting points within a typical aliphatic homologous series, large disparities in the melting points of closely similar compounds can arise because of differences in molecular shape, as illustrated with the polyols erythritol and pentaerythritol.

$$HOCH_2-\underset{\underset{OH}{|}}{C}H-\underset{\underset{OH}{|}}{C}H-CH_2OH \qquad\qquad HOCH_2-\underset{\underset{\underset{CH_2OH}{|}}{|}}{\overset{\overset{CH_2OH}{|}}{C}}-CH_2OH$$

erythritol, mp 121°C   pentaerythritol, mp 254°C

The boiling point of a liquid is much more directly related to the functional group and molecular size, since forces operating in the liquid state are less affected by symmetry and rigidity than those in a crystal. Within any straight-chain aliphatic homologous series, the boiling point increases in a regular way with increasing molecular weight, with increments between successive members becoming smaller, the longer the chain. Branching, particularly in the vicinity of a functional group, markedly lowers the boiling point within a set of isomers. The presence of an alicyclic or aromatic ring causes a significant increase in boiling point over that of an aliphatic compound having the same functional groups and number of carbon atoms. Table 46.1 contains representative data on the boiling points of a few simple monofunctional aliphatic alcohols. Within each group, the lowest boiling point is that of the most highly branched isomer, and the highest boiling point corresponds to the longest chains (*n*-alkanol).

Table 46.1 Boiling Point Ranges of Aliphatic Alcohols (°C)

TYPE	4-CARBON	6-CARBON	8-CARBON
Primary	108–116	148–156	183–194
Secondary	99	120–139	150–180
Tertiary	83	120–125	149–165

It is obvious that a relatively low boiling point for an unknown, say, below 100°C, greatly limits the number of possible compounds and may even define the structure uniquely with little other data. A boiling point in the range of 100 to 160°C, together with information on functional groups, provides a rough indication of molecular size. On the other hand, the number of possible compounds increases enormously in this range, and the boiling point cannot be used to pinpoint one or two candidates if all known compounds are admitted as possibilities. If the boiling point is above 180°C, the value as a means of narrowing possibilities is practically nil. Moreover, boiling points above this temperature are likely to be very inaccurate. Your instructor will indicate whether the boiling point should be determined or not.

PROCEDURES

1. Melting Points. For solid unknowns, the melting point is determined in the usual way, raising the temperature of an initial sample rapidly to get an approximate range, then repeating at a rate of 1°C to 2°C per minute in this region. If a sample is recrystallized, the melting point should be checked before and after recrystallization.

2. Boiling Points. If sufficient sample is available (at least 3 to 5 mL) the boiling point can be observed by distillation in a simple 10-mL distilling flask (Fig. 46.1). The flask is mounted at an angle on wire gauze so that excessive heating of the glass is avoided. A test tube fitted over the side arm and chilled in ice serves as a condenser and receiver. Be sure to add a boiling

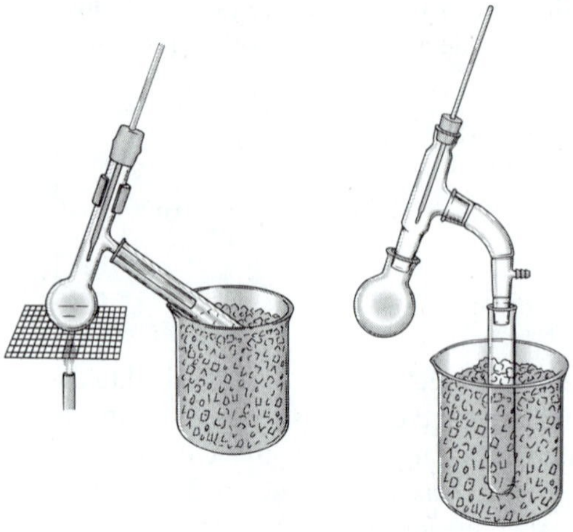

Figure 46.1   Distillation apparatus for unknowns.

stone; heat with a microburner. Care must be taken to distill slowly enough to permit thermal equilibrium to be reached. Walls of the flask above the thermometer should be wet with condensing vapor, and there should be a drop of condensate on the thermometer bulb. For strictly accurate results the barometric pressure should be taken into account.

An alternative procedure that is often more accurate and convenient is a **micro boiling point** determination. This method is described in Experiments F and G of the Experimental Section in Chapter 5. The determination should be performed on a freshly distilled sample, and it should be repeated at least once.

3. Specific Gravity. Specific Gravity is defined as the ratio of the mass of a substance compared to the mass of an equal volume of water. A pycnometer is used for the comparison of the masses of the two substances.

a. Fabrication of a Pycnometer. Make a simple pycnometer by sealing the end of a piece of 8-mm glass tubing and cutting it off so that it is approximately 3 inches long. Be sure to fire-polish the new end. Make a scribe mark with a triangular file about ½ inch from the open end. A 25-mL Erlenmeyer flask can be used to hold the tube upright. You now have a pycnometer that will be accurate to 3 figures. (A 10 × 75-mm test tube, properly scribed and held upright in a small Erlenmeyer flask, may also be used.)

b. Determination of the Specific Gravity of an Unknown Liquid.

1. Determine the mass of the pycnometer to the nearest milligram.
2. Add deionized water to the tube so that the bottom of the meniscus rests on the top of the scribe mark.
3. Determine the mass of the pycnometer and the water. To find the mass of the water subtract the mass of the empty pycnometer from the mass of the pycnometer and the water. (Optional: Assume the density of water is 1.000 gram/ mL and determine the volume of the pycnometer.)
4. Empty and dry the pycnometer. Check the mass to ensure the pycnometer is dry.
5. Add the unknown liquid to the scribe mark. Be sure there are no drops of liquid outside of the pycnometer or above the scribe mark.
6. Determine the mass of the pycnometer and the unknown liquid. To find the mass of the unknown liquid subtract the mass of the empty pycnometer from the mass of the pycnometer and unknown liquid.

$$\text{Specific gravity} = \frac{\text{Mass of unknown liquid}}{\text{Mass of water}}$$

4. Refractive Index. The refractive index is the ratio of the velocity of light in a vacuum to the velocity of light in a substance. The refractive index is a constant for a pure substance and can be used along with other methods for identification of substances. It may also be used as a measure of purity for a known substance. A **refractometer** is used to measure the index of refraction to a few parts per 10,000. A light of known wavelength is used (very often the "D" line of sodium, 5890 nm) because the index varies with the wavelength used. Moreover, the index of refraction decreases between 3.5 and 5.5×10^{-4} per degree increase in temperature, so the temperature at which the measurement was made needs to be specified. Note that there are no units.

The index of refraction of water at 20°C is 1.3330. It decreases 0.0001 per degree Celsius increase.

PROCEDURE

Determine the refractive index of water to be sure the refractometer is in calibration and that you have understood the directions. After drying the apparatus, determine the refractive index of your unknown. The refractive index values are listed in column 7(n_D), of Section 3 ("Physical Constants of Organic Compounds") of the *CRC Handbook of Chemistry and Physics*. Catalogs from chemical companies such as Aldrich also list the values.

B. SOLUBILITY CLASSIFICATION

A good deal can be learned about a compound from its solubility in a few media; the solubility classification complements spectral data and helps to determine the direction of further work. The solubility of the unknown is checked in water, dilute acid, dilute base, and concentrated sulfuric acid in that order, stopping when a positive result is obtained. If the compound is soluble in water, nothing is learned by testing in dilute acid or base; if it is soluble in any aqueous medium, it will also dissolve in or react with concentrated H_2SO_4.

A summary scheme of solubility classification is given in Chart 46.1.

1. Water. Solubility of an organic compound in water reveals the presence of ionic or "polar" groups that can be solvated or can participate in hydrogen bonding. The extent of solubilization depends, of course, on the "ratio" of functional group to carbon skeleton. Liquids containing hydroxyl, carboxyl, amino, or amide groups and no more than four or five carbon atoms are miscible with water in all proportions (indicated by "∞" in solubility tables) or have appreciable solubility. With two or more of these groups, a much larger molecule will be soluble in water.

With solids, crystal structure has a major influence, and solubility in any solvent, water or organic, is related to melting point as well as to the polarity of the molecule. Thus oxamide, $NH_2COCONH_2$, which has the unusually high melting point of 410°C, is only very slightly soluble in water.

In testing for solubility in water, approximately 50 mg of solid or 0.1 mL (2 drops) of liquid is added to 1 mL of water in a 10×75-mm test tube. Large hard crystals of a solid may dissolve slowly and should be powdered and stirred well. If a clear solution is obtained, or a major amount of the compound dissolves with the amounts specified, the compound is considered "soluble" in water.

If the unknown is soluble in water, the pH should be estimated with indicator paper to detect the presence of acidic or basic groups in the molecule. The solubility in ether should also be checked if the unknown is readily soluble in water. A multiplicity of polar groups, as in a polyol, may render the compound insoluble in ether. With a solid, solubility in water and

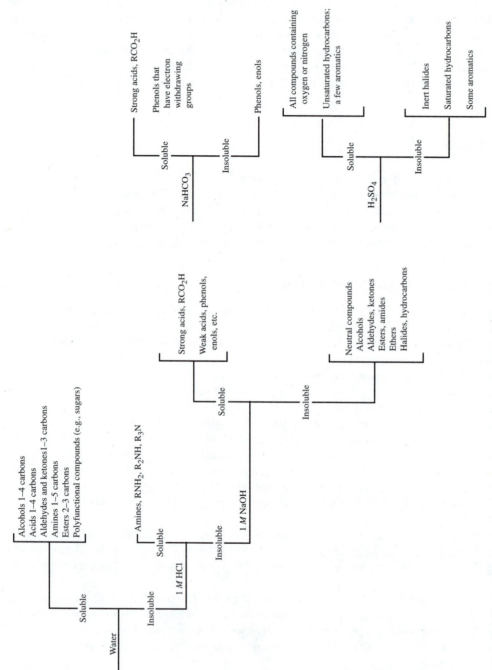

Chart 46.1 Summary scheme for solubility classification.

insolubility in ether may also suggest the possibility of an ionic salt, and this should be explored by treating the solution with acid or base, in which case the free acid or base may precipitate.

2. Aqueous Acid or Base. Compounds that can be converted to ionic species can be recognized by their solubility in water at certain pH values. Three major classes of compounds can be distinguished in this way: *strong acids, weak acids,* and *bases.* All compounds with $K_a >$ 10^{-12} (p$K_a <$ 12) are converted to anions to a significant extent in 1 M NaOH (pH 14). Acids with pK_a in the range of 3 to 7, including all carboxylic acids and nitrophenols, are also soluble in aqueous sodium bicarbonate (pH 8). Solubility in NaOH, but not NaHCO$_3$, indicates a weak acid such as a phenol, enol, or aliphatic nitro compound. Organic bases with K_b 10^{-3} to 10^{-10} are soluble in 1 M HCl by virtue of the conversion to cations. The only compounds that can be protonated in dilute aqueous acid are amines.

$$RCO_2H + HCO_3^- \rightarrow RCO_2^- + CO_2 + H_2O$$

$$R_3N + H_3O^+ \rightarrow R_3NH^+ + H_2O$$

In detecting basic or acidic properties by these solubility criteria, the important point is whether the unknown is significantly more soluble in aqueous acid or base than in water. If a test is doubtful, neutralization of the test solution should cause reprecipitation of an amine or an acid that is dissolved because of salt formation. One possible complication that must be kept in mind is the relatively low solubility of certain salts, particularly in the presence of an excess of the common ion. For example, it is possible to mistake the formation of a slightly soluble hydrochloride for insolubility of a solid amine.

3. Concentrated Sulfuric Acid. Virtually all compounds of moderate molecular size that contain a nitrogen or oxygen atom or a double or triple bond are protonated in 96% H$_2$SO$_4$ and therefore dissolve to some extent. The solution may become dark in color, heat up appreciably or polymer may separate; these reactions constitute a positive test, but slight darkening without actual solution may be due to trace impurities.

C. ANALYTICAL DATA

In certain cases you may be given additional information about an unknown in the form of an elemental analysis (% C, H, N, and so forth) or a molecular weight. The latter, if determined by measuring accurately a mass spectrum, can fix the molecular formula. Approximate values ($\pm$10%), such as would be obtained by freezing-point determinations, are useful only in conjunction with other data.

In using an elemental analysis, it must be recognized that methods of analysis are such that the results are accurate to only $\pm$0.3% (absolute) and that an empirical and not necessarily a molecular formula is obtained. As an example, consider a compound containing C, H, and possibly O, which analyzed for 74.02% C and 6.48% H. By difference it contains 19.5% O. The calculated molar proportions of C, H, and O are thus 5.06:5.27:1. Rounding to C$_5$H$_5$O$_1$, the theoretical analysis is 74.06% C and 6.21% H, in good agreement with that found. Since a compound containing only C, H, and O cannot have an odd number of hydrogen atoms, the molecular formula must be C$_{10}$H$_{10}$O$_2$ or some higher multiple.

EXAMPLE

A solid that melts at room temperature, Unknown A, boils at 224 to 226°C and is not soluble in water, 5% hydrochloric acid, or 5% sodium hydroxide; however, it is soluble in concentrated sulfuric acid. Elemental analysis does not indicate the presence of halogens, sulfur, or nitrogen.

Unknown A does not react with cold, dilute, aqueous KMnO$_4$ solution but does give a precipitate with 2,4-dinitrophenylhydrazine reagent. It gives a negative result with Tollens reagent.

Preparation of the semicarbazone and 2,4-dinitrophenylhydrazone gives derivatives that melt at 200 to 203°C and 250 to 256°C, respectively.

What compound found in the Tables of Derivatives (Tables 46.2–46.13) at the end of this chapter is this?

SOLUTION

Test	Indication
1. Insoluble in water	Not a low molecular weight compound or one with more than one polar functional group
2. Insoluble in HCl or NaOH	Not an acidic or basic functional group
3. Soluble in H_2SO_4	Alkene, alkyne, or oxygen-containing group
4. No reaction with $KMnO_4$	Not alkene or alkyne
5. Precipitate with 2,4-DNP	Aldehyde or ketone present
6. Negative Tollens test	Ketone, not aldehyde
7. From table of ketones at the end of this chapter, three with boiling points within 5° of 225°C are found: butyrophenone (230°C) isobutyrophenone (227°C) *para*-methylacetophenone (226°C)	
8. Comparison of derivatives' melting points indicates *para*-methylacetophenone	
9. For further confirmation the Iodoform Test might be tried	

ADDITIONAL DATA ON UNKNOWNS

After assessing the information obtained from physical properties and solubility, as illustrated in the preceding pages, it frequently will be possible to draw a tentative conclusion as to the functional groups present and proceed to selection and preparation of a derivative. In other cases, certain features will be established, but further data will be desirable to define the environment, or even the nature of the functional group, or to decide between two possible interpretations. Additional information can usually be obtained from one or more of the approaches discussed in the following paragraphs. More complete information on these and other procedures can be obtained from the references at the end of the chapter.

D. DETECTION OF OTHER ELEMENTS

One important item that may be needed is information on the presence of elements other than C, H, and O. A qualitative test for halogen and nitrogen should always be carried out. A common functional group containing nitrogen (amines) is usually revealed by the basicity, but a confirmatory test for nitrogen can be done readily.

Detection of N, Cl, Br, and I is accomplished by the total decomposition of the compound with hot metallic sodium, followed by detection of anions in the usual way. With rare exceptions, any compound containing a C—N bond will give cyanide ion under these conditions; excess carbon is usually converted to the amorphous element.

$$[C, H, O, N, X, S] \xrightarrow{Na} \xrightarrow{H_2O} C, HO^-, CN^-, X^-, S^=$$

In this and other tests that require observation of a positive or negative result, there is one all-important rule: *always run a control*. It is quite futile to attempt a conclusion from the reaction of an unknown when the observer does not know exactly the appearance of a positive and a negative result. Run a known compound first, on the scale that will be used for the unknown, to see the behavior and to insure that you are doing it properly. It may be equally important in some cases to run a *blank* in order to observe a negative result. If there is doubt about the result with a known, clear it up before turning to the unknown.

SAFETY NOTE

Metallic sodium is extremely reactive with water or hydroxylic solvents. Handle and transfer with a spatula; the metal is soft, and a small piece can be picked up by spearing it with the tip of a spatula. The sodium fusion should be carried out behind a shield or hood window, and always wear safety goggles.

PROCEDURE

1. Sodium Fusion.   Obtain a small piece of sodium (a cube about 4 mm on edge) and place in a clean *dry* 10 × 75-mm test tube. Handle the test tube with a holder or clamp, not with your fingers. Have ready a sample of approximately 100 mg of a solid unknown or 0.1 mL (2 drops) of a liquid. Heat the sodium over a flame until it melts and begins to glow red. Add the sample to the tube, making sure that it falls directly on the molten sodium and not on the wall of the test tube where it may be volatilized. With a very volatile liquid, add a second 0.1-mL sample. Heat the tube briefly again and allow it to cool. Add 1 mL of ethanol dropwise and stir thoroughly with a glass rod to dissolve the excess sodium. After bubbling (hydrogen evolution) has stopped, cautiously add, dropwise, distilled water. Heat the mixture to boiling, filter through paper, and rinse with distilled water. If the solution is brown in color (indicating incomplete decomposition in the fusion), add a little charcoal and filter again.

2. Test for N.   This test, and the alternate one described below, are quite sensitive and care must be taken that a false positive is not obtained because of nitrogen containing impurities. If there is doubt, a standard should be run. In a 13 × 100-mm test tube, mix 10 drops of a 0.1 M solution of *p*-nitrobenzaldehyde in 2-methoxyethanol (methyl cellosolve), 10 drops of a 0.1 M solution of *o*-dinitrobenzene in 2-methoxyethanol, and 1 drop of 0.50 M aqueous sodium hydroxide, and then add 1 drop of the sodium fusion solution. The formation of a deep blue-purple color indicates the presence of nitrogen, while a yellow or tan color constitutes a negative test.

3. Alternate Test for N.   To 1 mL of the solution, add 2 drops of saturated ferrous ammonium sulfate solution and 2 drops of 30% potassium fluoride solution. Boil the mixture for 30 seconds and acidify by dropwise addition of 30% sulfuric acid until the iron hydroxide just dissolves. The appearance of a brilliant blue precipitate of Prussian blue indicates the presence of cyanide ion and nitrogen in the compound.

4. Test for S.   Acidify 1 mL of the fusion solution with acetic acid and add several drops of 0.1 M lead(II) acetate solution. A black precipitate of lead sulfide indicates the presence of sulfur.

5. Test for Halogen.   Acidify a 2-mL portion of the alkaline fusion solution in a small beaker with 3 M nitric acid and boil gently for 3 to 5 minutes to expel any HCN that may be present. The appearance of a distinct white or yellow precipitate on addition of silver nitrate indicates the presence of halogen. A yellow color suggests bromide or iodide.

To differentiate the halogens in the event of a positive silver halide precipitate, acidify another 2- to 3-mL portion of the original solution and add approximately 0.3 mL of methylene chloride. Then add a few drops of fresh chlorine water, a few mg of calcium hypochlorite, or several drops of Clorox (be sure the pH remains acidic); stopper and shake the tube. A yellow-orange color in the CH_2Cl_2 layer indicates bromine; a violet color, iodine.

6. Alternate Test for Halogen.   A rapid test for chlorine, bromine, or iodine can be carried out on the original sample using the **Beilstein procedure,** based on the formation of a volatile copper halide when an organic halide is strongly heated with copper oxide. The value of the test is that it is easily carried out and is highly sensitive. To perform the test, a small loop of copper wire is heated to redness in a Bunsen burner flame until the flame is no longer colored. The wire is cooled and the now oxide-coated loop is immersed into a small amount of the solid or liquid to be tested. On reheating in the Bunsen flame, a blue-green flame indicates the presence of chlorine, bromine, or iodine (copper fluoride is nonvolatile).

Although the test is rapid, there are potential problems. Occasionally, the blue-green flame is fleeting and one must look quickly so as not to miss it. Further, highly volatile substances may vaporize before the copper halide forms, and such compounds as urea, quinoline, and pyridine derivatives can give a false positive because of the formation of copper cyanide. Thus, the test should always be confirmed by other methods such as those mentioned earlier.

E. EQUIVALENT WEIGHTS

There are many methods for the quantitative determination of various functional groups in organic compounds; these reveal the number of such groups in the molecule if the molecular formula is known. If the formula of the compound is not known, quantitative analysis of a particular group provides the *equivalent weight,* i.e., molecular weight ÷ the number of functional groups per molecule.

The simplest of all these quantitative methods is acid–base titration, which is used to determine the *neutralization equivalent* of acids. Practically any carboxylic acid can be titrated with standard base to a sharp end point with phenolphthalein, since the pH at complete neutralization for an acid of pK_a 3 to 6 is somewhat above 7. With care, the neutralization equivalent of a pure, dry acid can easily be determined with an accuracy of 1%, providing very useful information about an unknown acid. The neutralization equivalent of a monobasic acid is the molecular weight, whereas for dibasic or polybasic acids, it is some integral fraction of the molecular weight. The procedure given can also be used for the neutralization equivalent of amine salts, but it is generally difficult to obtain the latter sufficiently pure and dry for accurate results.

PROCEDURE

Weigh out a 150- to 200-mg sample of benzoic acid to ±1 mg on an analytical balance. Place the acid in a 125-mL Erlenmeyer flask with 50 mL of water, add 2 to 3 drops of phenolphthalein solution, and titrate to a pink end point with ~0.1 M NaOH from a 25- or 50-mL buret. From the volume of titrant and weight of benzoic acid (eq wt 122) used, calculate the exact molarity of the NaOH solution. As a check of your technique and the result obtained, titrate a second sample of benzoic or another known acid.

Repeat the process using the standardized base with a similar accurately measured quantity of the unknown acid (*in duplicate*), and calculate its equivalent weight. Liquid acids should be freshly distilled and weighed in a capped vial with minimum exposure to air, since they are generally hygroscopic. Solid unknown acids should be recrystallized and thoroughly dried. Acids that are very insoluble in water can be titrated in aqueous alcohol. If this is necessary, the standardization should be carried out with the same amount of added alcohol.

Other quantitative methods, including the saponification equivalent of an ester or quantitative hydrogenation of an olefinic compound, may be desirable in certain cases. Procedures for these or other special methods should be obtained from the literature or the references at the end of the chapter, and arrangements should be made with the instructor for necessary reagents and equipment.

F. MISCELLANEOUS CHEMICAL TESTS

A large variety of qualitative chemical tests for organic functional groups have been developed. A few of the simpler tests may be found useful in resolving ambiguities in other data. *In each case the test should be run on known compounds similar to those suspected to be the unknown to insure that the reagent is correctly prepared and so as to be able to recognize the appearance of a positive test.*

PROCEDURES

1. Chromic Acid Test for 1° and 2° Alcohols. This test, described in Chapter 18, is a reliable method for confirming the presence of a primary or secondary hydroxyl group (aliphatic aldehydes also react). A rapid color change to dark green constitutes a positive test. With amines or phenols, a brown color and a dark precipitate are usually seen.

2. Ferric Chloride Test for Phenols. The presence of phenolic or enolic hydroxyl groups in a compound is usually indicated by formation of a red or violet color due to an Fe(III) complex when treated with ferric chloride solution. Some phenols give negligible or very weak tests because of interference by substituents in the ring.

The test is carried out by adding 1 drop or a few crystals of the unknown to 1 mL of freshly prepared 1% $FeCl_3$ solution.

3. Iodoform Test. Methyl ketones ($RCOCH_3$) and also methyl carbinols ($RCHOHCH_3$) react rapidly with iodine under basic conditions to give a carboxylic acid and iodoform. (Also see Chapter 22.)

$$RCHOHCH_3 \xrightarrow[HO]{I_2} RCOCH_3 \xrightarrow[HO^-]{3I_2} RCOCl_3 \xrightarrow{HO^-} RCO_2 + CHI_3 \downarrow$$

To carry out the test, dissolve approximately 20 mg or 1 drop of the unknown in 0.5 mL of water (if insoluble in water, dissolve in 0.5 mL of methanol) and add 0.5 mL of 10% aqueous NaOH. Add dropwise, with shaking, a solution of KI_3* until the dark iodine color persists. Warm the solution slightly, and if the color fades, add more KI_3 until the color remains for 1 to 2 minutes at 50°C. Then add a drop or two of NaOH solution to remove excess iodine, and dilute with water. Iodoform, if present, will separate as a dense, pale yellow solid, mp 119 to 121°C.

4. 2,4-Dinitrophenylhydrazine Test for Aldehydes and Ketones. This reagent is commonly used for preparing dinitrophenylhydrazones of aldehydes and ketones and can also be used as a qualitative test for these functional groups to distinguish them from other carbonyl compounds, particularly esters, that do not react. The procedure used for preparing 2,4-dinitrophenylhydrazones (p. 220) is followed, and the formation of a red or orange precipitate constitutes a positive test. The test is also described in Chapter 22.

Since the reagent solution contains sulfuric acid, amines may give a heavy precipitate of the amine sulfate, which appears yellow in the solution and can be mistaken for a positive test.

5. Tollens Test for Aldehydes. Tollens reagent, a solution of silver ammonia complex $Ag(NH_3)_2{}^+$, provides a test (also described in Chapter 22) to distinguish between aldehydes and

---

* Prepare by dissolving 5 g KI and 2.5 g I_2 in 25 mL of water.

ketones. Aldehydes react by undergoing oxidation, and silver ion is reduced to form a metallic silver mirror.

$$RCHO + Ag(NH_3)_2^+ + 2HO^- \rightarrow RCO_2^- \, NH_4^+ + 2Ag \downarrow + H_2O + 3NH_3$$

Reagent Preparation: In a clean, dry test tube add 1.0 mL of 0.05 *M* silver nitrate. Add dropwise approximately 5 drops of 3.0 *M* sodium hydroxide until a gray precipitate of silver oxide (Ag$_2$O) appears. Mix the solution thoroughly and add another drop of the NaOH solution to see if further precipitate forms. Do not add excess NaOH. When the silver oxide precipitation is complete, add dropwise and with vigorous stirring, 2% aqueous ammonia until the silver oxide precipitate just disappears. DO NOT ADD EXCESS aqueous ammonia or the reagent will not work properly. Dilute the mixture (1:1) with deionized or distilled water.

SAFETY NOTES

1. This reagent must be prepared immediately before use because it can form silver fulminate if stored. All reagent residues and the silver mirror should be destroyed by adding nitric acid.
2. Silver nitrate can stain the skin black. Use gloves!
3. The residual silver ions can be precipitated as silver chloride by hydrochloric acid, filtered, and discarded as directed by your instructor.

6. Hinsberg Test for Amines. Primary and secondary amines react with benzenesulfonyl chloride in the presence of sodium hydroxide to give solid benzenesulfonamides. The sulfonamide from a simple primary amine is usually soluble in excess hydroxide solution because of the acidity of the —NHSO$_2$Ar group. Tertiary amines do not react under mild conditions.

$$RNH_2 + C_6H_5SO_2Cl \xrightarrow{HO^-} RNHSO_2C_6H_5 \xrightarrow{HO^-} [RNSO_2C_5H_4]^- Na^+$$
$$\text{soluble in aq NaOH}$$

$$R_2NH + C_6H_5SO_2Cl \xrightarrow{HO^-} R_2NSO_2C_6H_5$$
$$\text{insoluble in NaOH}$$

The test can be ambiguous if the amine is insoluble in water or is a solid, since it is necessary to distinguish between unreacted amine in excess, unreacted benzenesulfonyl chloride, and an insoluble sulfonamide.

To carry out the test (also described in Chapter 29), mix approximately 50 mg or 1 drop of amine and 0.2 g of benzenesulfonyl chloride in 4 mL of 10% aqueous NaOH solution, stopper the test tube and shake vigorously until the oily sulfonyl chloride has reacted (5 minutes or longer may be needed). If a clear or nearly clear solution is obtained, a primary amine is indicated; acidification should give a precipitate of RNHSO$_2$C$_6$H$_5$.

Formation of a significant amount of insoluble solid from a liquid amine indicates that the unknown is a secondary amine. To distinguish from the possibility of unreacted solid amine, acidify the mixture; an amine will be soluble.

G. DERIVATIVES

The final point in the identification is conversion of the unknown to a solid derivative whose melting point can be compared with a literature value. The derivative may be any compound that is formed in a reaction that is characteristic for the unknown. It is obviously desirable to choose, when possible, a derivative that can be obtained in good yield and readily isolated, and that has a melting point in the most convenient region, i.e., 80 to 180°C. In many cases, the derivative will be confirmatory evidence for a tentative conclusion, but occasionally it may be necessary to choose between two possible candidates; in this case, the derivative or reaction product from both compounds must be known.

In all cases the derivative should be recrystallized from a suitable solvent before measuring its melting point, since the melting point will be the basis for identification of the derivative. The product that precipitates or crystallizes from the reaction mixture may be quite impure, and an incorrect melting point is worse than none at all in this situation.

For acids, alcohols, amines, and carbonyl compounds, a variety of simple condensation products can serve as derivatives, and a number of these are given in the Tables of Derivatives at the end of this chapter (Tables 46.2–46.13). Procedures for the more useful of these standard derivatives are given below. Solid acids obtained by hydrolysis of esters, amides, or nitriles are usually satisfactory derivatives if the acid represents the major portion of the molecule. It may, however, be much better to characterize the entire molecule by a more specific reaction. For example, hydrolysis of acetanilide gives acetic acid and aniline, both liquids. Either the acid or amine or both could be isolated and converted to solid derivatives, but a much simpler and far more effective derivative is obtained by nitration to *p*-nitroacetanilide (see Chapter 27) or bromination to the *p*-bromo derivative.

A number of the unknowns may lend themselves to special derivatives that will be suggested by the tentative structure deduced from spectral and other data. Cyclization products can be readily obtained from many bifunctional compounds, and occasionally rearrangement or partial degradation will provide highly characteristic derivatives. In such cases, details for carrying out the reaction and isolation of the product should be obtained from the literature for the specific compound that is under consideration.

Note on Derivatives

In directions for preparing derivatives of unknowns, amounts of reagents are given for a compound of "average" molecular weight, i.e., 100–200 g/mol, and solvents suggested are for compounds with typical solubilities. For certain unknowns, these directions may not be optimal, and it may be desirable to modify the procedure according to the properties expected for the unknown in question. For experiments on a **microscale,** the quantities given can usually be carried out at one-half to one-tenth scale. Merely reduce the amounts of sample, reagents, and solvents by the required amounts.

SAFETY NOTE

In many of these procedures, highly reactive reagents, such as acid chlorides and isocyanates, or strong acids and bases, are used. Most solvents are flammable. Keep in mind the safety notes that you have seen throughout this book and the standard precautions. Always wear safety glasses. Pipet only with a bulb. Wash thoroughly with water if any chemicals come in contact with the skin. If in doubt, work in a hood.

1. Acids. The most satisfactory general derivatives of carboxylic acids are neutralization equivalents (Section E of this chapter) and amides, particularly the anilides and *p*-toluidides, which are prepared by the general sequence:

$$RCO_2H + SOCl_2 \rightarrow RCOCl + SO_2 + HCl$$
$$RCOCl + 2\ R'NH_2 \rightarrow RCONHR' + R'NH_3^+Cl^-$$

The second reaction requires the use of some base to combine with HCl. In the derivatization of an acid, an excess of the amine is usually used, but if the amine is the important component, some other base, usually aqueous NaOH, is used as the acid acceptor.

It should be noted that in the reaction with thionyl chloride, polyfunctional acids will often give cyclization or condensation products rather than an acid chloride. Thus, dibasic acids may form anhydrides, and α-acylamino acids give oxazolones (azlactones).

(an anhydride)

(an oxazolone)

PROCEDURES

a. Acid Chloride. A very suitable procedure is that given for naphthalene-1-acetic acid in Chapter 34. Cool a mixture of 1 g or 1 mL of the acid plus 1 mL of thionyl chloride in a test tube and add 0.2 mL of dimethylformamide. Allow the mixture to warm and stand 30 minutes, and then proceed with the next step.

b. Amides. For conversion to the anilide or *p*-toluidide, dilute the acid chloride with 5 to 10 mL of methylene chloride, and add it to a solution of 1.5 g of aniline or *p*-toluidine in 10 mL of methylene chloride. After mixing, allow the reaction to stand for a few minutes, add 10 mL of water and transfer the mixture to a separatory funnel. Add more solvent if necessary to dissolve all of the amide. At this point it is convenient to add sufficient ether to make the organic layer lighter than water. Wash the organic phase with dilute HCl until the aqueous layer is acidic, then with bicarbonate solution and finally water. Dry the organic phase (MgSO$_4$) and evaporate the solvent on the steam bath and recrystallize and collect the derivative.

Conversion to the amide may be preferable if the acid is a high-melting solid. In this case, add the acid chloride, without dilution, *dropwise* to a mixture of 10 mL of concentrated ammonium hydroxide and an equal volume of ice.

2. Alcohols. The most generally useful derivatives of alcohols are esters of substituted benzoic or carbamic acids; the carbamate esters are commonly called urethanes. The *p*-nitrobenzoates or 3,5-dinitrobenzoates are obtained by treatment of the alcohol with the acid chloride and pyridine, which serve as a catalyst and acid acceptor.

$$ArCOCl + ROH + C_5H_5N \rightarrow ArCO_2R + C_5H_5NH^+Cl^-$$

Urethanes are prepared from the alcohol and an aryl isocyanate;

$$ArN{=}C{=}O + ROH \rightarrow ArNHCO_2R$$

In any reactions with isocyanates, the following sequence leading to the diarylurea will occur if water is present.

$$ArN{=}C{=}O + H_2O \rightarrow [ArNHCO_2H] \rightarrow ArNH_2 + CO_2$$
$$ArN{=}C{=}O + ArNH_2 \rightarrow ArNHCONHAr$$

The urea is a high-melting insoluble compound (diphenylurea, mp 238°C) and can seriously interfere with the isolation of the desired derivative. Alcohols should be dry, and an excess of isocyanate must be avoided.

Chromic acid at room temperature converts primary alcohols to carboxylic acids and secondary alcohols to ketones. The latter transformation is often useful in dealing with an aliphatic alcohol, since derivatives of the ketone may be more reliable.

PROCEDURES

a. p-Nitro-and 3,5-Dinitrobenzoates. Dissolve 1 mL of the alcohol in 3 mL of pyridine and add 0.5 g of the acid chloride corresponding to the derivative desired. Warm the solution for a few minutes and pour into 10 mL of water. If a well-crystallized ester separates, this is collected; otherwise, extract the product with ether and wash the ether solution free of pyridine with dilute HCl, dry, and evaporate.

b. Urethanes. To approximately 0.1 mL or 0.1 g of the anhydrous alcohol, add about 0.1 mL of phenyl isocyanate (**CAUTION:** lachrymator) and warm the solution for a few minutes in a beaker of hot water (do not expose to steam bath vapors). Cool, add about 1 mL of high boiling petroleum ether, and heat to dissolve the product. If a very slightly soluble precipitate remains, it is probably the urea. This must be removed by filtration, followed by cooling the filtrate in ice, and scratching to induce crystallization.

3. Phenols. Nitrobenzoates or urethanes of phenols can be obtained by the same procedures described for alcohols. For a few phenols, only the benzoate esters are described: these should be prepared by the procedure given for eugenol benzoate (Chapter 10).

For a few phenols, derivatives can be obtained by reaction with chloroacetic acid to give the aryloxyacetic acid.

$$\text{ArOH} + \text{ClCH}_2\text{CO}_2\text{H} \xrightarrow{\text{NaOH}} \text{ArOCH}_2\text{CO}_2\text{Na} \xrightarrow{\text{H}_3\text{O}^+} \text{ArOCH}_2\text{CO}_2\text{H}$$

PROCEDURE

a. Aryloxyacetic Acids. To 1 g or 1 mL of the phenol add 5 mL of 30% NaOH solution and 1.5 g of chloroacetic acid. Stir and warm the mixture, and if solid is present, add water dropwise to obtain a clear solution. Heat the solution on the steam bath for 30 to 40 minutes, cool, and dilute with 10 mL of water. Acidify the solution (test paper) with 6 M HCl and extract the acid with ether. Wash the ether solution with a little water and then extract with saturated bicarbonate solution. Cautiously acidify the bicarbonate solution; stir to prevent loss from foaming. Collect the precipitate of aryloxyacetic acid and recrystallize from hot water.

4. Ethers. A variety of derivatives are possible for aryl ethers; most of these involve reactions of the aromatic ring rather than the ether linkage. Examples include bromination, chlorosulfonation followed by conversion to the sulfonamide, and nitration. For highly substituted aryl ethers, a more appropriate derivative may be the corresponding phenol obtained by cleavage of the ether linkage.

Enol ethers may be hydrolyzed to a carbonyl compound as described on page 364. No generally useful derivatives exist for saturated aliphatic ethers.

PROCEDURES

SAFETY NOTE

Bromine causes severe burns on contact with the skin. In severe cases, seek medical help. Avoid breathing the vapors of the bromination reagent and carry out the reaction in a hood if possible.

a. Bromination. Dissolve 0.2 g or 0.2 mL of the aryl ether in 2 mL of acetic acid, and add a solution of 10% bromine in acetic acid in 0.5-mL portions until the yellow color persists for at least a minute after mixing. Add 5 mL of water, mix well, and collect the precipitate by filtration. Rinse the solid with water, and recrystallize from aqueous ethanol or hexane.

b. Nitration. Aryl ethers can be nitrated under the conditions used for acetanilide in Chapter 27, adjusting the procedure to use 0.2 to 0.5 g or less.

c. Cleavage of Ethers. Place 0.2 g or 0.2 mL of the alkyl aryl ether, 2 mL of acetic acid, and 2 mL of concentrated hydriodic acid in a test tube. Add a boiling stone and heat in a steam bath *in the hood* for 1 hour. Cool the solution, and pour into 25 mL of water. Add, in small portions, sufficient solid NaHCO$_3$ to neutralize the solution (pH ~8), mixing well between additions. Transfer the mixture to a separatory funnel and extract the phenol with two 5-mL portions of methylene chloride or ether. Wash with water, dry, and evaporate the solvent to obtain the phenol.

5. Amines. A variety of amides and ureas are available as derivatives of 1° and 2° amines. Amides derived from secondary amines are often much more difficult to isolate because the R$_2$NCOAr structures are not hydrogen-bonded, and the compounds are therefore lower-melting and more soluble. The ureas are generally better derivatives.

PROCEDURES

a. Acetamides. Acetamides of aromatic amines are readily prepared by dissolving approximately 0.5 g of the amine in 1 to 2 mL of acetic anhydride and heating for a few minutes. Then add a few mL of water and warm the mixture until the excess acetic anhydride is destroyed (second liquid layer disappears). If the amide does not crystallize directly, extract with ether, wash the ether solution with bicarbonate solution, dry, and evaporate.

b. Benzamides. Mix the amine (0.5 g) with 10 mL of 10% sodium hydroxide solution and add 1 mL of benzoyl chloride. Shake or stir the mixture for 10 minutes and isolate the amide by collecting the solid and washing with water. In some cases it may be desirable to add a solvent such as methylene chloride and recover the amide from solution.

c. Substituted Ureas and Thioureas. The reaction of an amine with an isocyanate or isothiocyanate leads to the corresponding urea; the precautions about water mentioned under alcohols must be kept in mind with isocyanates. Isothiocyanates are much less reactive and the reaction with amines occurs on warming, but hydrolysis with water is negligible and an excess of the reagent can be removed by recrystallization.

$$\text{ArN}=\text{C}=\text{O} + \text{RNH}_2 \longrightarrow \text{ArNHCONHR}$$

$$\text{ArN}=\text{C}=\text{S} + \text{RNH}_2 \longrightarrow \text{ArNHCSNHR}$$

To 1 g of the amine, add 0.5 mL or less of the isocyanate or isothiocyanate; a few mL of methylene chloride or toluene can be used as diluent. Warm the solution if an isothiocyanate was used. Excess amine should be removed by washing the solution with dilute acid before isolating the urea.

6. Amides and Nitriles. Alkaline hydrolysis of these compounds leads to the acid salt and ammonia or the free amine.

$$\text{RCONHR}' + \text{H}_2\text{O} \xrightarrow{\text{NaOH}} \text{RCO}_2\text{Na} + \text{R}'\text{NH}_2$$

$$\text{RC}\equiv\text{N} + \text{H}_2\text{O} \xrightarrow{\text{NaOH}} \text{RCO}_2\text{Na} + \text{NH}_3$$

Amides or esters of carbamic acids give amines, alcohols, and CO_2.

$$CH_3NHCONH_2 + H_2O \xrightarrow{\text{NaOH}} CH_3NH_2 + NH_3 + CO_3^=$$

$$C_6H_5NHCO_2C_2H_5 + H_2O \xrightarrow{\text{NaOH}} C_6H_5NH_2 + C_2H_5OH + CO_3^=$$

With high-melting amides or nitriles of aromatic acids, the rate of hydrolysis and the solubility are low, and alcoholic alkali or hydrolysis in acid is recommended. Most nitrogen-containing compounds are quite soluble in 40 to 60% sulfuric acid and fairly high temperatures can be obtained by heating to reflux such solutions.

$$ArCONHR' + H_2O \xrightarrow{\text{H}_2\text{SO}_4} ArCO_2H + R'NH_3^+ \, HSO_4^-$$

An additional possibility for the characterization of *aromatic* nitriles is conversion to the amide. This reaction, which strictly speaking is not a hydrolysis but a hydration, is carried out by treatment of the nitrile with alkaline hydrogen peroxide; hydroperoxide ion (HO_2^-) is the specific reagent involved:

$$RC{\equiv}N + 2H_2O_2 \xrightarrow{\text{HO}^-} RCONH_2 + H_2O + O_2$$

The hydrolysis and hydration procedures given are generally useful only for amides or nitriles of aromatic acids or amides of aromatic amines. Low molecular weight aliphatic amides or nitriles cannot easily be converted to useful derivatives by hydrolysis, since isolation of the product is usually impractical. Reduction of these compounds to amines with sodium borohydride in the presence of cobalt(II) chloride is one alternative that can generally be used to obtain a derivative:

$$R'CONHR \xrightarrow[\text{CoCl}_2]{\text{NaBH}_4} R'CH_2NHR$$

$$RC{\equiv}N \xrightarrow[\text{CoCl}_2]{\text{NaBH}_4} RCH_2NH_2$$

PROCEDURES

a. Alkaline Hydrolysis. Add 1 mL of a liquid or 1 g of a solid amide or nitrile to 30 mL of 10% NaOH or KOH solution and heat the solution to reflux for 30 minutes. Test for the presence of a volatile amine or ammonia by holding a piece of moist indicator paper at the top of the condenser. Cool the solution to room temperature. If an insoluble amine is present, or if the presence of an alphatic amine of intermediate solubility is suspected from the odor, extract the solution with several 10- to 15-mL portions of ether, wash the ether solution, dry with $MgSO_4$ or K_2CO_3, and evaporate the solvent to obtain the amine. A low molecular weight aliphatic amine will probably be lost in this procedure because of its solubility in water and its volatility during removal of ether.

After removing the amine, if any, acidify the aqueous solution and collect the acid if it is a solid. A low molecular weight aliphatic acid can be isolated by thorough extraction, but will usually be of little value as a derivative.

b. Acid Hydrolysis. Slowly add 5 mL of concentrated sulfuric acid to 10 mL of water. To this solution add 1 g of the amide or nitrile (this procedure will generally be used with high-melting solids). Heat the solution to reflux for 30 to 60 minutes, then cool in ice and dilute with an equal volume of water. If a solid acid separates, this is collected, washed with water, and characterized.

The amine is recovered by making the aqueous solution basic by the addition of 10% NaOH. A heavy precipitate of inorganic salts usually separates, and more water must be added. The amine is then recovered as described in the section on alkaline hydrolysis.

c. Conversion of Nitrile to Amide. In a 100-mL round-bottom flask, place 1 mL or 1 g of the nitrile, 5 mL of 20% hydrogen peroxide, and 1 mL of 6 M (20%) NaOH solution. If the nitrile is insoluble, add 5 to 10 mL of ethanol. The reaction should occur exothermically, and cooling may be necessary at first. Keep the temperature at 40 to 50°C by cooling or gentle warming for 2 hours. Neutralize the solution and concentrate it by distillation to remove most of the alcohol. The amide will usually separate from the aqueous solution on chilling; if it does not, extract with methylene chloride, and isolate the amide by evaporating the solvent from the dried extract.

d. Reduction to Amines. In a 125-mL Erlenmeyer flask, dissolve 0.5 g of the nitrile or amide and 2 g of $CoCl_2 \cdot 6H_2O$ in 25 mL of methanol. In the hood, add 2 g of $NaBH_4$, in small portions, shaking vigorously after each addition. Hydrogen is evolved and a black precipitate usually separates. When the addition is completed, add 3 M HCl until the solution is acidic (5 to 10 mL) and heat the solution on the steam bath to evaporate the methanol. Make the residual aqueous solution basic with NaOH, extract the amine with ether, and prepare a derivative of the amine.

7. Aldehydes and Ketones. Semicarbazones and 2,4-dinitrophenylhydrazones are the most generally satisfactory derivatives of simple aldehydes and ketones:

$$RCOR' + NH_2NHCONH_2 \longrightarrow \begin{array}{c} R' \\ R \end{array}\!\!\!>\!C{=}NNHCONH_2$$

$$RCOR' + NH_2NHC_6H_3(NO_2)_2 \longrightarrow \begin{array}{c} R' \\ R \end{array}\!\!\!>\!C{=}NNHC_6H_3(NO_2)_2$$

PROCEDURES

(These derivatives are also described in Chapter 22.)

a. Semicarbazones. Dissolve 0.1 g of the unknown aldehyde or ketone in 1 mL of ethanol. Add 0.10 g of semicarbazide hydrochloride and 0.15 g of sodium acetate in 2 mL of water. Warm the mixture for 5 minutes on a steam bath, cool in an ice bath, and allow the semicarbazone to crystallize. Collect the crystals on a Hirsch funnel and recrystallize them from methanol. Determine the melting point.

b. 2,4-Dinitrophenylhydrazones. Dissolve approximately 0.2 g of the solid or liquid ketone or aldehyde in 1 mL of ethanol and add dropwise 3 mL of the 2,4-dinitrophenylhydrazine solution (1 g of 2,4-dinitrophenylhydrazine, 5 mL of concentrated sulfuric acid, 8 mL of water, and 25 mL of ethanol). The dinitrophenylhydrazone should precipitate from the solution. If a product does not precipitate, warm the solution briefly on the steam bath; this is necessary for sterically hindered carbonyl compounds. Isolate the derivative by filtration and recrystallize it from ethanol or aqueous ethanol.

8. Acetals, Ketals, and Enol Ethers. Acetals, ketals, and related carbonyl derivatives such as enol ethers, imines, and enamines may be hydrolyzed in warm dilute aqueous or alcoholic acid. The aldehyde or ketone that is liberated can be isolated by extraction and converted to a derivative as described in the preceding section. If the 2,4-dinitrophenylhydrazone is desired, generally it is sufficient to warm the unknown directly with the acidic 2,4-dinitrophenylhydrazine reagent; the hydrolysis takes place and the hydrazone is formed and can be isolated by filtration.

9. Esters. Hydrolysis (saponification) of an ester to the acid and alcohol is a very general reaction, but the choice of a procedure depends on the type of ester and the product that is to be isolated. For esters of aromatic acids the acid is nearly always the most practical derivative, and hydrolysis should be carried out in aqueous alcohol by the first procedure given; this procedure, of course, does not permit characterization of the alcohol portion.

With an ester of an aliphatic acid, isolation of the acid is usually impractical. In order to isolate the alcohol, the hydrolysis can be carried out in the high-boiling solvent diethylene glycol ($HOCH_2CH_2OCH_2CH_2OH$, bp 244°C); the alcohol from the ester is isolated by distillation.

Another approach for the derivatization of an aliphatic ester is the direct conversion to a substituted amide. An aromatic amine is converted to the more reactive amide anion by treatment with ethylmagnesium bromide, and the ester is then refluxed with this reagent.

$$ArNH_2 + C_2H_5MgBr \longrightarrow ArNH^- + MgBr^+ + C_2H_6$$

$$ArNH^- MgBr^+ + RCO_2R' \longrightarrow RCONHAr + R'OMgBr$$

PROCEDURES

a. Hydrolysis—Ethanolic Base. In a 50-mL round-bottom flask, place 5 mL of ethanol, 5 mL of water, 1 mL or 1 g of the ester, and 1 g of KOH. Add a boiling stone and heat the solution under reflux for 30 to 40 minutes. Cool, add 5 mL additional water, and arrange the condenser for distillation. Distill the solution until 5 to 6 mL is collected to remove alcohol, cool the solution, and acidify with 2 *M* HCl. Collect the acid or, if it is a liquid, extract with ether, dry, evaporate, and convert it to a derivative.

b. Hydrolysis—Diethylene Glycol. In a 10-mL distilling flask (Fig. 46.1), place 3 mL of diethylene glycol (bp 244°C), 0.5 g of KOH pellets, and 0.5 mL of water. Heat the mixture until the alkali has dissolved and then cool and add 1 to 2 mL of the ester. Heat again until the ester dissolves and then more strongly, distilling the alcohol into a cooled test tube. The potassium salt of the acid may separate as a solid during the reaction.

c. Conversion of Ester to Substituted Amide. In a 50-mL round-bottom flask, prepare a solution of ethylmagnesium bromide from 10 mL of anhydrous ether, 1 mL of ethyl bromide, and 0.2 g of magnesium (see Chapter 19). After the formation of the Grignard compound, briefly heat it to reflux to ensure completion of the reaction, add a solution of 1.5 mL of aniline or 1.5 g of *p*-toluidine in a small amount of dry ether. After reaction of the Grignard reagent is complete, add 1 mL of the ester and heat the mixture to reflux for 10 minutes. Cool, add 10 mL of 2 *M* HCl, and shake to extract unreacted amine, adding more ether if needed. Isolate the derivative by the usual procedure from the ether solution.

10. Alkyl Halides. Alkyl halides that contain no reactive functional groups can be derivatized by conversion to the Grignard compound and treatment of the latter with an aryl isocyanate to give a substituted amide of the homologous acid. This method is quite general for halides that can form Grignard compounds.

$$R{-}X \xrightarrow{\text{Mg}} RMgX \xrightarrow{\text{ArNCO}} R{-}\overset{\overset{\displaystyle O}{\displaystyle \|}}{C}{-}NHAr$$

PROCEDURE

Convert 1 to 2 mL of the halide (the sample must be anhydrous) to the Grignard compound in the usual way (see Chapter 19) with approximately 0.2 g of magnesium and 10 mL of anhydrous ether; a crystal of iodine can be added if needed. To this solution add, in small portions, a solution of 0.3 to 0.5 mL of the isocyanate (an excess must be avoided; **CAUTION:** lachrymator) in 5 mL of ether or methylene chloride. After hydrolysis with 2 *M* HCl, isolate the amide from the ether solution in the usual way.

11. Nitro Compounds. For aromatic nitro compounds, the oxidation of alkyl side chains or further nitration often provides satisfactory derivatives (see following section). Another general approach is the reduction to the amine, which can then be converted to an amide.

PROCEDURE

In a 100-mL round-bottom flask with reflux condenser, place 2 g of tin (granulated or mossy) and 1 g of the nitro compound. Then add 20 mL of 2 M HCl through the condenser in small portions with constant shaking. After warming the solution on the steam bath for 10 minutes, allow it to cool and slowly add 40% NaOH solution to liberate the amine, which can then be steam distilled or extracted with ether and further characterized by a derivative.

12. Aryl Halides and Aromatic Hydrocarbons. As noted in the discussions of phenol and aryl ether derivatives, the benzene ring can often serve as a satisfactory functional group for derivative preparation. For aryl halides and aromatic hydrocarbons, ring substitution may be the most practical reaction available. Nitration (Chapter 27) and chlorosulfonation (Chapter 38) are the two most generally useful reactions. In addition, aryl bromides can be converted to Grignard reagents and converted with CO_2 to benzoic acids (Chapter 19) or with phenyl isocyanate to the corresponding anilide as described for alkyl halides (10. above).

Alkylbenzenes can be degraded oxidatively to the corresponding benzoic acids. Other aliphatic side chains can also be oxidized, e.g., 2-phenylethyl chloride or phenylacetic acid to benzoic acid. The reaction is not generally useful for dialkylated or polyalkylated benzenes because of the inconveniently high melting points of the acids produced.

PROCEDURE

a. Side Chain Oxidation. In a 50-mL round-bottom flask place 2 g of $KMnO_4$, 25 mL of water, 5 mL of 10% NaOH, 0.5 g or 0.5 mL of the unknown, and a boiling stone. Heat gently to reflux for 1 hour or until the purple color has disappeared. Cool and acidify with dilute sulfuric acid. Reheat the mixture to boiling and add a few grains of sodium bisulfite to dissolve any tan manganese dioxide present. Cool the solution and collect the acid by filtration. Recrystallize from ethanol or aqueous ethanol.

REFERENCES

Bruno, T.J.; Sovronos, P. *CRC Handbook of Basic Tables for Chemical Analysis;* CRC Press: Boca Raton, FL, 1989.

Kemp, W. *Qualitative Organic Analysis;* McGraw-Hill: New York, 1979.

Dictionary of Organic Compounds; Pollock, J.R.A., and Stevens, R., Eds.; Oxford University Press: New York, 1965, and supplements.

Handbook of Chemistry and Physics, 80th ed.; Lide, D.R., Ed.; CRC Press: Boca Raton, FL, 2000.

Pasto, D.J.; Johnson, C.R. *Laboratory Text for Organic Chemistry;* Prentice Hall: Englewood Cliffs, NJ, 1979.

Pasto, D.J.; Johnson, C.R. *Organic Structure Determination;* Prentice Hall: Englewood Cliffs, N.J., 1969.

Properties of Organic Compounds on CD-ROM, Version 5.0; Lide, D.R., Milne, G.W.A., Eds.; CRC Press: Boca Raton, FL, 1995.

Rappoport, Z. *Handbook of Tables for Organic Compound Identification,* 3rd ed.; Chemical Rubber: Cleveland, 1967.

Shriner, R.L.; Fuson, R.C.; Curtin, D.Y.; Morrill, T.C. *The Systematic Identification of Organic Compounds,* 6th ed.; Wiley: New York, 1980.

TABLES OF DERIVATIVES

Table 46.2   Acids

| COMPOUND | MP, °C | BP, °C | Derivative, Mp, °C | | |
			AMIDE	ANILIDE	p-TOLUIDIDE
Acetic		118	82	114	147
Acetylanthranilic	185		171	167	
o-Anisic (methoxybenzoic)	100		128	131	
p-Anisic	184		162	169	186
Benzoic	122		128	163	158
o-Bromobenzoic	150		155	141	
Butanoic		163	115	95	72
Chloroacetic	63		118	134	120
o-Chlorobenzoic	140		139	118	131
m-Chlorobenzoic	158		134	122	
p-Chlorobenzoic	242		179	194	
4-Chloro-3-nitrobenzoic	182		156	131	
Cinnamic	133		147	153	168
Dichloroacetic		189	98	118	153
2,4-Dichlorobenzoic	158				
3,4-Dichlorobenzoic	208		133		
Diglycolic (oxydiacetic)	148		135 (mono)	118 (mono)	
3,4-Dimethoxybenzoic	182		164	154	
2,2-Dimethylsuccinic	140				
Diphenylacetic	146		167	180	
p-Ethoxybenzoic	198		202	169	
Glutaric	97		174	224	
Hippuric (benzoylglycine)	187		183	208	
p-Hydroxybenzoic	215		162	202	204
Itaconic	165 (d)		192 (di)		
p-Methoxyphenylacetic	85		189		
Methylpropanoic (isobutyric)		155	129	105	109
Methylsuccinic	115		165 (mono)		
Naphthalene-1-acetic	133		181	155	
1-Naphthoic	162		205	161	
m-Nitrobenzoic	140		142	155	162
p-Nitrobenzoic	241		201	217	
Phenylacetic	76		154	117	
2-Phenylbutanoic	42		86		
3-Phenylpropanoic	48		82	92	
5-Phenylpentanoic	60		109	90	
Phthalic	208 (d)		149		
Propionic		140	81	106	126
Salicylic (o-hydroxybenzoic)	158		139	136	156
Succinic	188		242 (mono)		
o-Toluic	105		142	125	144
m-Toluic	111		97	126	118
p-Toluic	177		158	140	160
3,4,5-Trimethoxybenzoic	170		176		

Table 46.3   Alcohols

| COMPOUND | BP, °C | MP, °C | Derivative, Mp, °C | |
			3,5-DINITROBENZOATE	PHENYLURETHANE
Benzyl alcohol	206		113	78
1-Butanol	117		64	57
2-Butanol	99		75	65
2-Buten-1-ol	122			
Cinnamyl alcohol	257	33	121	90
2-Chloroethanol	130			51
1-Chloro-2-propanol	127		83	
Cholesterol		148		168
Cyclohexanol	161		112	82
Cyclopentanol	141		115	132
Diphenylmethanol		69	141	140
Ethanol	78		93	52
2-Ethyl-1-butanol	149		52	
1-Heptanol	177		47	68
2-Heptanol	160		49	
1-Hexanol	156		58	42
Methanol	65		108	47
4-Methoxybenzyl alcohol	260	25		92
1-Methoxy-2-propanol	119		85	
4-Methylbenzyl alcohol		60	118	79
2-Methyl-1-butanol	129		70	31
3-Methyl-1-butanol	132		61	57
3-Methyl-2-butanol	113		76	68
2-Methyl-3-buten-2-ol	98			
2-Methyl-1-pentanol	148		51	
3-Methyl-2-pentanol	134		43	
4-Methyl-2-pentanol	132		65	143
2-Methyl-1-propanol	108		87	86
1-Octanol	192		61	74
2-Octanol	179		32	114
1-Pentanol	138		46	46
2-Pentanol	119		61	
3-Pentanol	116		101	48
2-Phenoxyethanol	237			
1-Phenylethanol	203		95	94
2-Phenylethanol	219		108	80
1-Phenyl-1-propanol	219			
2-Phenyl-2-propanol	202	34		
1-Propanol	97		74	51
2-Propanol	82		123	88
2-Propen-1-ol	97		48	70
2-Propyn-1-ol	115			

Table 46.4 Aldehydes

COMPOUND	BP, °C	MP, °C	Derivative, Mp, °C	
			SEMICARBAZONE	2,4-DINITRO-PHENYLHYDRAZONE
o-Anisaldehyde (o-methoxybenzaldehyde)	246	38	215	254
p-Anisaldehyde	247		210	254
Benzaldehyde	179		222	237
Butanal	75		106	123
2-Butenal	103		199	190
o-Chlorobenzaldehyde	208		225	207
p-Chlorobenzaldehyde	214	47	230	270 (d)
Cinnamaldehyde	252		215	255
Citral	228		164	116
Citronellal	206		82	77
2,5-Dimethoxybenzaldehyde		52	—	—
3,4-Dimethoxybenzaldehyde		44	177	263
p-Dimethylaminobenzaldehyde		74	222	325
o-Ethoxybenzaldehyde	248		219	
2-Ethylbutanal	116		99	134
Furfural	161		202	230
Heptanal	156		109	108
Hexanal	131		106	104
o-Hydroxybenzaldehyde (salicylaldehyde)	197		231	248
p-Hydroxybenzaldehyde		115	224	280 (d)
2-Methylbutanal	93		103	120
3-Methylbutanal	92		107	123
a-Methylcinnamaldehyde	270		208	
5-Methylfurfural	187		211	212
1-Naphthaldehyde	292	34	221	
p-Nitrobenzaldehyde		106	221	
o-Nitrobenzaldehyde		44	256	250
Phenylacetaldehyde	194		156	121
Piperonal (3,4-methylenedioxybenzaldehyde)	264	36	230	266
o-Tolualdehyde	200		212	195
p-Tolualdehyde	204		215	234

Table 46.5   Amides

COMPOUND	MP, °C
Acetamide	82
Acetanilide	114
Acetoacetanilide	85
o-Acetoacetanisidide	84
p-Acetoacetanisidide	115
o-Acetoacetotoluide	104
o-Acetotoluide	112
m-Acetotoluide	66
p-Acetotoluide	153
Benzamide	130
p-Bromoacetanilide	167
p-Bromobenzamide	155
o-Bromobenzanilide	141
m-Bromobenzanilide	136
p-Chloroacetanilide	179
p-Chloroacetoacetanilide	134
m-Chlorobenzamide	134
p-Chlorobenzanilide	194
Cinnamanilide	153
o-Ethoxybenzamide	133
m-Ethoxybenzamide	139
o-Methoxybenzamide	129
N-Methylacetanilide	102
o-Toluamide	142
m-Toluamide	97
p-Toluamide	158

Table 46.6   Amines

COMPOUND	BP, °C	MP, °C	Derivative, Mp, °C		
			ACETAMIDE	BENZAMIDE	PHENYLTHIOUREA
Aniline	183		114	160	154
Benzylamine	184		60	105	156
N-Benzylaniline		37			
o-Bromoaniline	229		99	116	146
p-Bromoaniline		66	167	204	148
n-Butylamine	77			42	65
iso-Butylamine	69			57	82
sec-Butylamine	63			76	101
tert-Butylamine	46			134	120
o-Chloroaniline	207		87	99	156
m-Chloroaniline	230		72	120	124
p-Chloroaniline		70	179	192	152
Cyclohexylamine	134		104	149	148
Di-*n*-butylamine	160				86
2,4-Dichloroaniline		63	145	117	
2,5-Dichloroaniline		50	132	120	
Diethylamine	55			42	34
Di-*n*-propylamine	110				69
Di-*iso*-propylamine	86				
p-Ethoxyaniline	250		137	173	136
Ethyl *p*-aminobenzoate		89	110	148	
N-Ethylaniline	205		54	60	
o-Ethylaniline	216		111	147	
n-Hexylamine	128			40	77
o-Methoxyaniline	225		87	84	
p-Methoxyaniline		58	128	155	154
2-Methoxy-5-methylaniline		50	110		
4-Methoxy-2-methylaniline		30	134		
N-Methylaniline	196		102	63	87
o-Nitroaniline		71	92	94	142
m-Nitroaniline		114	155	155	160
p-Nitroaniline		147	210	199	
α-Phenylethylamine	185		57	120	
β-Phenylethylamine	198		114	116	135
Piperidine	105			48	101
o-Toluidine	199		112	143	136
m-Toluidine	203		65	125	94
p-Toluidine		45	153	158	141

Table 46.7   Aromatic Halides and Hydrocarbons

COMPOUND	BP, °C	MP, °C	Nitration Product Positions	Nitration Product MP, °C	Carboxylic Acid MP, °C
Anthracene		216			
Biphenyl		70	4,4′	233	
Bromobenzene	157		2,4	75	
4-Bromobiphenyl		89			
p-Bromochlorobenzene		67	2	72	
o-Bromotoluene	181				147
p-Bromotoluene	185	28	2	47	251
Chlorobenzene	132				
o-Chlorotoluene	159				140
m-Chlorotoluene	162		4,6	91	158
p-Chlorotoluene	162		2	38	242
p-Cymene (isopropyltoluene)	175		2,6	54	
o-Dibromobenzene	224		4,5	114	
p-Dibromobenzene		89	2	84	
2,5-Dibromotoluene					157
o-Dichlorobenzene	179		4,5	110	
p-Dichlorobenzene		53	2	54	
2,4-Dichlorotoluene	195				160
2,6-Dichlorotoluene	199				130
Diphenylmethane		26			
Ethylbenzene	135				122
Fluorene		115			
Mesitylene	164		2,4	86	
1-Methylnaphthalene	240				162
2-Methylnaphthalene		32	1	81	
α-Methylstyrene	169				122
2-Phenylethyl chloride	190				122
Styrene	146				122
Toluene	111		2,4	70	122
o-Xylene	142		4,5	71	
m-Xylene	139		2,4	83	
p-Xylene	137		2,3,5	137	

Table 46.8   Esters

COMPOUND	BP, °C	MP, °C
Diethyl ethylmalonate	75 (5 mm)	
Diethyl glutarate	237	
Diethyl maleate	225	
Diethyl malonate	198	
Diethyl oxalate	185	
Diethyl phthalate	296	
Diethyl phenylmalonate	170 (14 mm)	16
Diethyl succinate	216	
Diethyl suberate	268	
Ethyl acetate	77	
Ethyl acetoacetate	181	
Ethyl *p*-anisate	270	
Ethyl benzoate	213	
Ethyl benzoylacetate	270	
Ethyl cinnamate	271	
Ethyl cyanoacetate	208	
Ethyl-*p*-hydroxybenzoate		116
Ethyl 2-methylacetoacetate	187	
Ethyl *p*-nitrobenzoate		57
Ethyl phenylacetate	229	
Ethyl propionate	98	
Ethyl salicylate	234	
Ethyl *p*-toluate	241	
Isopropenyl acetate	96	
Isopropyl acetate	91	
Isopropyl formate	68	
Isopropyl benzoate	218	
Isopropyl salicylate	255	
Methyl acetate	57	
Methyl acetoacetate	169	
Methyl *p*-anisate		49
Methyl benzoate	198	
Methyl *n*-butyrate	102	
Methyl *iso*-butyrate	92	
Methyl *o*-chlorobenzoate	230	
Methyl *m*-chlorobenzoate	231	
Methyl cinnamate		35
Methyl heptanoate	173	
Methyl hexanoate	150	
Methyl *p*-hydroxybenzoate		130
Methyl mandelate		57
Methyl *p*-nitrobenzoate		95
Methyl pentanoate	130	
Methyl phenylacetate	218	

(Table continued on next page.)

Table 46.8   Esters (*Continued*)

COMPOUND	BP, °C	MP, °C
Methyl propionate	79	
Methyl *o*-toluate	213	
Methyl *p*-toluate		30
Phenyl acetate	197	
Phenyl benzoate		69
Phenyl salicylate		42

Table 46.9   Ethers (Aryl)

COMPOUND	BP, °C	MP, °C	Derivative, Mp, °C NITRO
Anisole	154		87 (di)
o-Bromoanisole	218		106
p-Bromoanisole	223		88
o-Chloroanisole	195		95
p-Chloroanisole	200		98
o-Dimethoxybenzene	206		92 (dibromo)
m-Dimethoxybenzene	214		140 (dibromo)
p-Dimethoxybenzene		55	142 (dibromo)
o-Methylanisole	171		63 (bromo)
m-Methylanisole	177		91 (tri)
p-Methylanisole	176		
Phenetole (ethoxybenzene)	172		58

Table 46.10 Ketones

COMPOUND	BP, °C	MP, °C	SEMICARBAZONE	2,4-DINITRO-PHENYLHYDRAZONE
Acetone	56		187	126
2-Acetonaphthone		54	234	262
Acetophenone	200		198	250
Benzophenone		48	167	239
p-Bromoacetophenone		51	208	235
Butanone	80		146	117
Butyrophenone	230		187	190
Chloroacetone	119		164	125
p-Chloroacetophenone	232		201	231
p-Chloropropiophenone		36	176	
Cyclohexanone	156		167	162
Cyclopentanone	131		203	146
3,3-Dimethyl-2-butanone	106		158	125
2,4-Dimethyl-3-pentanone	125		160	95
Fluorenone		83		283
2-Heptanone	151		127	89
3-Heptanone	148		103	
4-Heptanone	145		133	75
Hexane-2,5-dione	188		220 (di)	255 (di)
2-Hexanone	129		122	110
5-Hexen-3-one	129		102	108
4-Hexen-3-one	139		157	
p-Hydroxypropiophenone		148		229
Isobutyrophenone	222		181	163
p-Methoxyacetophenone		38	197	220
p-Methoxypropiophenone		28		
p-Methylacetophenone	226	28	205	258
3-Methyl-2-butanone	94		113	120
2-Methylcyclohexanone	163		195	137
4-Methylcyclohexanone	169		199	130
5-Methyl-3-heptanone	160		102	
6-Methyl-3-heptanone	160		132	
5-Methyl-2-hexanone	145		147	95
Methylcyclohexyl ketone	180		177	140
4-Methyl-2-pentanone	119		135	95
4-Methyl-3-penten-2-one	130		164	203
m-Nitroacetophenone		81	257	228
p-Nitroacetophenone		80		
2-Octanone	173		123	58
2,4-Pentandione	139		122 (mono)	209
2-Pentanone	102		112	144
3-Pentanone	102		139	156

(Table continued on next page.)

Table 46.10   Ketones *(Continued)*

| COMPOUND | BP, °C | MP, °C | Derivative, Mp, °C | |
			SEMICARBAZONE	2,4-DINITRO-PHENYLHYDRAZONE
Phenylacetone	216		198	156
4-Phenyl-2-butanone	235		142	
4-Phenyl-3-buten-2-one		41	187	
Propiophenone	218		174	191

Table 46.11   Nitriles

COMPOUND	BP, °C	MP, °C
Acetonitrile	81	
Acrylonitrile	78	
Adiponitrile	295	
Benzonitrile	191	
p-Bromobenzonitrile		112
Butyronitrile	118	
Chloroacetonitrile	127	
o-Chlorobenzonitrile	232	47
m-Chlorobenzonitrile		41
p-Chlorobenzonitrile		92
o-Chlorophenylacetonitrile	242	24
p-Chlorophenylacetonitrile	265	30
Glutaronitrile	286	
Isobutyronitrile	108	
Malononitrile	219	
p-Methoxybenzonitrile		62
1-Naphthaleneacetonitrile		35
1-Naphthonitrile	299	35
2-Naphthonitrile	306	66
o-Nitrobenzonitrile		110
m-Nitrobenzonitrile		118
p-Nitrobenzonitrile		147
p-Nitrophenylacetonitrile		116
Phenylacetonitrile	234	
o-Tolunitrile	205	
m-Tolunitrile	212	
p-Tolunitrile	217	27

Table 46.12   Nitro Compounds

COMPOUND	BP, °C	MP, °C
o-Bromonitrobenzene	261	43
m-Bromonitrobenzene	256	54
p-Bromonitrobenzene		126
4-Bromo-3-nitrotoluene		33
o-Chloronitrobenzene	246	32
m-Chloronitrobenzene	235	44
p-Chloronitrobenzene		83
2,5-Dibromonitrobenzene		85
2,4-Dichloronitrobenzene		52
2,4-Dimethylnitrobenzene	238	
2,5-Dimethylnitrobenzene	234	
2,6-Dimethylnitrobenzene	226	15
2,4-Dinitroanisole		89
1,3-Dinitrobenzene		90
1,4-Dinitrobenzene		172
2,4-Dinitrobromobenzene		72
2,4-Dinitrochlorobenzene		52
2,4-Dinitrotoluene		70
2,6-Dinitrotoluene		66
o-Nitroanisole	265	
p-Nitroanisole		54
Nitrobenzene	210	
4-Nitrobiphenyl		114
o-Nitrotoluene	224	
m-Nitrotoluene	231	16

Table 46.13   Phenols

| COMPOUND | MP, °C | BP, °C | Derivative, Mp, °C | | |
			BENZOATE	3,5-DINITRO-BENZOATE	PHENYL-URETHANE
4-*t*-Butylphenol	100		81		
4-Chloro-3,5-dimethylphenol	115		[acetate, 48]		
o-Chlorophenol		176			121
p-Chlorophenol	43		88	186	148
o-Cresol (methylphenol)		190		138	142
m-Cresol		202	55	165	128
p-Cresol	36		70	189	146
2,4-Dichlorophenol	45		97	142	
3,5-Dichlorophenol	68		55		
2,4-Dimethylphenol	27	212	38	165	103
2,5-Dimethylphenol	75		61	137	161
2,6-Dimethylphenol	49			159	133
3,4-Dimethylphenol	62		59	182	120
3,5-Dimethylphenol	68			195	151
4-Ethylphenol	47		60	132	120
o-Hydroxyphenol (catechol)	104		84 (di)	152 (di)	169 (di)
m-Hydroxyphenol (resorcinol)	110		117 (di)	201 (di)	164 (di)
p-Hydroxyphenol (hydroquinone)	169		199 (di)	317 (di)	
2-Isopropyl-5-methylphenol (thymol)	51		33	103	107
2-Isopropylphenol		212	[aryloxyacetic acid, 133]		
4-Isopropylphenol	61		71		
2-Methoxyphenol	30	205	58	141	148
4-Methoxyphenol	56		87		137
4-Methyl-2-nitrophenol	34				
5-Methyl-2-nitrophenol	53				
1-Naphthol	94		56	217	177
2-Naphthol	122		107	210	155
o-Nitrophenol	45			155	
p-Nitrophenol	114		143	186	
Phenol	42	180	68	146	126

QUESTIONS

The CRC *Handbook of Chemistry and Physics,* Section 3, entitled "Physical Constants of Organic Compounds," lists, in a sequence near to the IUPAC nomenclature, approximately 22,000 compounds and their associated physical constants. Each compound is given a sequence or identification number. Another table, "Boiling Point Index," groups compounds according to their boiling points. These tables can be used to identify unknowns, or at least narrow the choices, if a compound's index of refraction and its density are known. Following are examples of the use of these tables along with experimental data:

- Carefully determine the boiling point, index of refraction, and density for an unknown.
- Consult the "Boiling Point Index" section of the CRC *Handbook* and write down the identification numbers for each of the compounds grouped within 3 to 5 degrees of the experimentally determined boiling point.
- Using the Identification Numbers, consult the body of Section 3, "Physical Constants of Organic Compounds," and match each of the compounds with the refractive index and density data.
- Write down the compounds that are the closest matches.

Very often one or two simple chemical tests can distinguish among the choices. While this procedure seems simple, it is dependent on careful determination of the experimental data for a pure sample.

1. Draw the structures of compounds that have the following physical constants:

	BOILING POINT	REFRACTIVE INDEX	DENSITY
a.	88–90	1.4922	1.579
b.	139	1.4972	0.8642
c.	216	1.460	0.904
d.	144	1.4237	0.8710

2. Compound A, C_8H_8, bp 144–146°C, is insoluble in water but soluble in concentrated H_2SO_4. It decolorizes Br_2 in CH_2Cl_2 and forms a brown precipitate with cold, dilute aqueous $KMnO_4$. Oxidation with hot concentrated $KMnO_4$ followed by acidification yields Compound B, a white precipitate, mp 120–122°C which has a neutralized, equivalent of 122 ± 2.
   a. Draw the structures of A and B.
   b. What is the structure of the product formed with bromine?

3. Compound C, $C_7H_{14}O$, bp 150–152, density 0.811, is insoluble in water but soluble in concentrated H_2SO_4. It forms a yellow precipitate, mp 122–124°C, upon addition of basic KI/I_2 solution. It forms a reddish-yellow precipitate (mp 88–90°C) with 2,4-dinitrophenylhydrazine reagent but does not react with Tollens reagent. It forms a semicarbazone (mp 126–128°C).
   a. What is the structure of C?
   b. What is the yellow precipitate?

4. Compound D, bp 161–163°C, is not soluble in water or cold, conc. H_2SO_4. It fails to react with cold dilute $KMnO_4$ but, in the presence of fluorescent lights, does eventually decolorize bromine in CH_2Cl_2 upon standing. The sodium fusion test reveals the presence of chloride. Heating with hot, conc. $KMnO_4$, followed by acidification gives a white precipitate, mp 240–242°C, and neutralization equivalent of 155 ± 2.
   a. What is the structure of D?
   b. What is the white precipitate?

5. Compound E, bp 174–176°C, is soluble in water. It does not react with Na metal or Br_2/CH_2Cl_2 but gives a silver mirror with Tollens reagent. It forms a precipitate with 2,4-dinitrophenylhydrazine. A semicarbazone derivative melts at 105–107°C, and the 2,4-dinitrophenylhydrazine melts at 122–124°C.
   a. What is the structure of E?

Appendix A

Selected Atomic Weights

Aluminum	26.98	Magnesium	24.31
Barium	137.34	Manganese	54.94
Boron	10.81	Mercury	200.59
Bromine	79.91	Nitrogen	14.01
Calcium	40.08	Oxygen	16.00
Carbon	12.01	Phosphorus	30.97
Chlorine	35.45	Potassium	39.10
Chromium	52.00	Selenium	78.96
Copper	63.54	Silicon	28.09
Fluorine	19.00	Silver	107.87
Hydrogen	1.008	Sodium	22.99
Iodine	126.90	Sulfur	32.06
Iron	55.85	Tin	118.69
Lithium	6.94	Zinc	65.37

Appendix B

Common Acids and Bases

	PERCENT BY WEIGHT	SP. GR.	MOLES/L	G/100 ML
Hydrochloric acid, conc.	37	1.19	12.0	44.0
10%	10	1.05	2.9	10.5
5%	5	1.02	1.4	5.1
1 M	3.6	1.02	1.0	3.6
Sulfuric acid, conc.	96	1.84	18.0	177.
Nitric acid, conc.	70	1.41	15.7	99.4
Acetic acid, glacial	99.7	1.05	17.4	105.
Ammonia (aq.), conc.	28	0.90	14.8	25.2
Sodium hydroxide, 10%	10	1.11	2.8	11.1
Sodium bicarbonate, 5%	5	1.04	0.62	5.2
Sodium carbonate, 5%	5	1.05	0.50	5.3

Appendix C

Common Organic Liquids

	BOILING POINT 760 MM (MP)	DENSITY, 20°C G/ML	MOLECULAR WEIGHT
Acetic acid	118° (16.6°)	1.05	60.0
Acetic anhydride	140°	1.08	102.1
Acetone	56°	0.79	58.1
Acetyl chloride	52°	1.10	78.5
Aniline	184°	1.02	93.1
Benzene	80° (5.5°)	0.88	78.1
Benzoyl chloride	197°	1.21	140.6
1-Butanol	118°	0.81	74.1
t-Butyl alcohol	82° (25.5°)	0.79	74.1
Carbon disulfide	45°	1.26	76.1
Carbon tetrachloride	77°	1.59	153.8
Chlorobenzene	132°	1.11	112.6
Chloroform	61°	1.49	119.4
Cyclohexane	81° (6.5°)	0.78	84.2
Cyclohexanol	161° (25°)	0.96	100.2
1,2-Dichloroethane	84°	1.26	99.0
Diethyl ether	35°	0.71	74.1
1,2-Dimethoxyethane (glyme)	85°	0.87	90.1
Dimethylformamide	153°	0.94	73.1
Ethanol	78°	0.79	46.1
Ethyl acetate	77°	0.90	88.1
Heptane	98°	0.68	100.2
Hexane	69°	0.66	86.2
Methanol	65°	0.79	32.0
Methyl acetate	57°	0.93	74.1
Methyl t-butyl ether	56°	0.74	88.2
Methylene chloride	40°	1.34	84.9
Nitrobenzene	211° (5.7°)	1.20	123.1
Pentane	36°	0.63	72.2
1-Propanol	97°	0.80	60.1
2-Propanol	82°	0.79	60.1
Pyridine	116°	0.98	79.1
Tetrahydrofuran	65°	0.89	72.1
Toluene	111°	0.87	92.1

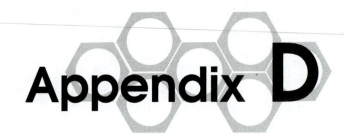

Appendix D

Representative Calculations

In order to complete many organic experiments, it is necessary to be familiar with one or more simple calculations. The value of carefully performed laboratory work may be lost if the calculations are done incorrectly. Among important kinds of problems are those dealing with relationships of weight, volume, and number of moles. The following examples illustrate the solutions to some typical problems.

1. The directions of an experiment call for using 20.0 mL of toluene (density = 0.87 g/mL; molecular weight = 92.1 g/mole). What weight of toluene and how many moles are involved?
   Solution:

   $$20.0 \text{ mL} \times 0.87 \text{ g/mL} = 17.4 \text{ g of toluene}$$

   $$17.4 \text{ g} \times \frac{1 \text{ mole toluene}}{92.1 \text{ g}} = 0.189 \text{ mole of toluene}$$

2. Suppose an experiment calls for 50.0 g of benzoyl chloride to be used. In this case, and frequently, it is more convenient to measure the material by volume rather than by weight if the density is known.
   Solution:

   $$\text{density of benzoyl chloride} = 1.21 \text{ g/mL}$$

   $$\frac{50.0 \text{ g}}{1.21 \text{ g/mL}} = 41.3 \text{ mL}$$

3. If a solution of nitric acid has a density of 1.41 g/mL and contains 70% HNO_3 by weight, what is its molarity?
   Solution:

   $$1000 \text{ mL} \times 1.41 \text{ g/mL} = 1410 \text{ g/liter}$$

   $$1410 \text{ g/liter} \times 70\% = 987 \text{ g } HNO_3/\text{liter}$$

   $$987 \text{ g } NHO_3/\text{liter} \times 1 \text{ mole}/63.0 \text{ g} = 15.7 \text{ mole/liter} = 15.7 \text{ } M$$

4. What volume of 15.7 M HNO$_3$ is needed to prepare 500 mL of 2.0 M HNO$_3$?
   Solution:

   $$M_1V_1 = M_2V_2 \quad 15.7\ M \times V_1 = 2.0\ M \times 500\text{mL}$$
   $$V_1 = 63.7\ \text{mL}$$

5. Consider the reaction of 2.50 g of aniline with 4.0 mL of acetic anhydride to form 2.92 g of acetanilide. What is the theoretical yield and the percentage yield of acetanilide in this preparation?
   Consider the reaction involved:

   Solution:

   $$2.50\ \text{g} \times \frac{1\ \text{mole}}{93.1\ \text{g}} = 0.0269\ \text{mole aniline}$$

   $$4.0\ \text{mL} \times 1.08\ \text{g/mL} = 4.3\ \text{g of acetic anhydride}$$

   $$4.3\text{g} \times \frac{1\text{mole}}{102\ \text{g}} = 0.042\ \text{mole acetic anhydride}$$

   so aniline is the limiting reagent
   theoretical yield:

   $$0.0269\ \text{mole aniline} \times \frac{1\ \text{mole acetanilide}}{1\ \text{mole aniline}} = 0.0269\ \text{mole acetanilide}$$

   $$0.0269\ \text{mole acetanilide} \times \frac{135\ \text{g}}{1\ \text{mole}} = 3.63\ \text{g}$$

   $$\text{percentage yield} = \frac{\text{actual yield}}{\text{theoretical yield}} \times 100$$

   $$= \frac{2.92\ \text{g}}{3.63\ \text{g}} \times 100 = 80.4\%$$

6. Suppose benzene is nitrated to give an 80% yield of nitrobenzene. The nitrobenzene is then reduced to give a 95% yield of aniline. The aniline is then reacted with acetic anhydride to give 90% yield of acetanilide. Calculate the overall yield of acetanilide from benzene.
   Solution:

   $$80\% \times 95\% \times 90\% = 68.4\ \text{percent yield overall}$$

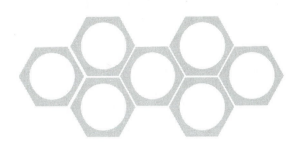

Index